W9-BCE-860

The America of 1750

Peter Kalm's Travels In North America

The English Version of
1770

Revised from the original Swedish and edited by

ADOLPH B. BENSON
Late Professor of German and Scandinavian in Yale University

with
A Translation of New Material from Kalm's Diary Notes

In Two Volumes
VOLUME II

New York
DOVER PUBLICATIONS, INC.

Published in Canada by General Publishing Company, Ltd., 30 Lesmill Road, Don Mills, Toronto, Ontario.
Published in the United Kingdom by Constable and Company, Ltd., 10 Orange Street, London, W.C. 2.

This Dover edition, first published in 1966, is an unabridged republication of the work originally published by Wilson-Erickson, Inc., in 1937.
The present edition contains two additional illustrations from Forster's English edition of 1770-71 (see plates facing pages 110 and 253). The map from the same English version is here reproduced approximately two and one-half times larger than in the 1937 edition.

Library of Congress Catalog Card Number: 66-19048

Manufactured in the United States of America
Dover Publications, Inc.
180 Varick Street
New York, N. Y. 10014

TABLE OF CONTENTS

LIST OF ILLUSTRATIONS

VOLUME II

PETER KALM'S TRAVELS IN NORTH AMERICA

VOLUME II

July the 24th

At Montreal. This morning I went from Prairie in a bateau to Montreal on the St. Lawrence River. The river is very rapid, but not very deep near Prairie, so that the boats cannot go higher than Montreal, except in spring with the high water, when they can come up to Prairie, but not further. The town of Montreal may be seen at Prairie and all the way down to it. On our arrival there we found a crowd of people at the gate of the town where we were to pass through. They were very desirous of seeing us, because they were informed that some Swedes were to come to town, people of whom they had heard something, but whom they had never seen; and we were assured by everybody, that we were the first Swedes that ever had come to Montreal. As soon as we had landed, the governor of the town sent a captain to me, who desired that I would follow him to the governor's house, where he introduced me to him in a room where the governor was with some friends. Baron Longueuil was as yet vice-governor, but he daily expected his promotion from France. He received me more civilly and generously than I can well describe, and showed me letters from the governor-general at Quebec, the Marquis de la Galissonnière, who mentioned that he had received orders from the French court to supply me with whatever I should want, as I was to travel in this country at the expense of his most Christian majesty. In short Governor Longueuil loaded me with greater favors than I could expect or even imagine, both during that stay and on my return from Quebec.

July the 25th

French Manners and Customs. The difference between the manners and customs of the French in Montreal and Canada, and those of the English in the American colonies, is as great as that between the manners of those two nations in Europe. The women in general are handsome here; they are well bred and virtuous, with an innocent and becoming freedom. They dress up very fine on Sundays; about the same as our Swedish women, and though on the other days they do not take much pains with other parts of their dress, yet they are very fond of adorning their heads. Their hair is always curled, powdered and ornamented with glittering bodkins and aigrettes. Every day but Sunday they wear a little neat jacket, and a short skirt which hardly reaches halfway down the leg, and sometimes not that far. And in this particular they seem to imitate the Indian women. The heels of their shoes are high and very narrow, and it is surprising how they can walk on them. In their domestic duties they greatly surpass the English women in the plantations, who indeed have taken the liberty of throwing all the burden of housekeeping upon their husbands, and sit in their chairs all day with folded arms.[1] The women in Canada on the contrary do not spare themselves, especially among the common people, where they are always in the fields, meadows, stables, etc. and do not dislike any work whatsoever. However, they seem rather remiss in regard to the cleaning of the utensils and apartments, for sometimes the floors, both in the town and country, are hardly cleaned once in six months, which is a disagreeable sight to one who comes from amongst the Dutch and English, where the constant scouring and scrubbing of the floors is reckoned as important as the exercise of religion itself. To prevent the thick dust, which is thus left on the floor from being noxious to the health, the women wet it several times a day, which lays the dust, and they repeat this as often as the dust is dry and begins to rise again. Upon the whole, however, they are not averse to the taking part in all the business of housekeeping, and I have with pleasure seen the daughters of the better sort of people and of the governor himself, not too finely dressed, going into kitchens and cellars to see that everything was done as it ought to be. And they also carry their sewing with them, even the governor's daughters.

The men are extremely civil and take their hats off to every person whom they meet in the streets. This is difficult for anyone whose duties demand that he be out doors often, especially in the

---

[1] It seems, that for the future, the fair sex in the English colonies in North America, will no longer deserve the reproaches Mr. Kalm stigmatizes them with repeatedly, since it is generally reported, that the ladies of late have vied one with another, in providing their families with linen, stockings, and homespun cloth of their own making, and that a general spirit of industry prevails among them at this present time.—F.

evening when every family sits outside their door, near the street. It is customary to return a visit the day after you have received one, even though one should have several scores to pay in one day.

Animals in Canada. I have been told by some Frenchmen, who had gone beaver hunting with the Indians to the northern parts of Canada, about fifty French miles from Hudson Bay, that the animals whose skins they endeavor to get, and which are there in great abundance, are beavers, wild cats, or lynx, and martens. These animals are the more valued, the further they are caught to the north, for their skins have better hair, and look better or worse the further they come from the north or south.

White partridges (*Perdrix blanches*) is the name which the French in Canada give to a kind of bird abounding during winter near Hudson Bay, and which undoubtedly are our *ptarmigans*, snow-hens (*Tetrao lagopus*). They are very plentiful at the time of a great frost, and when a considerable quantity of snow happens to fall. The greater the cold or snow, the greater the number of birds. They were described to me as having rough white feet, and being white all over, except for three or four black feathers in the tail; and they are considered very fine eating. From Edward's *Natural History of Birds*[1] it appears that the ptarmigans are common about Hudson Bay.

Hares are likewise said to be plentiful near Hudson Bay, and they are abundant even in Canada, where I have often seen and found them, corresponding perfectly with our Swedish hares. In summer they have a brownish gray and in winter a snowy white color, as with us.

The Trades. Mechanical trades, such as architecture, cabinet work, turning, and brick making, are not yet so advanced here as they ought to be, and the English in that particular outdo the French. The chief cause of this is that scarcely any other people than dismissed soldiers come to settle here, who have not had any opportunity of learning a mechanical trade, but have sometimes accidentally, and through necessity been obliged to do it. There are, however, some who have a good skill in mechanics, and I saw

[1] Page 72.—F.

a person here who made very good clocks and watches, though he had had but very little instruction.

July the 27th

The *common houseflies* (*Musca domestica*) were observed in this country about one hundred and fifty years ago, as I have been assured by several persons in this town, and in Quebec. All the Indians assert the same thing, and are of the opinion that the common flies first came over here with the Europeans and their ships, which were stranded on this coast. I shall not dispute this; however, I know, that while I was on the frontier between Saratoga and Crown Point, or Fort St. Frédéric, and sat down to rest or to eat, a number of our common flies always came and settled on me. It is therefore dubious whether they have not been longer in America than the term above-mentioned, or whether they have been imported from Europe. On the other hand, it may be urged that the flies were left in that wilderness at the time when Fort Anne was yet in a good condition, and when the English often travelled there and back again; not to mention that several Europeans, both before and after that time, had travelled through those places and carried the flies with them, which had been attracted by their provisions.

Wild cattle were abundant in the southern parts of Canada, and have been there since times immemorial. They were particularly plentiful in those parts where the Illinois Indians lived, which were nearly in the same latitude with Philadelphia; but further to the north they are seldom observed. I saw the skin of a wild ox to-day; it was as big as one of the largest ox hides in Europe, but had better hair. This was dark brown like that on a brown bearskin. That which was close to the skin is as soft as wool. This hide was not very thick and in general was not considered so valuable (in France) as a bearskin. In winter it is spread on the floor to keep the feet warm. Some of these wild cattle, as I am told, have a long and fine wool, as good if not better than sheep wool. They make stockings, cloth, gloves, and other pieces of worsted work of it, which look as well as if they were made of the best sheep wool.

The Indians employ it for several uses. The flesh is as good and fat as the best beef. Sometimes the hides are thick, and may be used as cowhides are in Europe. The wild cattle in general are said to be stronger and bigger than European cattle, and of a brownish red color. Their horns are short, though very thick close to the head. These and several other qualities, which they have in common with and in greater perfection than the domestic cattle, have induced some to endeavor to tame them, by which means they would obtain the advantages arising from their good hair, and, on account of their great strength, could employ them successfully in agriculture. With this view some have repeatedly gotten young wild calves and brought them up in Quebec and other places among the tame cattle, but they have usually died in three or four years time; and though they have seen people every day, they have always retained a natural ferocity. They have constantly been very shy, pricked up their ears at the sight of a man, and have trembled or run about, so that the art of taming them has not hitherto been successful. Some have been of the opinion that these cattle cannot bear the cold well, as they never go north of the place I mentioned, though the summers be very hot, even in those northern parts. They think that when the country about the Illinois is better peopled it will be more easy to tame these cattle, and that afterwards they may more easily be accustomed to the northerly climates.[1] The Indians and French in Canada, make use of the horns of these creatures to put gunpowder in. I have briefly mentioned the wild cattle in the former parts of this journey.[2]

The peace, which was concluded between France and England, was proclaimed to-day.[3] The soldiers were under arms, the artillery on the walls was fired off, and some salutes were given by the small firearms. All night fireworks were exhibited, and the whole town was illuminated. All the streets were crowded with people till late at night. The governor invited me to supper and to partake of the joy of the inhabitants. There were present a number of officers

---

[1] But by this means they would lose that superiority, which in their wild state they have over the tame cattle; as all the progenies of tamed animals degenerate from the excellence of their wild and free ancestors.—F.

[2] See pp. 110 and 150.

[3] The peace of Aix-la-Chapelle of October 18, 1748, ending King George's War.

and other persons of distinction, and they drank merrily far into the night.

July the 28th

On the Isle of Madeleine. This morning I accompanied the governor, Baron de Longueuil,[1] and his family to a little island called *Madeleine*, which is his own property. It lies in the St. Lawrence River directly opposite the town on the eastern side. The governor had here a very neat house, though it was not very large, a fine, extensive garden, and a yard. The river passes between the town and this island, and is very rapid. Near the town it is deep enough for large boats, but towards the island it grows more shallow, so that they are obliged to push the boats forward with poles. There was a mill on the island, turned by the mere force of the stream, without an additional milldam. In the mill I noticed that the stones did not consist of one single piece but were made of several pieces. The upper millstone was quite large, made of eight separate parts which were joined very close together and bound with a thick iron band. The lower stone was the same. The upper one had been imported from France but the other was native. The trough of the funnel through which the grain ran down was shaken in this manner: the upper part of the pinion-axle was joined above the millstone to a square piece of hard wood about four inches square. When this turned its four corners turned against the end of the hopper which projected out on one side, so the funnel was shaken and the grain ran down. The wheels and axle were made of white oak, but the cogs of the wheel and other parts of the machinery were made of the sugar maple or wild cherry. Still, the former was said to be the most in use, because it was considered hard wood, especially if it had grown in dry places.

Trees and Plants. The smooth sumach, or *Rhus glabra*, grows abundantly here. I have nowhere seen it so tall as in this place, where it had sometimes the height of eight yards and a proportional thickness.

[1] Charles le Moyne, second Baron de Longueuil, became governor of Montreal in 1749, and "commandant-general" of the Colony in 1752. See note by Marchand in his French edition, vol. 2, p. 48.

Sassafras is planted here, for it is never found wild in these parts, Fort Anne being the most northerly place where I have found it wild. Those shrubs which were on the island had been planted many years ago; however, they were only small shrubs, from two to three feet high, and scarcely that much. The reason is that the main stem is killed every winter almost down to the very root, and must produce new shoots every spring, as I have found from my own observations here; and so it appeared to be near the forts Anne, Nicholson, and Oswego. It will therefore be useless to attempt to plant sassafras in a very cold climate.

The *Mulberry trees* (*Morus rubra* L.) are likewise planted here. I saw four or five of them about five yards high, which the governor told me had been twenty years in this place and brought from more southerly parts, since they do not grow wild near Montreal. The most northerly place where I have found it growing wild is about twenty English miles north of Albany. I had this confirmed by the country people who live in that place, and who at the same time informed me that it was very scarce in the woods. When I came to Saratoga I inquired whether any of these mulberry trees had been found in that neighborhood, but everybody told me that they were never seen in those parts, and that the before-mentioned place twenty miles above Albany is the most northern point where they grow. The mulberry trees that were planted on this island succeed very well, though they are placed in poor soil. Their foliage was large and thick, but they did not bear any fruit this year. I was informed that they can bear a considerable degree of cold.

The *waterbeech* had been planted here in a shady place, and had grown to a great height. All the French hereabouts call it *cotonnier*.[1] It is never found wild near the St. Lawrence River, nor north of Fort St. Frédéric, where it is now very scarce.

The *red cedar* is called *cèdre rouge* by the French, and that was also planted in the governor's garden, whither it had been brought from more southern parts, for it is not to be found in the forests hereabouts. However, it grew very well here.

About half an hour after seven in the evening we left this pleas-

---

[1] Cotton tree. Mr. Kalm mentions before that this name is given to the *Asclepias Syriaca*. See page 387.

ant island, and an hour after our return the Baron de Longueuil received two agreeable pieces of news at once. The first was that his son who had been two years in France, had returned; and the second, that he had brought with him the royal patents for his father, by which he was appointed governor of Montreal and the country belonging to it.

People make use of *fans* here which are made of the tails of the wild turkeys. As soon as the birds are shot, their tails are spread like fans, and dried, by which means they keep their shape. The ladies and the men of distinction in town carry these fans, when they walk in the streets during the intense heat.

All the *grass* on the meadows round Montreal consists of a species of meadow grass, or the *Poa capillaris* L. This is a very slender grass which grows very close and succeeds even on the driest hills. It is however not rich in foliage; but the slender stalk is used for hay. We have numerous kinds of grasses in Sweden which are much more useful than this.

July the 30th

The *wild plum trees* grow in great abundance on the hills along the rivulets about the town. They were so loaded with fruit that the boughs were bent downwards by the weight. The fruit was not yet ripe, but when it comes to that perfection it has a red color and a fine taste, and preserves are sometimes made of it.

Black currants (*Ribes nigrum* L.) are plentiful in the same places, and its berries were ripe at this time. They are very small, and not by far so fine as those in Sweden.

Parsnips grow in great abundance on the rising banks of rivers, along the grain fields, and in other places. This led me to think that they were original natives of America, and not first brought over by the Europeans. But on my journey into the country of the Iroquois, where no European ever had a settlement, I never once saw it, though the soil was excellent; and from this it appears plain enough that it was transported hither from Europe, and is not originally an American plant. Therefore it is in vain sought for in any part of this continent except among the European settlements.

August the 1st

The governor-general of Canada commonly resides at Quebec, but occasionally he goes to Montreal, and generally spends the winter there. In summer he resides chiefly at Quebec on account of the king's ships which arrive there during that season and bring him letters which he must answer; besides, there is much other business at that time. During his residence in Montreal he lives in the castle, as it is called, which is a large house of stone, built by the former governor-general Vaudreuil, that still belongs to his family, who rents it to the king. General de la Galissonière is said to like Montreal better than Quebec, and indeed the location of the former is by far the more agreeable one.

Canada Money. Canada had scarcely any other money but paper currency. I hardly ever saw any coin, except French sols, consisting of copper, with a very small mixture of silver. They were quite thin by constant circulation, and were valued at a sol and a half. The bills were not printed but written. Their origin is as follows. The French king having found it very dangerous to send money for the pay of troops and other purposes over to Canada, on account of privateers, shipwrecks, and other accidents, ordered that instead of it the intendant, or king's steward at Quebec, or the commissary at Montreal, should write bills for the value of the sums which are due to the troops, and which he distributes to each soldier. On these bills is incribed that they bear the value of such or such a sum, till next October, and they are signed by the intendant or the commissary, and in the interval they bear the value of money. In the month of October, at a certain stated time, every one brings the bills in his possession to the intendant at Quebec, or the commissary at Montreal, who exchanges them for bills of exchange upon France, which are paid in lawful money, at the king's exchequer, as soon as they are presented. If the money is not yet wanted the bill may be kept till October of the following year, when it may be exchanged by one of those gentlemen for a bill upon France. The paper money can only be delivered in October and exchanged for bills upon France. They are of different values, and some do not exceed a livre, and perhaps some are still less. Towards autumn when the merchant ships come in from France, the merchants endeavor to get as many bills as they can and change them for bills upon the French treasury. These bills are partly printed, spaces being left for the name, sum, etc. But the first bill, or paper currency, is all written, and is therefore subject to be counterfeited, which has sometimes been done; but the great punishments which have been inflicted upon the authors of these forged bills, and which generally are capital, have deterred people from attempting it again; so that examples of this kind are very scarce at present. As there is a great want of small coin here, the buyers or sellers are frequently obliged to suffer a small loss and can pay no intermediate prices between one livre and two.[1] For example if I wanted to buy something for which the price was ten livres, and I did not have the bills for the exact amount, I was obliged to pay one or two livres extra. In this transaction the one who was anxious to buy or sell was generally the sufferer.

Wages. On the farms the wages for servants and day laborers was ordinarily a little less than in the cities. They commonly give one hundred and fifty livres a year to a faithful and diligent man servant, and to a maid servant of the same character one hundred livres. A journeyman to an artist gets three or four livres a day, and a common laboring man gets thirty or forty sols a day. The scarcity of laboring people occasions the wages to be so high; for almost everybody finds it easy to be a farmer in this uncultivated country where he can live well, and at so small an expense that he does not care to work for others.

Montreal is the second town in Canada, in regard to size and wealth; but it is the first on account of its fine location and mild climate. A short distance above the town the St. Lawrence River divides into several branches, and by that means forms several islands, among which the isle of Montreal is the greatest. It is ten French miles long, and near four broad, in its broadest part. The town of Montreal is built on the eastern side of the island, and close to one of the largest branches of the St. Lawrence River; and thus it has a very pleasant and advantageous location. The town has a square form, or rather it is a rectangular parallelogram, the long and

[1] The sol is the lowest coin in Canada, and is about the value of a penny in the English colonies. A livre, or franc, (for they are both the same) contains twenty sols; and three livres or francs make an écu or crown.—F.

eastern side of which extends along the great branch of the river. On the other side it is surrounded with excellent grain fields, charming meadows and delightful woods. It has the name of Montreal from a great mountain about half a mile westwards of the town, which lifts its head far above the woods. M. Cartier,[1] one of the first Frenchmen who surveyed Canada more accurately, named this mountain on his arrival on this island, in the year 1535, when he visited the mountain and the Indian town Hochelaga near it. The priests who, according to the Roman Catholic way, would call every place in this country after some saint or other, called Montreal, Ville Marie, but they have not been able to make this name general, for it has always kept its first name. It is pretty well fortified, and surrounded with a high and thick wall. On the east side it has the St. Lawrence River, and on all the other sides a deep ditch filled with water, which secures the inhabitants against all danger from sudden incursions of the enemy's troops. However, it cannot long stand a regular siege, because it requires a great garrison on account of its extent and because it consists chiefly of wooden houses. Here are several churches, of which I shall only mention that belonging to the friars of the order of St. Sulpicius, that of the Jesuits, that of the Franciscan Friars, that belonging to the nunnery, and that of the hospital. Of these the first is by far the finest, both in regard to its outward and inward ornaments, not only in this place but in all Canada. The priests of the seminary of St. Sulpicius have a fine large house, where they live together. The college of the Franciscan Friars is likewise spacious, and has good walls, but it is not so magnificent as the former. The college of the Jesuits is small, but well built. To each of these three buildings are annexed fine large gardens, for the amusement, health and use of the communities to which they belong. Some of the houses in the town are built of stone, but most of them are of timber, though very neatly built. Each of the better sort of houses has a door towards the street, with a seat on each side of it, for amusement and recreation in the morning and evening. The long streets are broad and straight and divided at right angles by the short ones: some are paved but most of them are very uneven. The gates of the town are numerous; on the east side of the town towards the

[1] Jacques Cartier (1494-about 1552 or later), famous French navigator.

river are five, two great and three lesser ones, and on the other side are likewise several. The governor-general of Canada, when he is at Montreal, resides in the so-called castle, which the government hires for that purpose of the family of Vaudreuil; but the governor of Montreal is obliged to buy or hire a house in town; though I was told that the government contributed towards paying the rent.

In the town is a *Nunnery*, and outside its walls half of another, for though the last was ready, it had not yet been confirmed by His Holiness the Pope. In the first they do not receive every girl that offers herself, for their parents must pay about five hundred éçus or crowns for them. Some indeed are admitted for three hundred éçus, but they are obliged to serve those who pay more than they. No poor girls are taken in.

The Hospital. The King has erected a hospital for sick soldiers here. The sick person is there provided with everything he wants, and the king pays twelve sols every day for his keep, attendants, etc. The surgeons are paid by the king. When an officer is brought to this hospital, who has fallen sick in the service of the crown, he receives victuals and attendance gratis; but if he has gotten a sickness in the execution of his private concerns, and comes to be cured here, he must pay it out of his own purse. When there is room enough in the hospital, they likewise take in some of the sick inhabitants of the town and country. They have the medicines and the attendance of the surgeons gratis, but must pay twelve sols per day for meat, etc.

Every Friday is a *market day* when the country people come to the town with provisions, and those who want them must supply themselves on that day because it is the only market day in the whole week. On that day, too, a number of Indians come to town to sell their goods and buy others.

The *declination of the magnetic needle* was here ten degrees and thirty-eight minutes, west. M. Gillion, one of the priests here, who had a particular taste for mathematics and astronomy, had drawn a meridian in the garden of the seminary which he said he had examined repeatedly by the sun and stars, and found it to be very exact. I compared my compass with it, taking care that no iron was near it, and found its declination just the same as that which I have before mentioned. According to M. Gillion's observations, the latitude of Montreal is forty-five degrees and twenty-seven minutes.

Temperature at Montreal. M. Pontarion, another priest, had made thermometrical observations in Montreal, from the beginning of this year, 1749. He made use of the Reaumur thermometer, which he placed sometimes in a window half open, and sometimes in one entirely open, and accordingly it will seldom mark the greatest degree of cold in the air. However, I shall give a short abstract of his observations for the winter months. In January the greatest cold was on the 18th day of the month, when the Reaumurian thermometer was 23° below the freezing point (—19¾°F.). The least degree of cold was on the 31st of the same month when it was just at the freezing point, but most of the days of this month it was from 12° to 15° below freezing. In February the greatest cold was on the 19th and 25th when the thermometer was 14° below, and the least was on the 3rd day of that month when it rose 8° above the freezing point (50° F.); but it was generally 11° below it. In March the greatest cold was on the 3rd when it was 10° below the freezing point (9.5° F.) and on the 22nd, 23rd, and 24th it was mildest, being 15° above it (65¾° F.): in general it was 4° below it (23° F.). In April the greatest degree of cold happened on the 7th, the thermometer being 5° below the freezing point (20¾° F.); the 25th was the mildest day, it being 20° above the freezing point (77° F.); but in general it was 12° above it (59° F.). These are the contents chiefly of M. Pontarion's observations during those months; but I found, by the manner he made his observations, that the cold had every day been from 4° to 6° (9 to 13½° F.) greater than he had marked it. He had likewise marked in his journal that the ice in the St. Lawrence River broke on the 3rd of April at Montreal, and only on the 20th day of that month at Quebec. On the 3rd of May some trees began to flower at Montreal, and on the 12th the hoarfrost was so great that the trees were covered with it as with snow. The ice in the river close to this town is over a French foot thick every winter, and sometimes it is two feet, as I was informed by all whom I consulted on that head.

Several of the friars here told me that the summers had been remarkably longer in Canada since its cultivation than they used to be before; it begins earlier and ends later. The winters, on the other hand, are much shorter, but the friars were of the opinion that they were as severe as formerly, though they were not of the same duration, and the summer at present was no hotter than it used to be. The coldest winds at Montreal are those from the north and northwest.

August the 2nd

En Route for Quebec. Early this morning we left Montreal and went in a bateau on our journey to Quebec in company with the second major of Montreal, M. de Sermonville. We went down the St. Lawrence River, which was here pretty broad on our left; on the northwest side was the isle of Montreal, and on the right a number of other isles, and the shore. The isle of Montreal was closely inhabited along the river; it was very flat and the rising land near the shore consisted of pure earth and was between three or four yards high. The woods were cut down along the riverside for the distance of an English mile. The dwelling houses were built of wood or stone indiscriminately, and whitewashed on the outside. The other buildings, such as barns, stables, etc. were all of wood. The ground next to the river was turned either into grain fields or meadows. Now and then we perceived churches on both sides of the river, the steeples of which were generally on that side of the church which looked towards the river, because they are not obliged here to put the steeples on the west end of the churches as in Sweden. Within six French miles of Montreal we saw several islands of different sizes in the river, and most of them were inhabited. Those without houses were sometimes turned into grain fields, but generally into grazing land. We saw no mountains, hills, rocks or stones to-day, the country being flat throughout, and consisting of pure earth.

All the *farms in Canada* stand separate from one another, so that each farmer has his possessions entirely separate from those of his neighbor. Each church, it is true, has a little village near it; but that consists chiefly of the parsonage, a school for the boys and girls of the place, and of the houses of tradesmen, but rarely of farmhouses; and if that was the case, their fields were still separated. The farmhouses hereabouts are generally all built along the rising banks of the river, either close to the water or at some distance from it, and about three or four arpens from each other. To some farms are annexed small orchards but they are in general without them; however, almost every farmer has a kitchen-garden.

Peach Trees. I have been told by all those who have made journeys to the southern parts of Canada and to the Mississippi River that the woods there abound with peach trees which bear excellent fruit, and that the Indians of those parts say that the trees have been there since time immemorial.

The *farmhouses* are generally built of stone, but sometimes of timber, and have three or four rooms. The windows, seldom of glass, are most frequently of paper. They have iron stoves in one of the rooms and fireplaces in the rest, always without dampers. The roofs are covered with boards, and the crevices and chinks are filled up with clay. Other farm buildings are covered with straw. The fences are like our common ones.

Road Shrines. There are several crosses put up by the roadside, which is parallel to the shores of the river. These crosses are very common in Canada, and are put up to excite devotion in the travellers. They are made of wood, five or six yards high, and proportionally broad. In that side which faces the road is a square hole, in which they place an image of our Savior, the Cross, or of the Holy Virgin with the Child in her arms, and before that they put a piece of glass, to prevent its being spoiled by the weather. Everyone who passes by crosses himself, raises his hat or does some other bit of reverence. Those crosses which are not far from churches, are very much adorned, and they put up about them all the instruments which they think the Jews employed in crucifying our Savior, such as a hammer, tongs, nails, a flask of vinegar, and perhaps many more than were really used. A figure of the cock, which crowed when St. Peter denied our Lord, is commonly put at the top of the cross.

The country on both sides was very delightful to-day, and the fine state of its cultivation added greatly to the beauty of the scene. It could really be called a village, beginning at Montreal and ending at Quebec, which is a distance of more than one hundred and eighty miles, for the farmhouses are never above five arpens and sometimes but three apart, a few places excepted. The prospect is exceedingly beautiful when the river flows on for several miles in a straight line, because it then shortens the distances between the houses, and makes them form one continued village.

Women's Dress. All the women in the country without exception, wear caps of some kind or other. Their jackets are short and so are their skirts, which scarcely reach down to the middle of their legs. Their shoes are often like those of the Finnish women, but are sometimes provided with heels. They have a silver cross hanging down on the breast. In general they are very industrious. However I saw some, who, like the English women in the colonies, did nothing but prattle all day. When they have anything to do within doors, they (especially the girls) commonly sing songs in which the words *amour* and *coeur* are very frequent. In the country it is usual that when the husband receives a visit from persons of rank and dines with them, his wife stands behind and serves him, but in the town the ladies are more distinguished, and would willingly assume an equal if not a superior position to their husbands. When they go out of doors they wear long cloaks, which cover all their other clothes and are either grey, brown or blue. Men sometimes make use of them when they are obliged to walk in the rain. The women have the advantage of being in a *déshabillé* under these cloaks, without anybody's perceiving it.

We sometimes saw *windmills* near the farms. They were generally built of stone, with a roof of boards, which together with its wings could be turned to the wind.

The breadth of the river was not always the same to-day; in the narrowest place it was about a quarter of an English mile broad; in other parts it was nearly two English miles. The shore was sometimes high and steep, and sometimes low or sloping.

At three o'clock this afternoon we passed by a river [the Chambly] which empties into the St. Lawrence and comes from Lake Champlain, in the middle of which is a large island. The large boats which go between Montreal and Quebec pass on the southeast side of this island because it is deeper there; but some boats prefer the northwest side because it is nearer and yet deep enough for them. Besides this island there are several more hereabouts which are all inhabited. Somewhat further on, the country on both sides of the river is uninhabited near the shore till we come to the Lac St. Pierre, because it is so low as to be overflowed at certain times of the year. But still further inland, where the country is more elevated, it was said to be just as well populated as those places that we had passed previously to-day.

Lac St. Pierre is a part of the St. Lawrence River which is so broad that we could hardly see anything but sky and water before us, and I was everywhere told that it is seven French miles long and three broad. From the middle of this lake, as it is called, you see a large high country in the west which appears above the woods. In the lake are many places covered with a kind of rush, or *Scirpus palustris* L. There are no houses in sight on either side of the lake, because the land is rather too low there, and in spring the water rises so high that they can go with boats between the trees. However, at some distance from the shores where the ground is higher, the farms are close together. We saw no islands in the lake this afternoon, but the next day we came upon some.

Late in the evening we left Lac St. Pierre and rowed up a little river called Rivière de Loup, in order to come to a house where we might pass the night. Having rowed about an English mile we found the country inhabited on both sides of the river. Its shores are high but the country in general is flat. We passed the night in a farmhouse. The territory of Montreal extends to this place, but here begins the jurisdiction of the governor of Trois Rivières, which they reckon eight miles from here.

August the 3rd

At five o'clock in the morning we set out again, and first rowed down the little river till we came into the Lac St. Pierre, down which we proceeded. After we had gone a good way, we perceived a high chain of mountains in the northwest, which were very much elevated above the low, flat country. The northwest shore of the lake was now in general very closely inhabited; but on the southeast side we saw no houses, and only a country covered with woods, which is sometimes said to be under water, but behind which there are, as I am told, a great number of farms. Towards the end of the lake, the river went into its proper bounds again, being not even a mile and a half broad, and afterwards it grew still narrower.

From the end of Lac St. Pierre to Trois Rivières, they reckon three French miles, and at about eleven o'clock in the morning we arrived at the latter place, where we attended divine service.

Trois Rivières is a little market town which had the appearance of a large village. It is, however, numbered among the three great towns of Canada, which are Quebec, Montreal, and Trois Rivières. It is said to lie in the middle between the two first, and is thirty French miles distant from each. The town is built on the north side of the St. Lawrence River on a flat, elevated sandbar and its location is very pleasant. On one side the river passes by, and it is here an English mile and a half broad. On the other side are fine grain fields, though the soil is very sandy. In the town are two churches of stone, a nunnery, and a house for the friars of the order of St. Francis. This town is likewise the seat of the third governor in Canada, whose house is also of stone. Most of the other houses are of timber, a single story high, tolerably well built, and stand very much apart. The streets are crooked. The shore here consists of sand, and the rising grounds along it are pretty high. When the wind is very violent here, it raises the sand, and blows it about the streets, making it very troublesome to walk in them. The nuns, who are about twenty in number, are very ingenious in all kinds of needlework. This town formerly flourished more than any other in Canada, for the Indians brought their goods to it from all sides; but since that time they have gone to Montreal and Quebec, and to the English, on account of their wars with the Iroquois, or Five Nations, and for several other reasons, so that this town is at present very much reduced by it. Its present inhabitants live chiefly by agriculture, though the neighboring ironworks may serve in some measure to support them. About an English mile below the town a great river flows into the St. Lawrence River, but first divides into three branches, so that it appears as if three rivers emptied themselves there. This has given occasion to call the river and this town, Trois Rivières (the Three Rivers).

The tide goes about a French mile above Trois Rivières, though it is so trifling as to be hardly observable. But about the equinoxes, and at the new moons and full moons in spring and autumn, the difference between the highest and lowest water is two feet. Accordingly the tide in this river goes very far up, for from the above-mentioned place to the sea they reckon about a hundred and fifty French miles.[1]

While my companions were resting I went on horseback to view the ironworks. The country which I passed through was pretty high, sandy, and generally flat. I saw neither stones nor mountains here.

The *ironworks* which is the only one in this country, lies three miles to the west of Trois Rivières. Here are two great forges, besides two lesser ones under the same roof. The bellows were made of wood, and everything else is as in the Swedish forges. The melting furnaces stand close to the forges, and are the same as ours. The ore is gotten two French miles and a half from the ironworks, and is carried thither on sledges in the winter. It is a kind of moor ore, which lies in veins within six inches or a foot from the surface of the ground. Each vein is from six to eighteen inches deep, and below it is a white sand. The veins are surrounded with this sand on both sides and covered at the top with a thin earth. The ore is pretty rich and lies in the veins in loose lumps the size of two fists, though there are a few which are nearly eighteen inches thick. These lumps are full of holes, which are filled with ochre. The ore is so soft that it may be crushed betwixt the fingers. They make use of a gray limestone which is quarried in the neighborhood for promoting the smelting of the ore. To that purpose they likewise employ a clay marle, which is found near this place. Charcoal is to be had in great abundance here, because all the country round this place is covered with woods, which have never been disturbed except by storms and old age. The charcoal from evergreen trees, that is, from the fir, is best for the forge, but that of deciduous trees is best for the smelting oven. The iron which is here made was to me described as soft, pliable, and tough, and is said to have the quality of not being attacked by rust as easily as other iron. In this point there appears a great difference between the Spanish iron and this, in shipbuilding. This smeltery was first founded in 1737, by private persons, who afterwards ceded it to the king. They cast cannon and mortars here of different sizes, iron stoves which are in use all over Canada, kettles, etc. not to mention the bars which are made here. They have likewise tried to make steel here,

[1] By the "sea" here is meant the Gulf of St. Lawrence, not the Atlantic Ocean.

(From Swedish original)

Type of Fence

but cannot bring it to any great perfection because they are unacquainted with the best manner of preparing it. Here are many officers and overseers who have very good houses, built on purpose for them. It is agreed on all sides that the revenues of the ironworks do not pay the expenses which the king must every year have for maintaining them. They lay the fault on the bad state of the population and say that the inhabitants in the country are few, and that these have enough to do in attending to their agriculture, and that it therefore costs large sums to get a sufficient number of workmen. But however plausible this may appear, yet it is surprising that the king should be a loser in carrying on this work, for the ore is easily broken, very near the furnaces, and it is very fusible. The iron is good, and can be very conveniently transported over the country. These are, moreover, the only ironworks in the country from which everybody must supply himself with iron tools and what other iron he wants. But the officers and workmen belonging to the smeltery appear to be in very affluent circumstances. A river runs down from the ironworks into the St. Lawrence River, by which all the iron can be sent in boats throughout the country at a low rate.—In the evening I returned again to Trois Rivières.

[At this point Kalm introduces a long description of a prevalent type of Canadian fence. It does not seem necessary to include the verbal details here, and Forster and all his imitating translators omit it entirely, but as a middle course, and to illustrate Kalm's curious interest and versatility (as well as his sense of completeness) we are reproducing the drawing of the fence from the original.]

August the 4th

At the dawn of day we left this place and went on towards Quebec. We found the land on the north side of the river somewhat elevated, sandy, and closely inhabited along the shore. The southeast shore, we were told, is equally well inhabited, but the woods along that shore prevented our seeing the houses which are built further back in the country, the land close to the river being so low as to be subject to annual inundations. Near Trois Rivières, the river grows somewhat narrow but it enlarges again as soon as you come a little below that place, and has the breadth of more than two English miles.

Prayers. The French in their travels generally read a *Kyrie eleison* every morning before they started off. Every time when I followed them on water they chose the *Kyrie eleison,* which is found in their prayer books for Saturday and is almost wholly directed to the Holy Virgin. Almost every word of it is in Latin and although women, common people, and in fact most of the higher classes in Canada hardly understood a word of it, the whole morning prayer and the benediction were in this language. If there were women in the company the foremost of them was elected to read this litany in a very loud voice and to enumerate all the titles of honor which in it are attributed to the Holy Virgin. But in the absence of women it was done by one of the most distinguished of the men. At the end of every sentence the others present would answer *Ora pro nobis.* When priests were present they conducted the service. The women knew this Latin litany perfectly, so that they didn't miss a word. Some of the titles of the Holy Virgin were *Mater divinae gratiae, Virgo potens, Virgo clemens, Rosa mystica, Domus aurea, Regina angelorum,* etc. At everyone of these expressions everyone exclaimed *Ora pro nobis.* It was both strange and amusing to see and hear how eagerly the women and soldiers said their prayers in Latin and did not themselves understand a word of what they said. When all the prayers were ended the soldiers cried *Vive le Roi!* and that is about all they understood of the prayer proceedings. I have noticed in the papal service that it is directed almost entirely toward the external; the heart representing the internal is seldom touched. It all seems to be a ceremony. In the meantime the people are very faithful in these observances, because everyone tries by these means to put God under some obligation and intends by it to make himself deserving of some reward.

As we went on we saw several churches of stone, and often very well built ones. The shores of the river are closely inhabited for about three quarters of an English mile back in the country, but beyond that the woods and the wilderness increase. All the rivulets joining the St. Lawrence River are likewise well inhabited on both sides. I observed throughout Canada that the cultivated lands lie only along the St. Lawrence River and near the other rivers in the country, the environs of towns excepted, round which the country is all cultivated and inhabited within the distance of twelve or eighteen English miles. The great islands in the river are likewise inhabited.

The shores of the river now became higher, more oblique and steep; however, they consisted chiefly of earth. Here and there some rivers or large brooks flow into the St. Lawrence River, among which one of the largest is the Rivière Puante, which unites on the southeast side with the St. Lawrence, about two French miles below Trois Rivières, and has on its banks, a little way from its mouth, a town called Becancourt which is wholly inhabited by Abenakee Indians who have been converted to the Roman Catholic religion, and have Jesuits among them. At a great distance, on the northwest side of the river we saw a chain of very high mountains, running from north to south, elevated above the rest of the country, which is quite flat here without any remarkable hills.

Here were several limekilns along the river; and the limestone employed in them is quarried in the neighboring hills. It is compact and gray, and the lime it yields is quite white.

The fields here are generally sown with wheat [in this country they ate only wheat bread] oats, corn and peas. Pumpkins and watermelons are planted in abundance near the farms.

A humming bird (*Trochilus colubris*) flew among the bushes in a place where we landed to-day. The French call it *Oiseau mouche,* and say it is pretty common in Canada; and I have seen it since several times at Quebec.

About five o'clock in the afternoon we were obliged to take our night's lodgings on shore, the wind blowing very strong against us and being attended with rain. I found that the nearer we came to Quebec, the more open and free from woods was the country. The place where we passed the night is twelve French miles from Quebec.

Fish Traps. They have a very peculiar method of catching fish near the shore here. They place hedges along the shore, made of twisted oziers, so close that no fish can get through them, and from one foot to a yard high, according to the different depth of the water. For this purpose they choose places where the water runs off during the ebb, and leaves the hedges quite dry. Within this inclosure they place several weels, or wickerwork fish traps, in the form of cylinders, but broader at the base. They are placed upright, and are about a yard high and two feet and a half wide: on one side near the bottom is an entrance for the fishes, made of twigs, and sometimes of yarn made into a net. Opposite to this entrance, on the other side of the weel, facing the lower part of the river, is another entrance, like the first, and leading to a box of boards about four feet long, two deep and two broad. Near each of the weels is a hedge, leading obliquely to the long hedge, and making an acute angle with it. This hedge is made in order to lead the fish directly into the trap, and it is placed on that end of the long hedge which points towards the upper part of the river. When the tide comes up the river, the fish, and chiefly the eels, go up with it along the river side; when the water begins to ebb, the fish likewise go down the river, and meeting with the hedges, they swim along them, till they come through the weels into the boxes of boards [or eelpots], at the top of which there is a hole with a cover, through which the fish can be taken out. This apparatus is made chiefly for catching eels. In some places hereabouts they place nets instead of the hedges of twigs.

The shores of the river consisted no longer of pure loam; but of a species of slate. They are very steep and nearly perpendicular here, and the slate of which they consist is black, with a brown cast. The slate is divisible into thin shivers, no thicker than the blade of a knife. This slate moulders as soon as it is exposed to the open air, and the shore is covered with fine grains of sand which are nothing but particles of such mouldered slate. Some of the strata run horizontally, others obliquely, dipping to the south and rising to the north, and sometimes the contrary way. Sometimes they form bendings like large semicircles: sometimes a perpendicular line cuts off the strata to the depth of two feet, and the slates on both sides of the line form a perpendicular and smooth wall. In some places hereabouts, they find amongst the slate a stratum about four inches thick of a gray, compact but pretty soft limestone, of which the Indians for many centuries have made tobacco pipes and the French still make them.[1]

[1] This limestone seems to be a marle, or rather a kind of stone-marle for there is a whitish kind of it in the Krim Tartary, and near Stiva or Thebes, in Greece, which

AUGUST THE 5TH

Approaching Quebec. This morning we continued our journey rowing against the wind. The appearance of the shores was the same as yesterday; they were high, pretty steep or perpendicular, and consisted of the black slate before described. The country at the top was a plain without eminences, and was closely inhabited along the river for about the space of an English mile and a half inland. There are no islands in this part of the river, but several stone places, perceptible at low water only, which have several times proved fatal to travellers. The breadth of the river varies; in some parts it was a little more than three quarters of a mile, in others half a mile, and in some over two miles. The inhabitants made use of the same method of catching eels along the shores here as that which I described yesterday. In many places they make use of nets instead of the wicker traps.

Insects. Bedbugs (*Cimex lectularius*) abound in Canada; and I met with them in every place where I lodged, both in the towns and country, and the people know of no other remedy for them than patience.

The crickets (*Gryllus domesticus*) are also abundant in Canada, especially in the country where these disagreeable guests lodge in the chimneys; nor are they uncommon in the towns. They stay here both summer and winter, and frequently cut clothes in pieces for a pastime.

The cockroaches (*Blatta orientalis*) have never been found in the houses here.

Landing at Quebec. The shores of the river grow more sloping as you come nearer to Quebec. To the northward appears a high ridge of mountains. About two French miles and a half from Quebec the river becomes very narrow, the shores being within the reach of a musketshot from each other. The country on both sides was sloping, hilly, covered with trees, and had many small rocks. The shore was stony. About four o'clock in the afternoon we happily arrived at Quebec. The city does not appear till one is close to

---

is employed by the Turks and Tartars for making heads of pipes, and that from the first place is called Keffekil, and in the latter, Sea-Scum; it may be very easily cut, but grows harder in time.—F.

it, the view being intercepted by a high mountain on the south side. However, a part of the fortifications appears at a good distance, being situated on the same mountain. As soon as the soldiers, who were with us, saw Quebec, they called out that all those who had never been there before should be ducked, if they did not pay something to release themselves. This custom even the governor-general of Canada was obliged to submit to on his first journey to Quebec. We did not care when we came in sight of this town to be exempted from this old custom, which is very advantageous to the rowers, as it enables them to spend a merry evening on their arrival at Quebec after their troublesome labor.

Immediately after my arrival, the officer who had accompanied me from Montreal led me to the palace of the then vice-governor-general of Canada, the Marquis de la Galissonnière, a nobleman of uncommon qualities, who behaved towards me with extraordinary goodness during the time he stayed in this country.[1] He had already ordered some apartments to be got ready for me, and took care to provide me with everything I wanted, besides honoring me so far as to invite me to his table almost every day I was in town.

August the 6th

Quebec, the chief city in Canada, lies on the western shore of the St. Lawrence River, close to the water's edge on a neck of land bounded by that river on the east side, and by the St. Charles River on the north. The mountain, on which the town is built rises still higher on the south side and behind it begin great pastures. The same mountain also extends a good way westward. The city is divided into the lower and the upper section.[2] The lower lies on the river east of the upper. The neck of land I mentioned before, was formed by the dirt and filth, which had from time to time been accumulated there, and by a cliff which projects out at that point, not by any gradual diminution of the water. The upper city lies above the other on a high hill and takes up five or six times the space of the lower, though it is not quite so populous. The mountain on which the upper city is located reached above the houses of the lower city.

[1] The Marquis returned to France on September 24, 1749.

[2] *La haute Ville et la basse Ville.*

Notwithstanding, the latter are three or four stories high, and the view from the palace of the lower city (part of which is immediately under it) is amazing. There is only one easy way of getting to the upper city, and that is where a part of the mountain has been blown away. This road is very steep, although it is serpentine. However, people go up and down it in carriages and with wagons. All the other roads up the mountain are so steep that it is very difficult to climb to the top on them. Most of the merchants live in the lower city, where the houses are built very close together. The streets in it are narrow, very rough, and almost always wet. There is likewise a church and a small marketplace. The upper city is inhabited by people of quality, by several persons belonging to the different offices, by tradesmen and others. In this part are the chief buildings of the town, among which the following are worthy of particular notice.

I. The *Palace* is situated on the south or steepest side of the mountain, just above the lower city. It is not properly a palace but a large building of stone, two stories high, extending north and south. On the west side of it is a courtyard, surrounded partly by a wall and partly by houses. On the east side, or towards the river, is a gallery as long as the whole building, and about two fathoms broad, paved with smooth flags and protected on the outside by iron railings from which the city and the river exhibit a charming view. This gallery serves as a very agreeable walk after dinner, and those who come to speak with the governor-general wait here till he is at leisure. The palace is the lodging of the governor-general of Canada, and a number of soldiers stand guard before it, both at the gate and in the courtyard. When the governor or the bishop comes in or goes out they must present arms and beat the drum. The governor-general has his own chapel where he offers prayers. However, he often goes to mass at the church of the Recollets,[1] which is very near the palace.

II. The *churches* in this town are seven or eight in number and are all built of stone.

1. The *Cathedral* is on the right hand, coming from the lower to the upper city, somewhat beyond the bishop's house. The people were at present employed in ornamenting it. The organ had just

[1] A kind of Franciscan friars, called *Ordo St. Francisci strictioris observantiæ*.

been removed because of the improvements being made on it. On its west side was a round steeple of two stories, the lower one of which contained bells. The pulpit and some other parts within the church were of gilt. The seats were excellent.

2. The Jesuits' Church is built in the form of a cross and has a round steeple. This is the only temple that has a clock, and I shall mention it more particularly below.

3. The Recollets Church, or the Temple of the Barefooted Friars, is opposite the gate of the palace on the west side, looks well, and has a pretty high pointed steeple, with a compartment below for the bells.

4. The Church of the Ursulines has a round spire.

5. The Church of the Hospital.

6. The Bishop's Chapel.

7. The church in the lower city was built in 1690, after the town had been delivered from the English, and is called *Notre Dame de la Victoire*. It has a small steeple in the middle of the roof, square at the bottom and round at the top.

8. The little Chapel of the Governor-General may likewise be ranked amongst these churches.

III. The Bishop's House is the first on the right hand coming from the lower to the upper town. It is a fine large building, surrounded by an extensive courtyard and kitchen garden on one side, and by a wall on the other.

IV. The College of the Jesuits, which I shall describe more particularly, has a much more noble appearance in regard to its size and architecture than the Palace itself, and would be proper for a palace if it had a more advantageous location. It is about four times as large as the Palace, and is the finest building in town. It stands on the north side of a market, on the south side of which is the Cathedral.

V. The House of the Recollets lies to the west, near the Palace and directly over against it, and consists of a spacious building with a large orchard and kitchen garden. The house is two stories high. In each story is a narrow gallery with rooms and halls on one or both sides.

VI. The Hotel de Dieu, where the sick are taken care of, will be described more minutely later. The nuns that serve the sick are of the Augustine order.

VII. *Le Seminaire* or the house of the clergy is a large building on the northeast side of the cathedral. Here is on one side a spacious court, and on the other, towards the river, a large orchard with a kitchen garden. Of all the buildings in the town none has so fine a view as that in the garden belonging to this house, which lies on the high shore and commands a good distance down the river. The Jesuits on the other hand have the worst, and hardly any view at all from their college; nor have the Recollets any fine views from their home. In this building all the clergy of Quebec lodge with their superior. They have large pieces of land in several parts of Canada presented to them by the government, from which they derive a bountiful income so that they can live exceedingly well.

VIII. The Convent of the Ursuline Nuns will be described later.

These are the chief public buildings in town, but to the northwest, just before the town is:

IX. The house of the intendant, a public building whose size makes it fit for a palace. It is covered with tin, and stands in a second lower town, situated southward upon the St. Charles River. It has a large and fine garden on its north side. In this house all the deliberations concerning this province are held, and the gentlemen who have the management of the police and the civil power meet here and the intendant (mayor) generally presides. In affairs of great consequence the governor-general is also present. On one side of this building is the storehouse of the crown, and on the other the prison.

Most of the houses in Quebec are built of stone, and in the upper city they are generally but one story high, the public buildings excepted. I saw a few wooden houses in the town, but these may not be rebuilt when decayed. The houses and churches in the city are not built of bricks, but of the black "lime-slate" or [calcareous schist] of which the mountain consists and whereon Quebec stands. When these strata of slate are quarried deep in the mountain, they look very compact at first, and appear to have no fragments or *lamellæ* at all; but after being exposed a while to the air they separate into thin leaves. This slate is soft and easily cut, and the city walls as well as the garden ones consist chiefly of this material. The roofs of the public buildings are covered with common slate, which is brought from France because there is none in Canada.[1]

The *slated roofs* have for some years withstood the changes of air and weather without suffering any damage. The private houses have roofs of boards which are laid parallel to the spars, and sometimes to the eaves, or sometimes obliquely. The corners of houses are made of a gray small-grained limestone, which has a strong smell like stink-stone, and the windows are generally enchased with it. This limestone is more useful in those places than the lime-slate which always shivers in the air. The outsides of the houses are generally whitewashed. The windows are placed on the inner side of the walls; for they sometimes have double windows in winter. The middle roof has two, or at most three spars, covered with boards only. The rooms are warmed in winter by small iron stoves, which are removed in summer. There were no dampers anywhere. The floors are very dirty in every house and have all the appearance of being cleaned but once every year.

The Powder Magazine stands on the summit of the mountain on which the city is built, and south of the palace.

The streets in the upper city have a sufficient breadth, but are very rough on account of the rock on which it lies, and this renders them very disagreeable and troublesome, both to foot passengers and carriages. The black lime slabs basset out everywhere into sharp angles, which cut the shoes into pieces. The streets cross each other at all angles and are very crooked.

The many great orchards and kitchen gardens near the house of the Jesuits and other public and private buildings make the town appear very large, though the number of houses it contains is not very great. Its extént from south to north is said to be about six hundred toises,[2] and from the shore of the river along the lower town to the western wall between three hundred and fifty, and four hundred toises. It must be here observed that this space is not yet wholly inhabited, for on the west and south side, along the town

[1] In his French version of Kalm's *Travels,* Marchand points out that common slate is now found in several places in Canada, especially in the eastern sections. See vol. 2, p. 78, note.

[2] Fathoms.

walls, are large pieces of land without any buildings on them, and destined to be built upon in future time when the number of inhabitants will have increased in Quebec.

The bishop whose see is in the city is the only bishop in Canada. His diocese extends to Louisiana on the Mexican Gulf in the south and to the South Sea on the west.

No bishop, the pope excepted, ever had a more extensive diocese. But his spiritual flock is very small some distance from Quebec, and his sheep are often many hundred miles distant from each other.

Quebec as a Seaport. Quebec is the only seaport and trading town in all Canada, and from there all the produce of the country is exported. The port is below the town on the river, which is there about a quarter of a French mile broad, twenty-five fathoms deep, and its bottom is very good for anchoring. The ships are secured from all storms in this port; however, the northeast wind is the worst, because the town is more exposed in a storm from this direction. When I arrived here I counted thirteen large and small vessels, but in the evening before I left Quebec I counted twenty-three, and they expected more to come in. But it is to be remarked that no other ships than French ones can come into the port, though they may come from any place in France, or even from the French possessions in the West Indies. All foreign goods which are found in Montreal and other parts of Canada must first come from here. Similarly the French merchants from Montreal, after having spent six months among various Indian nations in order to purchase skins of beasts and furs, return about the end of August and go down to Quebec in September or October to sell their goods. The privilege of selling the imported goods should have vastly enriched the merchants of Quebec; but this is contradicted by others, who allow that there are a few in affluent circumstances, but that the majority possess no more than is absolutely necessary for their bare subsistence, and that several are very much in debt, which they say is owing to their luxury and vanity and to the fact that no one wanted to be poorer than the other. The merchants dress very finely, are extravagant in their repasts, and their ladies are every day in full dress and as much adorned as if they were to go to court.

The town is surrounded on almost all sides by a high wall, and especially towards the land. It was not quite completed when I was there and they were very busy finishing it. It is built of the above lamellated black limestone and of dark gray sandstone. For the corners of the gates they have employed gray limestone. They have not made any walls towards the water side, but nature seems to have worked for them by placing a rock there which it is impossible to ascend. All the rising land thereabouts is likewise so well planted with cannon that it seems impossible for an enemy's ships or boats to come to the town without running into imminent danger of being sunk. On the land side the town is likewise guarded by high mountains so that nature and art have combined to fortify it.

History of Quebec. Quebec was founded by its former governor, Samuel de Champlain, in the year 1608. We are informed by history that its rise was very slow. In 1629 towards the end of July it was taken by two Englishmen Lewis and Thomas Kirke[1] by capitulation and surrendered to them by the above-mentioned Champlain. At that time, Canada and Quebec were wholly destitute of provisions, so that they looked upon the English more as their deliverers than their enemies. The above-mentioned Kirkes were the brothers of the English admiral David Kirke, who lay with his fleet somewhat lower in the river. In the year 1632 the French got the town of Quebec and all Canada returned to them by the treaty of St. Germain-en-Laye. It is remarkable that the French were doubtful whether they should reclaim Canada from the English or leave it to them. The greater part were of the opinion that to keep it would be of no advantage to France, because the country was so cold, its expenses far exceeded the income, and France could not people so extensive a country without weakening herself, as Spain had done before; furthermore, that it was better to keep the people in France and employ them in all sorts of manufactures, which would oblige the other European powers who have colonies in America to bring their raw materials to French ports and take French manufactured goods in return. Those on the other hand who had more foresighted views knew that the climate was not so unfavorable as it had been represented. They likewise believed that that which caused the expenses was a fault of the Company, because it did not manage the country well. They would not have many people sent over at once,

[1] The capture of Quebec is generally credited to Sir David Kirke (1597-1656), the brother of Lewis and Thomas. See Kalm's account below.

but a few at a time, so that France might not feel it. They hoped that this colony would in future times make France powerful, for its inhabitants would become more and more acquainted with the herring, whale and cod fisheries, and likewise with the capturing of seals, and that by this means Canada would become a school for training seamen. They further mentioned the several sorts of furs, the conversion of the Indians, the shipbuilding, and the various uses of the extensive woods; and lastly, that it would be a considerable advantage to France even though it should reap no other benefit than to hinder by this means the progress of the English in America and of their increasing power, which would otherwise become insupportable to France, not to mention several other reasons. Time has shown that these reasons were the result of mature judgment and that they laid the foundation for the rise of France. It were to be wished that we had been of the same opinion in Sweden at a time when we were actually in possession of New Sweden, the finest and best province in all North America, or when we were yet in a condition to get possession of it. Wisdom and foresight do not only look upon the present times but also look into the future.

In the year 1663, at the beginning of February, the great earthquake was felt in Quebec and a great part of Canada, and there are still some vestiges of its effects at that time; however, no lives were lost.

On the 16th of October, 1690, Quebec was besieged by the English general, William Phipps,[1] who was obliged to retire in disgrace a few days afterward with great loss. The English have tried several times to repair their losses, but the St. Lawrence River has always been a very good defense for this country. An enemy, and one that is not acquainted with this river, cannot ascend without being ruined; for in the neighborhood of Quebec it abounds with hidden rocks, and has strong currents in some places so that the channel follows an extremely serpentine course. [In the last war, however, the Englishmen (1759) became masters of Quebec and a part of Canada.]

The name of Quebec is said to be derived from a Norman word, on account of its situation on a neck or point of land. For when one

[1] Sir William Phipps (1651-1695), later became governor of Massachusetts. He was in 1694 summoned to England to answer charges of inaction and failure in his campaigns against the French and Indians, but died before proceedings were undertaken.

comes up the river by l'Isle d'Orleans, that part of the St. Lawrence River which lies above the town does not come in sight and it appears as if the St. Charles River which lies just before were a continuation of the St. Lawrence. But on advancing further the true course of the river comes within sight, and has at first a great similarity to the mouth of a river or a great bay. This has given occasion for a sailor, who saw it unexpectedly, to cry out in his provincial dialect *Qué bec,*[1] that is, "what a point of land!" and from this it is thought the city obtained its name. Others derive it from the Algonquin word *Quebégo* or *Québec* signifying that which grows narrow because the river becomes narrower as it comes nearer to the town. The pronunciation of "Quebec" by the French was "Kebäk" almost without any accent; they gave each syllable a quantitative value. The word Canada was pronounced both by the French and English with the accent on the first syllable.

The St. Lawrence River is exactly a quarter of a French mile, or about three quarters of an English mile broad at Quebec. The salt water never comes up to the town and therefore the inhabitants can make use of the water in the river for their kitchens, drinking purposes, etc. All accounts agree that notwithstanding the breadth of this river and the violence of its course, especially during ebb, it is covered with ice during the whole winter, which is strong enough for walking, and a carriage may go over it. It is said to happen frequently that when the river has been open in May there are such cold nights in this month that it freezes again and will bear walking over. This is a clear proof of the intenseness of the cold here, especially when one considers what I shall mention immediately about the ebbing and flowing of the tide in this river. The greatest breadth of the river at its mouth is computed to be twenty-six French miles [or about seventy-five English miles], though the boundary between the sea and the river cannot well be ascertained, as the latter gradually looses itself in and unites with the former. The greatest part of the water contained in the numerous lakes of Canada, four or five of which are like large seas, is forced to empty into the ocean by means of this river alone. The navigation up this river from the sea is rendered very dangerous by the strength of the current, and by the number of sandbars which often arise in places where they never

[1] Meaning *Quel bec.*—F.

were before. The English encountered this formation of new sandbars once or twice when they intended to conquer Canada. Hence the French have good reasons to look upon the river as a barrier to Canada.[1]

The tide goes far beyond Quebec in the St. Lawrence River, as I have mentioned above. The difference between high and low water is generally between fifteen and sixteen feet, French measure; but with the new and full moon, and when the wind is likewise favorable, the difference is seventeen or eighteen feet, which is indeed considerable.

August the 7th

Ginseng is the current French name in Canada of a plant, the root of which has a very great value in China.[2] It has been growing since time immemorial in the Chinese Tartary and in Korea, where it is annually collected and brought to China. Father Du Halde[3] says it is the most precious and the most useful of all the plants in eastern Tartary, and attracts every year a number of people into the deserts of that country. The Mantchou-Tartars call it Orhota, that is, the most noble or the queen of plants.[4] The Tartars and Chinese praise it very much, and ascribe to it the power of curing several dangerous diseases and that of restoring to the body new strength and supplying the loss caused by the exertion of the mental and physical faculties. An ounce of ginseng brings the surprising price of seven or eight ounces of silver at Peking. When the French botanists in Canada first saw a picture of it, they remembered to have seen a similar plant in this country. They were confirmed in their conjecture by considering that several settlements in Canada lie in the same

[1] The river St. Lawrence, was no more a barrier to the victorious British fleets in the last war, nor were the fortifications of Quebec capable to withstand the gallant attacks of their land army, which disappointed the good Frenchmen in Canada of their too sanguine expectations and at present they are rather happy at this change of fortune, which has made them subjects of the British sceptre, whose mild influence they at present enjoy.—F.

[2] Botanists know this plant by the descriptive name of *Panex quinque folium, foliis ternatis quinatis* L.

[3] Jean-Baptiste Du Halde (1647-1743, French geographer. In 1735 he published in four volumes his *Description Géographique, historique, chronologique, politique et physique de l'Empire de la Chine et de la Tartarie chinoise.*

[4] Peter Osbeck's *Voyage to China,* Vol. I, p. 223.

latitude as those parts of the Chinese Tartary, and China, where the true ginseng grows wild. They succeeded in their attempt and found the same plant wild and abundant in several parts of North America, both in the French and English plantations, in level parts of the woods. It is fond of shade, and of a deep rich earth, and of land which is neither wet nor high. It is not common everywhere, for sometimes one may search the woods for the space of several miles without finding a single plant of it, but in those spots where it grows it is always found in great abundance. It flowers in May and June and its berries are ripe at the end of August. It bears transplanting very well, and will soon thrive in its new ground. Some people here who have gathered the berries and put them into their kitchen gardens told me that they lay one or two years in the ground without coming up. The Iroquois call the ginseng roots *Garangtoging* which it is said signifies a child, the roots bearing a faint resemblance to one: but others are of the opinion that they mean the thigh and leg by it, and the roots look very much like that. The French use this root for curing asthma, as a stomachic, and promoting fertility in women. The trade which is carried on with it here is very brisk, for they gather great quantities of it and send them to France, whence they are brought to China and sold there to great advantage.[1] It is said that the merchants in France met with amazing success in this trade at the first outset, but by continuing to send the ginseng over to China its price has fallen considerably there and consequently in France and Canada; however, they still find some profit in it. In the summer of 1748 a pound of ginseng was sold for six francs or livres at Quebec; but its common price here is one hundred sols or five livres. During my stay in Canada all the merchants at Quebec and Montreal received orders from their correspondents in France to send over a quantity of ginseng, there being an uncommon demand for it this summer. The roots were accordingly collected in Canada with all possible haste.[2] The Indians especially travelled about the

[1] Mr. Osbeck seems to doubt whether the Europeans reap any advantages from the ginseng trade or not, because the Chinese do not value the Canada roots so much as those of the Chinese Tartary and therefore the former bear scarce half the price of the latter. See Osbeck's *Voyage to China*, Vol. I, p. 223.—F.

[2] Marchand in the previously quoted *Voyage de Kalm en Amérique,* Vol. 2, p. 88, note 2, states that the ginseng trade in Canada was ruined by the merchants who, in order to hasten their profits, employed artificial heat to dry the plants instead

country in order to collect as much as they could and to sell it to the merchants at Montreal. The Indians in the neighborhood of this town were likewise so much taken up with this business that the French farmers were not able during that time to hire a single Indian, as they commonly do to help them in the harvest. Many people feared lest by continuing for several successive years to collect these plants without leaving one or two in each place to propagate their species, there would soon be very few of them left, which I think is very likely to happen, for by all accounts they formerly grew in abundance round Montreal, but at present there is not a single plant of it to be found, so effectually have they been rooted out. This obliged the Indians this summer to go far within the English boundaries to collect these roots. From the merchants in Montreal one received 40 francs a minot (39 liters) of these fresh roots. After the Indians have sold the fresh product to the merchants, the latter must take a great deal of pains with them. They are spread on the floor to dry, which commonly requires two months or more, according as the season is wet or dry. During that time they must be turned once or twice every day, lest they should spoil or moulder. Ginseng has never been found far north of Montreal. The father superior of the clergy here and several other people assured me that the Chinese value the Canada ginseng as much as the Tartarian [1] and that no one ever had been entirely acquainted with the Chinese method of preparing it. However, it is thought that amongst other preparations they dip the roots in a decoction of the leaves of ginseng. The roots prepared by the Chinese are almost transparent and look like horn inside, and the roots which are fit for use must be heavy, solid or compact inside.

"Maiden Hair". The plants which throughout Canada bear the name of *Herba capillaris* is likewise one of those with which a great trade is carried on in Canada. The English in their plantations call it "maiden hair"; it grows in all their North American colonies, which I travelled through and likewise in the southern parts of Canada; but I never found it near Quebec. It grows in the woods

---

of following the slow and natural method of dessication; that is, failure was due to greed and the method of preparation rather than to any deficiency in the quality of the Canadian plant itself.

[1] This is directly opposite to Mr. Osbeck's assertion. See the preceding page, note 1.—F.

in shady places and in a good soil.[1] Several people in Albany and Canada assured me that its leaves were very much used instead of tea, in consumption, cough, and all kinds of pectoral diseases. This they have learnt from the Indians who have made use of the plant for these purposes since ancient times. This American maiden hair is reckoned preferable in surgery to that which we have in Europe [2] and therefore they send a great quantity of it to France every year. The price varies and is regulated according to the grade of the plant, the care in preparing it, and the quantity which is to be gotten. For if it is brought to Quebec in great abundance, the price falls, and on the contrary it rises when the quantity gathered is but small. Usually the price at Quebec is between five and fifteen sols a pound. The Indians went into the woods about this time and travelled far above Montreal in quest of this plant.

Kitchen herbs succeed very well here. White cabbage is very fine but sometimes suffers greatly from worms. Onions (*Allium cepa*) are very much in use here together with other species of leeks. They likewise plant several species of pumpkins, melons, lettuce, wild chiccory or wild endive (*Cichorium intybus*), several kinds of peas, beans, Turkish beans, carrots, and cucumbers. They have plenty of red beets, horseradish, and common radishes, thyme, and marjoram. Turnips are sown in abundance and used chiefly in winter. Parsnips are sometimes eaten, though not very commonly. Few people took notice of potatoes, and neither the common (*Solanum tuberosum*) nor the Bermuda ones (*Convolvulus batatas*) were planted in Canada; only a few had any artichokes. When the French here are asked why they do not plant potatoes, they answer that they do not like them and they laugh at the English who are so fond of them. Throughout all North America the root cabbage [3] (*Brassica gongylodes* L.) is unknown to the Swedes, English, Dutch, Irish, Germans, and French. Those who have been employed in sowing and planting kitchen herbs in Canada and have had some experience in

---

[1] It is the *Adiantum pedatum* of L. *sp. pl.* p. 1557. Cornutus, in his *Canadens. plant. historia*, p. 7, calls it *Adiantum Americanum*, and gives together with the description, a figure of it, p. 6.—F.

[2] *Adiantum Capillus Veneris*. True Maiden Hair.

[3] This is a kind of cabbage with large round eatable roots, which grow out above the ground, wherein it differs from the turnip cabbage (*Brassica Napobrassica*) whose roots grow in the ground. Both are common in Germany, and the former likewise in Italy.

gardening told me that they were obliged to send for fresh seeds from France every year, because they commonly loose their strength here in the third generation and do not produce such plants as would equal the original ones in quality.

Antiquities. The Europeans have never been able to find any alphabetical characters, much less writings or books among the Indians, who have inhabited North America since time immemorial, and comprise several nations and dialects. These Indians have therefore lived in the greatest ignorance and darkness for several centuries; they are totally unacquainted with the state of their country before the arrival of the Europeans, and all their knowledge of it consists of vague traditions and mere fables. It is not certain whether any other nations possessed America before the present Indian inhabitants came into it, or whether any other nations visited this part of the globe, before Columbus discovered it.[1] It is equally unknown whether the Christian religion was ever preached here in former times. I conversed with several Jesuits who undertook long journeys in this extensive country and asked them whether they had met with any marks that there had formerly been any Christians among the Indians who lived here, but they all answered they had not found any. The Indians have always been as ignorant of architecture and manual labor as of science and writing. In vain does one seek for well-built towns and houses, artificially built fortifications, high towers and pillars, and such like among them, which the Old World can show from the most ancient times. Their dwelling places are wretched huts of bark, exposed on all sides to wind and rain. All their masonry work consists in placing a few gray stones on the ground round their fireplace to prevent the firebrands from spreading too far in their hut, or rather to mark out the space intended for the fireplace in it. Travelers do not enjoy a tenth part of the pleasure in traversing these countries which they must receive on their journeys through our old countries where they, almost every day, meet with some vestige or other of antiquity. Now an ancient celebrated town presents itself to view; here the remains of an old castle; there a field where many centuries ago the most powerful and the most

[1] Kalm here seems to have forgotten, temporarily, the well-known fact, which he mentions elsewhere, that the Norsemen visited North America "long before Columbus's time" (see pp. 202 and 443). Since Kalm wrote his *Travels,* a vast amount of literature has appeared on this subject, and it seems unnecessary to elaborate on the topic here.

skilful generals and the greatest kings fought a bloody battle; now the native spot and residence of some great or learned man. In such places the mind is delighted in various ways, and represents all past occurrences in living color to itself. We can enjoy none of these pleasures in America. The history of the country can be traced no further than from the arrival of the Europeans, for everything that happened before that period, is more like fiction or a dream than anything that really happened. In later times there have, however, been found a few marks of antiquity, from which it may be conjectured that North America was formerly inhabited by a nation more versed in science and more civilized than that which the Europeans found on their arrival here; or that a great military expedition was undertaken to this continent from the known parts of the world.

This is confirmed by an account which I received from Mr. de Verandrier (or de Vérendrye)[1] who commanded the expedition to the "South Sea" (Pacific Ocean) in person, of which I shall presently give an account. I have heard it repeated by others who were eyewitnesses of everything that happened on that occasion. Some years before I came to Canada, the then governor-general, Chevalier de Beauharnois,[2] gave Mr. de Vérendrye an order to go from Canada, with a number of people, on an expedition across North America to the South Sea in order to examine how far those two places were distant from each other and to find out what advantages might accrue to Canada or Louisiana, from a communication with that ocean. They set out on horseback from Montreal and went as far west as they could on account of the lakes, rivers and mountains which fell in their way. As they came far into the country beyond many nations they sometimes met with large tracts of land, free from wood, but covered with a kind of very tall grass for the space of some days' journey. Many of these fields were everywhere covered

[1] Probably Pierre Gaultier de Varenne, Sieur de La Vérendrye (1685-1749), noted Canadian pioneer and explorer. A condensed account of his expedition to the West is given in the *Dictionnaire Générale du Canada* under his name. He died soon after this conversation took place. He had several sons, some of whom accompanied him on his expeditions. One of these, Louis Joseph de La V. (1717-1761), is undoubtedly the one to whom Marchand refers in his *op. cit.*, 2, 94, note 1, as Lieut. de La Verandrière.

[2] Charles de la Boische, Marquis de Beauharnois (1670-1749), was appointed the fifteenth governor of New France in 1726 and was recalled in 1747. See *Dictionnaire Générale du Canada.*

with furrows, as if they had once been plowed and sown. It is to be noticed that the nations which now inhabit North America could not cultivate the land in this manner, because they never made use of horses, oxen, plows, or any instruments of husbandry, nor had they ever seen a plow before the Europeans came to them. In two or three places, at a considerable distance from each other, our travellers met with impressions of the feet of grown people and children in a rock but this seems to have been no more than a *Lusus Naturæ.* When they came to the west where, to the best of their knowledge, no Frenchmen, or European, had ever been, they found in one place in the woods, and again on a large plain, great pillars of stone, leaning upon each other. The pillars consisted of one single stone each, and the Frenchmen could not but suppose that they had been erected by human hands. Sometimes they have found such stones laid upon one another, and, as it were, formed into a wall. In some of those places where they found such stones, they could not find any other sort. They have not been able to discover any characters or writing upon any of these, though they have made a very careful search for them. At last they met with a large stone, like a pillar, and in it a smaller stone was fixed, which was covered on both sides with unknown characters. This stone, which was about a foot of French measure in length, and between four or five inches broad, they broke loose, and carried to Canada with them, whence it was sent to France to the secretary of state, the Count of Maurepas. What became of it afterwards is unknown to them, but they think it is yet preserved in his collection. Several of the Jesuits, who had seen and handled this stone in Canada, unanimously affirm that the letters on it are the same as those in books containing accounts of Tataria, are called Tatarian characters,[1] and that on com-

[1] This account seems to be highly probable, for we find in Marco Polo that Kublai-Khan, one of the successors of Gengbizkhan, after the conquest of the southern part of China, sent ships out to conquer the kingdom of Japan, or, as they call it, Nipangri, but in a terrible storm the whole fleet was cast away, and nothing was ever heard of the men of the fleet. It seems that some of these ships were cast to the shores, opposite the great American lakes, between forty and fifty degrees north latitude, and there probably erected these monuments, and were the ancestors of some nations, who are called Mozemlecks, and have some degree of civilization. Another part of this fleet it seems reached the country opposite Mexico, and there founded the Mexican empire, which, according to their own records, as preserved by the Spaniards, and in their painted annals, in Purchas's *Pilgrimage*, are very recent; so that they can scarcely remember any more than seven princes before Motezuma

paring both together they found them just alike. Notwithstanding the questions which the French on the South Sea expedition asked the people there concerning the time when and by whom those pillars were erected, what their traditions and sentiments concerning them were, who had written the characters, what was meant by them, what kind of letters they were, in what language they were written, and other circumstances; yet they could never get the least explanation, the Indians being as ignorant of all those things as the French themselves. All they could say was that these stones had been in those places since ancient times. The places where the pillars stood were near nine hundred French miles westward of Montreal. The chief intention of this journey, *viz.* to come to the South Sea, and to examine its distance from Canada, was never attained on this occasion. For the people sent out for that purpose were induced to take part in a war between some of the most distant Indian nations, in which some of the French were taken prisoners, and the rest obliged to return. Among the last and most westerly Indians they were with, they heard that the South Sea was but a few days journey off; that they (the Indians) often traded with the Spaniards on that coast and sometimes also they went to Hudson

---

II. who was reigning when the Spaniards arrived there, 1519, under Fernanado Cortez; consequently the first of these princes, supposing each had a reign of thirty-three years and four months, and adding to it the sixteen years of Motezuma, began to reign in the year 1270, when Kublai Khan, the conqueror of all China and of Japan, was on the throne, and in whose time happened, I believe, the first abortive expedition to Japan, which I mentioned above, and probably furnished North America with civilized inhabitants. There is, if I am not mistaken, a great similarity between the figures of the Mexican idols, and those which are usual among the Tartars, who embrace the doctrines and religion of the Dalaï-Lama, whose religion Kublai-Khan first introduced among the Monguls, or Moguls. The savage Indians of North America, it seems, have another origin, and are probably descended from the Yukagbiri and Tchucktchi, inhabitants of the most easterly and northerly part of Asia, where, according to the accounts of the Russians, there is but a small traject to America. The ferocity of these nations, similar to that of the American, their way of painting, their fondness of inebriating liquors, (which the Yukaghiri prepare from poisonous and inebriating mushrooms, bought of the Russians) and many other things, show them plainly to be of the same origin. The Esquimaux seem to be the same nation with the inhabitants of Greenland, the Samoyedes and Lapponians. South America, and especially Peru, is probably peopled from the great unknown south continent, which is very near America, civilized, and full of inhabitants of various colors: who therefore might very easily be cast on the America continent in boats, or proas.—F.

I have let this note by Forster remain to show the character of the speculation on such a topic in the eighteenth century.—Ed.

Bay, to trade with the English. Some of these savages had houses which were made of earth. Many nations had never seen any Frenchmen; they were commonly clad in skins, but many were naked.

All those who had made long journeys in Canada to the south, but chiefly westward, agreed that there were many great plains destitute of trees, where the land was furrowed, as if it had been plowed. In what manner this happened, no one knew, for the grain fields of a great village or town of the Indians are scarce above four or six of our acres in extent; whereas those furrowed plains sometimes continued for several days' journey, except now and then a small smooth spot, and here and there some rising grounds.

I could not hear of any more relics of antiquity in Canada, notwithstanding my careful inquiries after them. In the continuation of my journey for the year 1750[1] I shall find an opportunity of speaking of two other remarkable curiosities. Our Swedish scholar, Mr. George Westmann, A.M., has clearly and circumstantially shown, that our Scandinavians, chiefly the northern ones, long before Columbus's time, have undertaken voyages to North America; see his dissertation on that subject, which he read at Åbo in 1747 for obtaining his degree.[2]

August the 8th

A Convent. This morning I visited the largest nunnery in Quebec. Men are prohibited from visiting under very heavy punishments, except in some rooms, divided by iron rails, where the men and women that do not belong to the convent stand without and the nuns within the rails and converse with each other. But to increase the many favors which the French nation heaped upon me as a Swede, the governor-general got the bishop's leave for me to enter the convent and see its construction. The bishop alone has the power of granting this favor to men, but he does it very sparingly. The royal physician and a surgeon, are, however, at liberty to go in as often as they think proper. Mr. Gauthier,[3] a man of great knowl-

[1] This part was not published until 1929. See Introduction.

[2] Westmann's thesis was published in Åbo, 1757, and bore the title of *Itinera priscorum Scandianorum in Americam*. It was written under Kalm's direction. See Bibliography, item 25.

[3] See page 375, note 2.

edge in medicine and botany, was then the royal physician here, and accompanied me to the convent. We first saw the hospital, which I shall presently describe, and then entered the convent which forms a part of the hospital. It is a great building of stone, three stories high, divided on the inside into long corridors on both sides of which are cells, halls and rooms. The cells of the nuns are in the highest story, on both sides of the long corridor. They are only small, not painted on the inside, but hung with paper pictures of saints and of our Savior on the Cross. A bed with curtains and good bed clothes, a little narrow desk, and a chair or two, comprise the whole furniture of a cell. They have no fires in winter, and the nuns are forced to lie in the cold cells. In the long hall is a stove which is heated in winter, and as all the rooms are left open some warmth can by this means come into them. In the middle story are the rooms where they pass the day together. One of these is the work room. This is large, finely painted and decorated, and has an iron stove. Here nuns were at their needlework, embroidering, gilding and making flowers of silk, which bear a great similarity to the natural ones. In a word, they were all employed in such fine and delicate work as was suitable to ladies of their rank in life. In another hall they assemble to hold their deliberations. Another apartment contains those who are indisposed, but such as are more dangerously ill have rooms to themselves. The novices or newcomers are taught and instructed in another hall. Another is destined for their refectory, or dining room, in which are tables on all sides. On one side of it is a small desk, on which is laid a French book concerning the life of those saints who are mentioned in the New Testament. When they dine, all are silent. One of the eldest enters the pulpit and reads a part of the book before mentioned and when they have gone through it, they read some other religious book. During the meal they sit on that side of the table which is turned towards the wall. In almost every room is a gilt table on which are placed candles, the pictures of our Savior on the Cross and of some saint. Before these tables they say their prayers. On one side is the church and near it a large room divided from the church by rails, so that the nuns can only look into it. In this room they remain during divine service. The priest is in the church where the nuns hand him his sacerdotal garments through a hole, for they are not allowed to go into the vestry or to be in the same room with the priest. There are still several other rooms and halls here, the use of which I do not remember. The lowest story contains a kitchen, bake house, several butteries, etc. In the garrets they keep their grain and dry their linen. In the middle story is a balcony on the outside, almost round the whole building, where the nuns are allowed to take the air. The view from the convent is very fine on every side. The river, the fields and the meadows beyond the town appear there to great advantage. On one side of the convent is a large garden, in which the nuns are at liberty to walk about. It belongs to the convent, and is surrounded with a high wall. There is a quantity of all sorts of vegetables in it and a number of apple, cherry and white walnut trees and red currant bushes. This convent, they say, contains about fifty nuns, most of them advanced in years, scarcely any being under forty years of age. At this time there were two young ladies among them who were being instructed in those things which belong to the knowledge of nuns. They are not allowed to become nuns immediately after their entrance, but must pass through a noviciate of two or three years in order to learn whether they will be constant. For during that time it is in their power to leave the convent, if a monastic life does not suit their inclinations. But as soon as they are received among the nuns and have taken their vows, they are obliged to continue their whole life in it. If they appear willing to change their mode of life, they are locked up in a room from which they can never get out. The nuns of this convent never go further from it than to the hospital, which lies near it and even constitutes a part of it. They go there to attend the sick and to take care of them. Upon my leaving, the abbess asked me if I was satisfied with their institution, whereupon I told them that their convent was beautiful enough, though their mode of living was much circumscribed. Thereupon she told me that she and her sisters would heartily ask God to make me a good Roman Catholic. I answered her that I was far more anxious to be and remain a good Christian, and that as a recompense for their honors and prayers I would not fail earnestly to ask God that they too might remain good Christians, because that would be the highest degree of a true religion that a mortal could find. Thereupon she smilingly bade me farewell. I was told by several people here, some of which were ladies, that none of the nuns went into a convent till she had attained an age in which she had small hopes of ever getting a husband. The nuns of all the three convents in Quebec looked very old, by which it seems that there is some foundation for this assertion. All agree here that the men are much less numerous in Canada than the women, for the men die on their voyages. Many go to the West Indies and either settle or die there; many are killed in battles, etc. Hence there seems to be a necessity of some women going into convents.

The hospital, as I have before mentioned, forms a part of the convent. It consists of two large halls, and some rooms near the apothecary's shop. In the halls are two rows of beds on each side. The beds next to the wall are furnished with curtains, the outward ones are without them. In each bed are fine bed clothes with clean double sheets. As soon as a sick person has left his bed, it is made again to keep the hospital in cleanliness and order. The beds are two or three yards distant, and near each is a small table. There are good iron stoves, and fine windows in this hall. The nuns attend the sick people, and bring them food and other necessaries. Besides them there are some men who attend, and a surgeon. The royal physician is likewise obliged to come hither once or twice every day, look after everything and give prescriptions. They commonly receive sick soldiers into this hospital, who are very numerous in July and August, when the king's ships arrive, and in time of war. But at other times, when no great number of soldiers are sick, other poor people can take their places, as far as the number of empty beds will reach. The king provides everything here that is requisite for the sick persons, *viz.* provisions, medicines, fuel, etc. Those who are very ill are put into separate rooms, in order that the noise in the great hall may not be troublesome to them.

The civility of the inhabitants here is more refined than that of the Dutch and English in the settlements belonging to Great Britain. On the street they raised their hat only to acquaintances and to those of the upper classes. Young men often kept their hats on inside where there were women, but most of them, especially the older ones, took them off. The English, on the other hand, do not idle their time away in dressing as the French do here. The ladies, especially, dress and powder their hair every day, and put their locks in papers every night. This idle custom had not been introduced in the English settlements. The gentlemen generally wear their own hair, but some have wigs, and there were a few so distinguished that they had a queue. People of rank are accustomed to wear lace-trimmed clothes and all the crown officers carry swords. All the gentlemen, even those of rank, the governor-general excepted, when they go into town on a day that looks like rain, carry their cloaks on their left arm. Acquaintances of either sex, who have not seen each other for some time, on meeting again salute with mutual kisses.

Canadian Plants. The plants which I have collected in Canada and which I have partly described, I pass over as I have done before, that I may not tire the patience of my readers by a tedious enumeration. If I should crowd my journal with my daily botanical observations, and descriptions of animals, birds, insects, ores, and like curiosities, it would be swelled to six or ten times its present size. I therefore spare all these things, consisting chiefly of dry descriptions of natural curiosities, for a *Flora Canadensis* or a similar work.[1] The same I must say in regard to the observations I have made in medicine. I have carefully collected all I could on this journey, concerning the medicinal use of the American plants and the cures some of which they reckon infallible in more than one place. But medicine not being my principal study (though from my youth I always was fond of it) I may probably have omitted remarkable circumstances in my accounts of medicines and cures, though one cannot be too accurate in such remarks, or at least one would not find them as they ought to be. This will excuse me for avoiding as much as possible to mention such things as belong to that field and are above my knowledge. Concerning the Canadian plants, I can here add that the further you go northward, the more you find the plants are the same as the Swedish ones: thus, on the north side of Quebec, a fourth part of the plants, if not more, are the same as the wild plants in Sweden. A few plants and trees which have a particular quality or are applied to some particular use, shall, however, be mentioned in a few words later.

The reindeer moss (*Lichen rangiferinus*) grows plentifully in the woods round Quebec. M. Gauthier, and several other gentlemen, told me that the French, on their long journeys through the woods, on account of their fur trade with the Indians, sometimes boil this moss and drink the decoction for want of better food, when their provisions are at an end, and they say it is very nutritive.

[1] This work was never completed or published, though the essential results of Kalm's investigations were utilized by other botanists, chiefly Linné.

Several Frenchmen who have been in the Terra Labrador, where there are many reindeer (which the French and Indians here call cariboux), related that all the land there is in most places covered with this reindeer moss, so that the ground looks white as snow.

August the 10th

The Jesuit College. To-day I dined with the Jesuits. A few days before, I paid my visit to them, and the next day their president and another Father Jesuit called on me and invited me to dine with them to-day. I attended divine service in their church, which is a part of their house. It is very fine within, though it has no seats, for everyone is obliged to kneel down during the service. Above the church is a small steeple with a clock. The building the Jesuits live in is magnificently built, and looks exceedingly fine, both without and within, which makes it similar to a fine palace. It consists of stone, is three stories high, exclusive of the garret, is covered with slate and built in a square form like the new [royal] palace at Stockholm, including a large court. Its size is such that three hundred families would find room enough in it, though at present there were not above twenty Jesuits in it. Sometimes there is a much greater number of them, especially when those return who have been sent as missionaries into the country. There is a long corridor along all the sides of the square, in every story, on both sides of which are either cells, halls, or other apartments for the friars. There are also their library, apothecary shop, etc. Everything is very well regulated, and the Jesuits are well accommodated here. On the outside is their college which is on two sides surrounded with great orchards and kitchen gardens, in which they have fine walks. A part of the trees here are the remains of the forest which stood here when the French began to build this town. They have besides planted a number of fruit trees, and the garden is stocked with all sorts of plants for use in the kitchen. The Jesuits dine together in a great hall. There are tables placed all round it along the walls, and seats between the tables and the walls, but not on the other side. Near one wall is a pulpit at which one of the fathers stands during the meal in order to read some religious book, but this day it was omitted, all the time being employed in conversation. They dine very well, and their dishes are as numerous as at the greatest feasts.

In this spacious building you do not see a single woman; all are "fathers", or "brothers", the latter of which are young men, brought up to be Jesuits. They prepare the meal and bring it upon the table, for common servants are not admitted.

The Jesuits in this country were dressed as follows; they had their own hair, but it was cut short; at the top of the head they were shaved so that it was only a bare spot. The older ones had hoods of black cloth, the younger went bare-headed inside the house but sometimes had a hat on. All shaved their beard, like all other Frenchmen in Canada. The tie was black and often was but the collar of the coat: the latter was long, black and reached to the shoes. It buttoned tight to the body with the buttons in front and was besides bound about the waist with a black band. It was so covered with buttons down the entire front that only these and the shoes could be seen. Often slippers were used in place of shoes. Sometimes they wore a tight-fitting jacket which reached to the knees. They had no collars like clergymen nor did their shirt sleeves show. The oldest had black caps, shaped like a cone, at the top was a black tassel which they also wore on their calotte.

Clergymen in Canada. Besides the bishop there are three kinds of clergymen in Canada; *viz.* Jesuits, priests and recollets. The Jesuits are without doubt the most important; therefore they commonly say here, by way of proverb, that a hatchet is sufficient to cut out a recollet; a priest can be made with a pair of scissors, but a Jesuit requires a paint-brush [1] to show how much he surpasses the others. The Jesuits are usually very learned, studious and civil and agreeable in company. In their whole deportment there is something pleasing. It is no wonder therefore that they captivate the minds of people. They seldom speak of religious matters, and if it happens they generally avoid disputes. They are very ready to do anyone a service, and when they see that their assistance is wanted they hardly give one time to speak of it, falling to work immediately to bring about what is required of them. Their conversation is very entertaining and learned, so that one cannot be tired of their company. Among all the Jesuits I have conversed with in Canada, I have not found one who was not possessed of these qualities in a very eminent degree.

[1] *Pour faire un recolet il faut une hachette, pour un prêtre un ciseau, mais pour un Jesuite il faut un pinceau.*—F.

They have large possessions in this country which the French king gave them. At Montreal they have likewise a fine church and a neat little house with a small but pretty garden next to it. They do not care to become pastors of a congregation in the town or country; but leave their places, together with the emoluments arising from them, to the priests. All their business here is to convert the heathens; and with that view their missionaries are scattered over every part of this country. Near every town and village peopled by converted Indians are one or two Jesuits, who take great care that they may not return to paganism but live as Christians ought to do. Thus there are Jesuits with the converted Indians in Tadoussac, Lorette, Becancourt, St. François, Sault St. Louis, and all over Canada. There are likewise Jesuit missionaries with those who are not converted; so that there is commonly a Jesuit in every large village belonging to the Indians, whom he endeavors on all occasions to convert. In winter he goes on their great hunts where he is frequently obliged to suffer all imaginable inconveniences such as walking in the snow all day, lying in the open air all winter, being out both in good and bad weather, the Indians not fearing any kind of weather, and lying in the Indian huts, which often swarm with fleas and other vermin, etc. The Jesuits undergo all these hardships both for the sake of converting the Indians and also for political reasons. The Jesuits are of great use to their king, for they are frequently able to persuade the Indians to break their treaty with the English, to make war upon them, to bring their furs to the French and not to permit the English to come amongst them. But there is some danger attending these attempts, for when the Indians are drunk they sometimes kill the missionaries who live with them, calling them spies, or excusing themselves by saying that the brandy had killed them. These are accordingly the chief occupations of the Jesuits here. They do not go to visit the sick in the town, they do not hear confessions, and attend no funerals. I have never seen them go in processions in remembrance of the Virgin Mary and other saints. They seldom go into a house in order to get food, and if they are invited they do not like to stay except they be on a journey. Everybody sees that they are, as it were, selected from the people on account of their superior genius and qualities. They are here held a most cunning set of people, who generally succeed in their undertakings and surpass all others in acuteness of understanding. I have therefore several times observed that they have enemies in Canada. They never receive any others into their society but persons of very promising parts, so that there are no blockheads among them. On the other hand the priests take any kind of people into their order they can find; and in the choice of monks, they are even less careful. The Jesuits who live here have all come from France, and many of them return thither again after a stay of a few years here. Some (five or six of whom are yet alive) who were born in Canada, went over to France and were received among the Jesuits there, but none of them ever came back to Canada. I know not what political reason hindered them. During my stay in Quebec one of the priests with the bishop's leave, gave up his priesthood and became a Jesuit. The other priests were very ill pleased with this, because it seemed as if he looked upon their condition as too lowly for himself. Those congregations in the country that pay rents to the Jesuits, have, however, divine service performed by priests, who are appointed by the bishop; and the land rent belongs only to the Jesuits. Neither the priests nor the Jesuits carry on any trade with furs and skins, leaving that entirely to the merchants.

This afternoon I visited the building called the Seminary, where all the priests live in common. They have a great house built of stone with corridors in it, and rooms on each side. It is several stories high, and close to it is a fine garden full of all sorts of fruit trees and vegetables, and it is divided by walks. The view from here is the finest in Quebec. The priests of the Seminary are not much inferior to the Jesuits in civility, and therefore I spent my time very agreeably in their company.

The priests are the second and most numerous class of the clergy in this country, for most of the churches, both in towns and villages (the Indian converts excepted) are served by priests. A few of them are likewise missionaries. In Canada are two seminaries: one in Quebec, the other in Montreal. The priests of the seminary in Montreal are of the order of St. Sulpicius, and supply only the congregation on the isle of Montreal and the town of the same name. At all the other churches in Canada, the priests belonging to the Quebec seminary officiate. The former, or those of the order of St. Sulpicius, all come from France, and I was assured that they never allow a native of Canada to come among them. In the seminary at Quebec, the natives of Canada constitute the greater part. In order to fit the children of this country for orders, there are schools at Quebec and St. Joachim where the youths are taught Latin and instructed in the knowledge of those things and sciences, which have a more immediate connection with the business they are intended for. However, they are not very particular in their choice, and people of a middling capacity are often received among them. They do not seem to have made great progress in Latin; for notwithstanding the service is read in that language, and they read their Latin breviary and other books, every day, yet most of them find it very difficult to speak it. All the priests in the Quebec seminary are consecrated by the Bishop of Canada. The dress of the priests differed from that worn by the Jesuits in that the former wore collars, either yellow or light blue. When they travelled they always brought with them their breviary or prayer book in a little skin bag, which they hung about the neck or on their arm, and read so often in it, that it seemed to us they had been prescribed to read certain sections in it every day. Both the seminaries receive great revenues from the king; that in Quebec has above thirty thousand livres a year. All the country on the west side of the St. Lawrence River from the town of Quebec to St. Paul Bay belongs to this seminary, besides their other possessions in the country. They lease the land to the settlers for a certain rent, which, if it be annually paid according to their agreement, the children or heirs of the settlers may remain in an undisturbed possession of the lands. A piece of land three arpens [1] broad and thirty, forty, or fifty arpens long, pays annually an eçu [2] and a couple of chickens, or some other additional trifle. In such places as have convenient waterfalls they have built watermills, or sawmills, from which they annually get considerable sums. The seminary of Montreal possesses the whole ground on which that town stands, together with the whole Isle of Montreal. I have been assured that the ground rent of the town and isle is computed at seventy thousand livres, besides what they get for saying masses, baptizing, holding confessions, attending at marriages and funerals, etc. All the revenues of ground rent belong to the seminaries alone, and the priests in the country have

[1] A Canadian arpent, linear measure, as noted before, is said to be 12 rods. A square arpent is a little less than an acre. Its value varied greatly.

[2] A French coin, value about one English crown.—F.

no share in them. But as the seminary in Montreal, consisting only of sixteen priests, has greater revenues than it can expend, a large sum of money is annually sent over to France, to the chief seminary there. The land rents belonging to the Quebec seminary are employed for the use of the priests in it, and for the maintenance of a number of young people who are brought up to take the orders. The priests who live in the country parishes, get the tithes from their congregation, together with the extra pay on visiting the sick, etc. In small congregations the king gives the priests an additional sum. When a priest in the country grows old, and has done good services, he is sometimes allowed to come into the seminary in town and spend the rest of his days there. The seminaries are allowed to place the priests on their own estates; but the other places are appointed by the bishop.

The recollets or Barefooted Monks are the third class of clergymen in Canada. They have a fine large dwelling house here, and a fine church, where they officiate. Near it is a large and fine garden, which they cultivate with great application. In Montreal and Trois Rivières they are lodged almost in the same manner as here. They do not endeavor to choose the best fellows amongst them, but take all they can get. They do not torment their brains with much learning and I have been assured that after they have put on their monastic habit they do not study to increase their knowledge but forget even what little they knew before. Their dress consists of a long black frock of coarse cloth extending down to the heels. On the back of it near the collar hangs a hood like a bag fastened to it, which they pull over the head in bad weather. It is like the hoods which our women now use on their cloaks, a custom which probably first came from the monks. On their head they have small calottes; the hair is cut very short, reaching only to the ears. Around the waist they wear a narrow hemp rope which encircles the body several times. In the summer they go barelegged, with wooden shoes, but in winter they have stockings. They wear no linen shirts but a woolen coat next to the body. When they walk in processions they put a black mantle outside of their frock, extending down to the waist. At night they generally lie on mats or some other hard mattresses. However, I have sometimes seen good beds in the cells of some of them. They have no possessions here, having made vows of poverty, and live chiefly on the alms which people give them. To this purpose, the young monks, or brothers, go into the houses with a bag, and beg what they want. They have no congregations in the country, but sometimes go among the Indians as missionaries. In each fort, which contains forty men, the king keeps one of these monks who officiates there instead of a priest. The king gives him lodging, provisions, servants, and all he wants; besides two hundred livres a year. Half of it he sends to the community he belongs to; the other half he reserves for his own use. On board the king's ships are generally no other priests than these friars, who are therefore looked upon as people belonging to the king. When one of the chief priests in the country dies, and his place cannot immediately be filled, they send one of these friars there to officiate while the place is vacant. Some of these monks come from France and some are natives of Canada. There are no other monks in Canada besides these, except now and then one of the order of St. Augustine or some other, who come with one of the king's ships, but goes off with it again.

August the 11th

A Convent. This morning I took a walk out of town, with the royal physician M. Gauthier, in order to collect plants and to see a nunnery at some distance from Quebec. This cloister, which is built very magnificently of stone, lies in a pleasant spot surrounded with grain fields, meadows, and woods, and from which Quebec and the St. Lawrence may be seen. A hospital for poor old people, cripples, etc. makes up part of the cloister and is divided up into two halls, one for men, the other for women. The nuns attend both sexes, with this difference however, that they only prepare the meal for the men and bring it in to them, give them medicine, clear the table when they have eaten, leaving the rest for male servants. But in the hall where the women are, they do all the work that is to be done. The regulation in the hospital was the same as in that at Quebec. To show me a particular favor, the bishop, at the desire of the Marquis de la Galissonnière, governor-general of Canada, granted me leave to see this nunnery also, where no man is allowed to enter without his leave, which is an honor he seldom confers on anybody. The abbess led me and M. Gauthier through all the apartments, accompanied by a great number of nuns. Most of the nuns here are of noble families and one was the daughter of a governor. She had a grand air. Many of them are old, but there are likewise some very young ones among them, who looked very well. They all seemed to be more polite than those in the other nunnery. Their rooms are the same as in the last place except for some additional furniture in their cells. The beds are hung with blue curtains; there are a couple of small bureaux, a table between them, and some pictures on the walls. There are however no stoves in any cell. But those halls and rooms in which they are assembled together, and in which the sick ones lie, are supplied with an iron stove. I did not find out the number of nuns here, but I saw a great number of them. Here are likewise some probationers preparing for their reception as nuns. A number of little girls are sent hither by their parents to be instructed by the nuns in the principles of Christian religion and in all sorts of ladies' work, and when they are through their parents take them home again. The convent at a distance looks like a palace, and, as I am told, was founded by a bishop who they say is buried in a part of the church.

We botanized till dinner time in the neighboring meadows, and then returned to the convent to dine with a venerable old father recollet, who officiated here as a priest. The dishes were all prepared by nuns, and as numerous and various as on the tables of great men. There were likewise several sorts of wine and among the many dainties served at the end of a meal were these: white Canadian walnuts coated with sugar, pears and apples with syrup, apples preserved in spirits of wine, small sugared lemons from the West Indies, strawberry preserves and angelica roots. The revenues of this convent are said to be considerable. At the top of the building is a small belfry. Considering the large tracts of land which the king has given in Canada to convents, Jesuits, priests, and several families of rank, it seems he must have very little left for himself.

Our *common raspberries* are so plentiful here on the hills near grain fields, rivers and brooks, that the branches look quite red on account of the number of berries on them. They are ripe about this time and eaten as a dessert after dinner. They are served either with or without fresh milk and powdered sugar. Sometimes they are kept through the winter in glass jars with syrup.

The *mountain ash*, or sorb tree (*Sorbus aucuparia*) is pretty common in the woods hereabouts.

The *northeast wind* is considered the most piercing of all here. Many prominent people assured me that this wind when it is very violent in winter pierces through walls of a moderate thickness, so that the whole wall on the inside of the house is covered with snow or a thick hoarfrost, and that a candle placed near a thinner wall is almost blown out by the wind which continually comes through. This wind damages the houses which are built of stone, and forces the owners to repair them very frequently on the northeast side. The north and northeast winds are consequently considered very cold here. In summer the north wind is generally attended with rain.

The difference of climate between Quebec and Montreal is by all said to be very great. The wind and weather of Montreal are often entirely different from what they are at Quebec. The winter there is not nearly so cold as in the latter place. Several sorts of fine pears will grow near Montreal, but are far from succeeding at Quebec, where the frost frequently kills them. Quebec has generally more rainy weather, spring begins later and winter sooner than at Montreal, where all sorts of fruits ripen a week or two earlier than at Quebec.

AUGUST THE 12TH

Mixed Blood. This afternoon I and my servant went out of town, to stay in the country for a couple of days that I might have more leisure to examine the plants which grow in the woods here, and the nature of the country. In order to proceed the better, the governor-general had sent for an Indian from Lorette to show us the way and teach us what use they make of the wild plants hereabouts. This Indian was an Englishman by birth, taken by the Indians thirty years ago when he was a boy and adopted by them according to their custom in the place of a relation of theirs killed by the enemy. Since that time he had constantly stayed with them, become a Roman Catholic and married an Indian woman. He dressed like an Indian, spoke English and French and many of the Indian dialects. In the wars between the French and English in this country, the French Indians made many prisoners of both sexes in the English plantations, adopted them afterwards, and married them to people of the Indian nations. Hence the Indian blood in Canada is very much mixed with European blood, and a large number of the Indians now living owe their origin to Europe. It is also remarkable that a great number of the people they had taken during the war and incorporated with their nations, especially the young people, did not choose to return to their native country, though their parents and nearest relations came to them and endeavored to persuade them to, and though it was in their power to do so. The free life led by the Indians pleased them better than that of their European relations; they dressed like the Indians and regulated all their affairs in their way. It is therefore difficult to distinguish them, except by their color, which is somewhat whiter than that of the Indians. There are likewise examples of some Frenchmen going amongst the Indians and following their mode of life. There is on the contrary scarcely one instance of an Indian adopting the European customs; for those who were taken prisoners in the war always endeavored to return to their own people again, even after several years of captivity, though they enjoyed all the privileges that were ever possessed by the Europeans in America.

Geological Formations. The lands which we passed over were everywhere laid out into grain fields, meadows, or pastures. Almost all around us were presented to our view farms and farmhouses and excellent fields and grazing land. Near the town the land is pretty flat and intersected now and then by a clear rivulet. The roads are very good, broad, and lined with ditches on each side in low ground. Further from the town the land rises higher and higher, and consists as it were of terraces, one above another. This rising ground is, however, pretty smooth, chiefly without stones and covered with rich earth. Under that is the black lime stratum which is so common hereabouts, and is broken into small lamellæ and corroded by the air. Some of the strata were horizontal, others perpendicular. I have likewise found such perpendicular strata of lime lamellæ in other places in the neighborhood of Quebec. All the hills are cultivated and some are adorned with fine churches, houses and crop-bearing fields. The meadows are commonly in the valleys, though some are in higher places. Soon after we had a fine view from one of these hills. Quebec appeared very plainly to the eastward, and the St. Lawrence River could likewise be seen. Further away, on the southeast side of that river, appeared a long chain of high mountains running generally parallel to it, though many miles distant from it. To the west again, at some distance from the rising lands where we were, the hills changed into a long chain of very high mountains, lying very close to each other and running parallel likewise to the river, that is, nearly from south to north. These high mountains consisted of grey rock composed of several kinds of [conglomerate] stones which I shall mention later. These mountains seemed to prove that the lime slate strata were of as ancient a date as the gray rock and not formed in later times; for the amazingly large rocks lay on the top of the mountains, which consisted of black slate.

Grass and Meadows. The high meadows in Canada are excellent and by far preferable to the meadows round Philadelphia and in the other English colonies. The further I advanced northward here the finer were the meadows, and the turf upon them was better and closer. Almost all the grass here is of two kinds, *viz.* a species of the narrow leaved meadow grass (*Poa angustifolia* L.), its spikes containing either three or four flowers which are so exceedingly small that the plant might easily be taken for a bent grass (*Agrostis* L.), and its seeds have several small downy hairs at the bottom. The other plant, which grows in the meadows, is the white clover.[1] These two plants form the hay in the meadows. They stand close and thick together, and the meadow grass (*Poa*) is pretty tall, but has very thin stalks. At the root of the meadow grass the ground is covered with white clover, so that one cannot wish for finer meadows than are found here. Almost all have been formerly tilled fields, as appears from the furrows on the ground which still remain. They can be mown but once every summer, as spring commences very late.

Haymaking. Farmers were now busy making hay and getting it in and I was told they had begun about a week ago. The scythes are like our Swedish ones; the men mow and the women rake. The hay was prepared in much the same way as with us, but the tools are a little different. The head of the rake is smaller, has tines on both sides and is a little heavier. The hay is raked into rows; and they also use a kind of wooden fork for both pitching and raking. In so doing, however, a good deal of the hay is left on the field

[1] *Trifolium repens* L.

since this does not rake as clean as an ordinary rake. There were no hillocks on these meadows. The hay is taken away in four-wheeled carts drawn by either horses or oxen. The oxen are hitched in such a way as to pull with their horns instead of their shoulders. Some of the hay barns were out on the fields. They have haystacks near most of their meadows, and on the wet ones they make use of conic haystacks. Their grass lots are usually without fences, the cattle being in the pastures on the other side of the woods and cow-herds take care of them where they are necessary.

The *grain fields* are pretty large. I saw no ditches anywhere, though they seemed to be needed in some places. They are divided into ridges, of the breadth of two or three yards broad, between the shallow furrows. The perpendicular height of the middle of the ridge, from the level to the ground is near one foot. All the grain is summer sown, for as the cold in winter destroys the grain which lies in the ground, it is never sown in autumn. I found white wheat most common in the fields. There are likewise large fields with peas, oats, in some places summer rye, and now and then barley. Near almost every farm I found cabbages, pumpkins, and melons. The fields are not always sown, but lie fallow every two years. The fallow fields not being plowed in summer the weeds grow without restraint in them and the cattle are allowed to roam over them all season.[1]

There was a superabundance of fences around here, since every farm was isolated and the fields divided into small pastures. It will be difficult to obtain material for these fences when the woods are used up; in the future they will probably have to use hedges for their enclosures. It is a stroke of good fortune though, that there is a large amount of cockspur hawthorn growing in the neighborhood. Happy they who will think of it in time! (Here follows a description of a palisade type of fence which it seems unnecessary to reproduce).

The houses in the country are built of stone or wood. The stone

[1] Here follows, in the original, an account of the fences made use of near Quebec, which is intended only for the Swedes, but not for a nation that has made such progress in agriculture and husbandry as the English.—F.

This is a rather typical case of the patronizing air of the English translator toward Kalm and Sweden. While some of the matter is superfluous to us, a part has been reintroduced here.—Ed.

houses are not of bricks, as there is not yet any considerable quantity of brick made here. People therefore take what stones they can find in the neighborhood, especially the black slate. This is quite compact when quarried, but shatters when exposed to the air; however, this is of little consequence as the stones stick fast in the wall, and do not fall apart. For want of it they sometimes make their buildings of limestone or sandstone, and sometimes of gray stone. The walls of such houses are commonly two feet thick and seldom thinner. The people here can have lime everywhere in this neighborhood. The greater part of the houses in the country are built of wood, and sometimes plastered over on the outside. The chinks in the walls are filled with clay instead of moss. The houses are seldom above one story high. The windows are always set in the inner part of the wall, never in the outer, unless double windows are used. The panes are set with putty and not lead. In the city glass is used for the windows for the most part, but further inland they use paper. In opening the windows they use hooks as with us. The floors are sometimes of wood and sometimes of clay. The ceiling consists generally of loose boards without any filling, so that much of the internal heat is wasted. In every room is either a chimney or a stove, or both. The stoves have the form of an oblong square; some are entirely of iron, about two feet and a half long, one foot and a half or two feet high, and near a foot and a half broad. These iron stoves are all cast at the ironworks at Trois Rivières. Some are made of bricks or stones, not much larger than the iron stoves, but covered at the top with an iron plate. The smoke from the stoves is conveyed up the chimney by an iron pipe in which there are no dampers, so that a good deal of their heat is lost. In summer the stoves are removed. The roofs are always very steep, either of the Italian type or with gables. They are made of long boards, laid horizontally, the upper overlapping the lower. Wooden shingles are not used since they are too liable to catch fire, for which reason they are forbidden in Quebec. Barns have thatched roofs, very high and steep. The dwelling houses generally have three rooms. The baking oven is built separately outside the house, either of brick or stone, and covered with clay. Brick ovens, however, are rare.

This evening we arrived at Lorette, where we lodged with the Jesuits.

AUGUST THE 13TH

Botanizing. In the morning we continued our journey through the woods to the high mountains, in order to see what scarce plants and curiosities we could get there. The ground was flat at first and covered with a thick wood all round, except in marshy places. Nearly half the plants which are to be found here grow in the woods and morasses in Sweden.

We saw *wild cherry trees* here of two kinds which are probably mere varieties, though they differ in several respects. Both are pretty common in Canada and both have red berries. One kind which is called *cerisier* by the French tastes like our Alpine cherries and their acid contracts the mouth and cheeks. The berries of the other sort have an agreeable sourness and a pleasant taste.

The three-leaved hellebore (*Helleborus trifolius*) grows in great quantities in the woods, and in many places it covers the ground by itself. However, it commonly chooses mossy places that are not very wet, and the wood sorrel (*Oxalis acetosella* L.), with the mountain enchanter's nightshade (*Circæa alpina* L.) are its companions. Its seeds were not yet ripe and most of the stalks had no seeds at all. This plant is called *Tissavoyanne jaune* by the French in Canada. Its leaves and stalks are used by the Indians for giving a fine yellow color to several kinds of work which they make of prepared skins. The French who have learnt this from them, dye wool and other things yellow with this plant.

We climbed with a great deal of difficulty to the top of one of the highest mountains here, and I was vexed to find nothing at its summit but what I had seen in other parts of Canada before. We had not even the pleasure of a view, because the trees with which the mountain is covered obstructed it. The trees that grow here are a kind of hornbeam, or *Carpinus ostrya* L., the American elm, the red maple, the sugar maple, that kind of maple which cures scorched wounds (which I have not yet described), the beech, the common birch tree, the sugar birch (*Betula nigra* L.) the mountain ash, the Canada pine, called *perusse*, the mealy tree with dentated leaves (*Viburnum dentatum* L.), the ash, the cherry tree (*Cerisier*) just before described and the berry-bearing yew (*Taxus baccata*).

The gnats in this wood were more numerous than we could have wished. Their bite caused such swelling of the skin that it was difficult to shave, and the Jesuits at Lorette said the best preservative against their attacks was to rub the face and naked parts of the body with grease. Cold water they reckoned the best remedy against the bite, when the wounded places were washed with it immediately afterward.

At night we returned to Lorette, having accurately examined the plants of note which we found to-day.

August the 14th

Lorette is a village, three French miles to the west of Quebec, inhabited chiefly by Indians of the Huron nation converted to the Roman Catholic religion. The village lies near a little river which tumbles over a rock there with a great noise and turns a sawmill and a flourmill. When the Jesuits who are now with them, arrived among them, they lived in their usual huts, which are made like those of the Laplanders. They have since laid aside this custom and built all their houses after the French fashion. In each house are two rooms, *viz.* their bedroom and the kitchen. In one room is a small oven of stone, covered on top with an iron plate. Their beds are near the wall, and they put no other clothes on them than those which they are dressed in. Their other furniture and utensils look equally wretched. There is a fine little church here, with a steeple and a bell. The steeple is raised pretty high and covered with white tin plates. They pretend that there is some similarity between this church in its shape and plan and the Santa Casa at Loretto in Italy, whence this village has gotten its name. Close to the church is a house built of stone for the clergymen, two Jesuits, who constantly live here. The divine service is as regularly attended here as in any other Roman Catholic church, and it is a pleasure to hear the vocal skill and pleasant voices of the Indians, especially of the women, when singing all sorts of hymns in their own language. The Indians dress chiefly like the other adjacent Indian nations; the men, however, like to wear waistcoats, or jackets, like the French. The women keep exactly to the Indian dress. It is certain that these Indians and their ancestors long ago, on being converted to the Christian religion, made a vow to God never to drink strong liquors. This vow they have kept pretty inviolable hitherto, so that one seldom sees one of them drunk, though brandy and other strong liquors are goods which other Indians prefer to life itself.

The Indians of Lorette. These Indians have made the French their patterns in several things besides the houses. They all plant corn; and some have small fields of wheat and rye. Some of them keep cows. They plant our common sun-flower (*Helianthus annuus*), in their corn fields and mix the seeds of it into their sagamite or corn soup. The corn which they plant here is of the small sort, which ripens sooner than the other. Its kernels are smaller but give more and better flour in proportion. It commonly ripens here at the middle, sometimes however at the end, of August. The mills belong to the Jesuits who get paid for everything they grind.

The Swedish winter wheat and winter rye has been tried in Canada, to see how well it would succeed; for Canadians employ nothing but spring wheat or rye, since it has been found that French wheat and rye die here in winter, if it be sown in autumn. Dr. Sarrazin[1] therefore (as I was told by the eldest of the two Jesuits here) got a small quantity of winter wheat and rye from Sweden. It was sown in autumn, not hurt by the winter, and gave good results. The ears were not so large as those of the Canadian grain, but weighed nearly twice as much, and gave a greater quantity of finer flour, than the summer variety. Nobody could tell me why the experiments have not been continued. They cannot, I am told, bake such white bread here of the summer grain as they can in France of their winter wheat. Many people assured me that all the spring wheat now used here came from Sweden or Norway; for the French, on their arrival, found the winters in Canada too severe for the French winter seed, and their summer variety did not always ripen on account of the shortness of the summer. Therefore they began to look upon Canada as little better than a useless country where nobody could live, till they fell upon the idea of getting their spring grain from the most northern parts of Europe, which has succeeded very well.

To-day I returned to Quebec making botanical observations on the way.

---

[1] See pp. 390, note.

AUGUST THE 15TH

The New Governor Arrives. The new governor-general of all Canada, the Marquis de la Jonquière,[1] arrived last night in the river before Quebec; but it being late he reserved his public entrance for to-day. He had left France on the second of June, but could not reach Quebec before this time on account of the difficulty which great ships find in passing the sandbars in the St. Lawrence River. The ships cannot venture to go up without a fair wind, being forced to sail in many windings and frequently in a very narrow channel. To-day was another great feast on account of the Ascension of the Virgin Mary which is very highly celebrated in Roman Catholic countries. This day was accordingly doubly remarkable both on account of the holiday and of the arrival of the new governor-general, who is always received with great pomp, as he is really a viceroy here.

About eight o'clock the chief people in town assembled at the house of Mr. de Vaudreuil,[2] who had lately been nominated governor of Trois Rivières and lived in the lower town, and whose father had likewise been governor-general of Canada. Thither came likewise the Marquis de la Galissonnière, who had till now been governor-general, and was to sail for France at the first opportunity. He was accompanied by all the people belonging to the government. I was likewise invited to see this festivity. At half an hour after eight the new governor-general went from the ship into a barge covered with red cloth. A signal with cannons was given from the ramparts for all the bells in the town to be set ringing. All the people of distinction went down to the shore to salute the governor who, on alighting from the barge, was received by the Marquis de la Galissonnière. After they had saluted each other, the commandant of the town addressed the new governor in a very

[1] Pierre-Jacques de Taffanel, Marquis de la Jonquière (1685-1752). See long article on him in the *Dictionnaire Générale du Canada,* which quotes Kalm on Jonquière's personal appearance. See *infra,* p. 465.

[2] Pierre de Rigaud, Marquis de Vaudreuil-Cavagnol (1698-1778), last governor of New France (1755-1760), had been appointed governor of the Three Rivers in 1733, and was governor of Louisiana from 1742 to 1755. Obviously he did not spend much time in Louisiana if he still made his home in Trois Rivières. Or was the candidate "lately" nominated for the governorship of Three Rivers one of the other eleven Vaudreuil brothers, of whom Pierre was one? It was the latter who surrendered all Canada to the British in 1760.

elegant speech which he answered very concisely. Then all the cannon on the ramparts gave a general salute. The whole street up to the cathedral was lined with men in arms, chiefly drawn from among the burgesses. The governor then walked towards the cathedral, dressed in a suit of red, with an abundance of gold lace. His servants went before him in green, carrying firearms on their shoulders. On his arrival at the cathedral he was received by the Bishop of Canada and the whole clergy assembled. The bishop was arrayed in his pontifical robes, and had a long gilt tiara on his head, and a great crozier of massy silver in his hand. After the bishop had addressed a short speech to the governor-general, a priest brought a silver crucifix on a long stick (two priests with lighted tapers in their hands, going on each side of it) to be kissed by the governor. The bishop and the priests then went through the long aisle up to the choir. The servants of the governor followed with their hats on, and arms on their shoulders. At last came the governor and his suite and after them a crowd of people. At the beginning of the choir the governor and the General de la Galissonnière, stopped before a chair covered with red cloth and stood there during the whole time of the celebration of the mass, which was celebrated by the bishop himself. From the church he went to the palace where the gentlemen of note in the town afterwards went to pay their respects to him. The members of the different orders with their respective superiors likewise came to him to testify their joy on account of his happy arrival. Among the numbers that came to visit him none stayed to dine but those that were invited beforehand, among which I had the honor to be one. The entertainment lasted very long, and was as elegant as the occasion required.

The governor-general, Marquis de la Jonquière, was very tall, and at that time above sixty years old. He had fought a desperate naval battle with the English in the last war, but had been obliged to surrender, the English being, as it was told, vastly superior in the number of ships and men. On this occasion he was wounded by a ball which entered one side of his shoulder and came out at the other, so that he walked slightly bent over. He was very complaisant but knew how to preserve his dignity when he distributed favors.

Early Refrigeration. Many of the gentlemen present at this entertainment asserted that the following method had been successfully employed to keep wine, beer, or water cool during the summer. The wine or other liquor is bottled; the bottles are well corked, hung up in the air and wrapped in wet cloth. This cools the wine in the bottles notwithstanding it was quite warm before. After a little while the cloths are again made wet, with the coldest water that is to be had and this continued. The wine, or other liquor, in the bottles is then always colder than the water with which the cloths are made wet. And though the bottles should be hung up in the sunshine, the above way of proceeding will always have the same effect.[1]

A Catholic Procession. The procession in memory of the Ascension of the Holy Virgin, which was held to-day in Quebec, was in its way quite magnificent. Believing that She ascended on this day, the Catholics marched from one church to the other through the whole city. The people flocked to see the procession, as though they had never seen it before. It was said that they were always anxious to behold such sights. First came two boys who were constantly ringing bells, followed by a man carrying a banner with a painting of Christ on the Cross on one side and one of Mary, Joseph and the Christ-Child between them on the other. Then came another man bearing a painted wooden image of the Savior on the Cross. Both the banner and the image were on long poles. Then followed the Recollets, or Mendicant Friars, dressed in the costume described above, walking far apart and two abreast; and since they affected poverty they carried aloft a plain wooden cross. Next, borne on a pole, part of which was of silver, came a silver image of the crucified Savior. Then followed several pairs of boys, about ten or twelve years old, clad in red tunics with white cottas over them and wearing red, cone-shaped caps. Then came some more

---

[1] It has been observed by several experiments, that any [container of] liquid dipped into another liquor, and then exposed into the air for evaporation, will get a remarkable degree of cold; the quicker the evaporation succeeds, after repeated dippings, the greater the cold. Therefore spirit of wine evaporating quicker than water, cools more than water; and spirit of sal ammoniac, made by quick-lime, being still more volatile than spirit of wine, its cooling quality is still greater. The evaporation succeeds better by moving the vessel containing the liquor, by exposing it to the air, and by blowing upon it, or using a pair of bellows. See de Marian, *Dissertation sur la Glace*, Prof. Richman in Nov. *Comment. Petrop. ad an.* 1747, & 1748. p. 284, and Dr. Cullen in the *Edinburgh physical and literary Essays and Observations,* Vol. II. p. 145.—F.

boys with black tunics and caps, followed by the priests. Some of the latter wore white chasubles, others long silk mantles of various colors, with black, cone-shaped hoods on their heads and bluish bands. A priest swinging a censer followed. Then came two priests bearing a silver image of the Virgin enclosed in a silver-plated shrine, surrounded on all sides by men carrying lanterns with wax-candles, and after came the most distinguished clergymen in their long robes. The bishop in full dress with his official crosier ended the religious group of the procession. After him marched the governor-general's private guard with guns on their shoulders, followed by the governor himself, de la Jonquière, and General de la Galissonnière, walking abreast. Last came a number of distinguished residents and a large group of common people. As the procession passed the castle the soldiers presented arms in salute, the drums were beaten, and the cannon roared from the forts, as was customary on such occasions. Those who stood nearest where the procession passed kneeled as the image of the Holy Virgin was carried by; but at the figure of Christ they remained standing. Onlookers who watched at some distance did not kneel. Thus the procession went on, the priests singing as they marched.[1]

August the 16th

The occidental *Arbor vitæ* (*Thuja occidentalis* L.), is a tree which grows very plentiful in Canada but not much further south. The most southerly place I have seen it in is a place a little to the south of Saratoga in the province of New York, and likewise near Casses [Cassius?] in the same province, which places are in forty-two degrees and ten minutes north latitude. Mr. Bartram, however, informed me that he had found a single tree of this kind in Virginia, near the falls in the James River. Doctor Colden likewise asserted that he had seen it in many places round his seat Coldingham, which lies between New York and Albany, about forty-one degrees thirty minutes north latitude. The French, all over Canada, call it *Cèdre blanc*. The English and Dutch in Albany also call it the white cedar. The English in Virginia have called a *Thuya* which grows with them a juniper. The places and the soil where it grows

[1] This passage, like several others, is omitted by Forster and subsequent translators, who did not use the original.

best are not always alike; however, it generally succeeds in ground where its roots have sufficient moisture. It seems to prefer swamps, marshes, and other wet places to all others, and there it grows pretty tall. Stony hills, and places where a number of stones lie together, covered with several kinds of mosses (*Lichen, Bryum, Hypnum*), seemed to be the next in order where it grows. When the seashores are hilly and covered with mossy stones, the *Thuya* seldom fails to grow on them. It is likewise seen now and then on the hills near rivers and other high grounds, which are covered with earth or mould; but it is to be observed that such places commonly have a sourish water in them, or receive moisture from the higher localities. I have however seen it growing in some pretty dry places; but there it never grows to any great size. It is found frequently in the clefts of mountains, but cannot grow to any remarkable height or thickness. The tallest trees I have found in the woods in Canada are from thirty to thirty-six feet high. A barked tree of exactly twelve inches diameter had ninety-two rings round the trunk; another of one foot and four inches in diameter had one hundred and forty-two rings.

Thuya Wood. The Canadians generally use this tree for the following purposes. Since it is thought the most durable wood in Canada, and that which best withstands rotting, so as to remain undamaged for over a man's age, fences of all kinds are scarcely made of any other than this wood. All the posts which are driven into the ground are made of the Thuya wood. The palisades round the forts in Canada are likewise made of the same material. The beams in the houses are made of it; and the thin narrow pieces of wood which form both the ribs and the bottom of the bark boats commonly used here are taken from this wood, because it is pliant enough for the purpose, especially while it is fresh. It is also very light. The Thuya wood is reckoned one of the best for the use in limekilns. Its branches are used all over Canada for brooms, and the twigs and leaves of it being naturally bent together seem to be very proper for the purpose. The Indians make such brooms and bring them to the towns to sell, nor do I remember having seen any made of any other wood. The fresh branches have a peculiar, agreeable scent which is pretty strongly smelled in houses where they make use of brooms of this kind.

Thuya in Medicine. This Thuya is used for several medicinal purposes. The commandant of Fort St. Frédéric, M. de Lusignan, could never sufficiently praise its excellence for rheumatic pains. He told me he had often seen it tried with remarkably good success upon several persons in the following manner. The fresh leaves are pounded in a mortar and mixed with hog's grease or any other grease. This is boiled together till it becomes a salve, which is spread on linen and applied to the part where the pain is. The salve gives certain relief in a short time. Against violent pains which move up and down in the thighs and sometimes spread all over the body, they recommend the following remedy. Take of the leaves of a kind of polypody[1] four-fifths, and of the cones of the Thuya one-fifth, both reduced to a coarse powder by themselves, and mixed together afterwards. Then pour milkwarm water on it so as to make a poultice, spread it on linen, and wrap it round the body: but as the poultice burns like fire, they commonly lay a cloth between it and the body, otherwise it would burn and scorch the skin. I have heard this remedy praised beyond measure, by people who said they had experienced its good effects. Among these was a woman who said that she had applied such a poultice for three days, after which her severe pain passed away entirely. An Iroquois Indian told me that a decoction of Thuya leaves was used as a remedy for a cough. In the neighborhood of Saratoga, they use this decoction in the intermittent fevers.

The Thuya tree keeps its leaves, and is green all winter. Its seeds are ripe towards the end of September, old style. The fourth of October of this year, 1749, some of the cones, especially those which stood much exposed to the heat of the sun, had already dropped their seeds, and all the other cones were opening in order to shed them. This tree has, in common with many other American trees, the quality of growing plentifully in marshes and thick woods, which may with certainty be called its native habitat. However, there is scarcely a single Thuya tree in those places which bears seeds; if, on the other hand, a tree accidentally stands on the outside of a wood, on the seashore, or in a field where the air can freely get to it, it is always full of seeds. I have found this to be the case with the Thuya on innumerable occasions. It is the same with the sugar maple, the maple which is good for healing scorched wounds, the white fir tree, the pine called *pérusse*, the mulberry tree, the sas-

[1] *Polypodium fronde pinnaia, pinnis alternis ad basin superne appendiculatis.*—F.

safras and several others. In England this tree is everywhere called *arbor vitæ* by the farmers.

August the 17th

The Ursuline Convent. To-day I went to see the nunnery of the Ursulines, which is located on nearly the same plan as the two other nunneries. It lies in the town and has a very fine church. The nuns are renowned for their piety and they go abroad less than any others. Men are not allowed to go into this convent either, except by the special license of the bishop, which is given as a great favor. The royal physician and the surgeon are alone entitled to go in as often as they please to visit the sick. At the desire of the Marquis de la Galissonière the bishop granted me leave to visit this nunnery together with the royal physician M. Gauthier. On our arrival we were received by the abbess, who was attended by a great number of nuns, for the most part old ones. We saw the church; and it being Sunday, we found some nuns on every side of it kneeling by themselves and saying prayers. As soon as we came into the church, the abbess and the nuns with her dropped on their knees and so did M. Gauthier and myself. We then went to an apartment or small chapel dedicated to the Virgin Mary, at the entrance of which they all fell on their knees again. In several places which we visited we found images and paintings and candles which burned in front of some of these. The nuns explained that these pictures and images were not kept here for worship, because God alone is to be worshipped, but only to arouse piety through them. We afterwards saw the kitchen, the dining hall and the apartment they work in, which is large and fine. They do all sorts of neat work there, gild pictures, make artificial flowers, etc. The dining hall is similar to those in the other two nunneries. Under the tables are small drawers for each nun to keep her napkin, knife and fork, and other things in. Their cells are small, and each nun has one to herself. The walls are not painted; a little bed, a table with a drawer, a crucifix and pictures of saints on it, and a chair, constitute the whole furniture of a cell. We were then led into a room full of young ladies about twelve years old and below that age, sent hither by their parents to be instructed in reading and in matters of religion. They are allowed to go to visit their relations once a day, but must not stay away long. When they have learned reading and have received instructions in religion they return to their parents again. Near the nunnery is a fine garden which is surrounded with a high wall. It belongs to this institution and is stocked with all sorts of kitchen herbs and fruit trees. When the nuns are at work, or during dinner, everything is silent in the rooms, unless some one of them reads to the others; but after dinner they have leave to take a walk for an hour or two in the garden, or to divert themselves within doors. After we had seen everything remarkable here we took our leave.

A Mineral Spring. About a quarter of a Swedish mile to the west of Quebec is a well of mineral waters which contains a great deal of iron ochre and has a pretty strong taste. M. Gauthier said that he had prescribed it with success in costive cases and like diseases.

Snakes. I have been assured that there are no snakes in the woods and fields round Quebec whose bite is poisonous, so that one can safely walk in the grass. I have never found any that endeavored to bite, and all were frightened. In the southern parts of Canada it is not advisable to be off one's guard.

A very small species of black ant (*Formica nigra* L.) live in ant-hills, in high grounds and in woods; they look exactly like our Swedish ants, but are much smaller.

August the 21st

Indians. To-day there were representatives of three Indian nations in this country with the governor-general, *viz.* Huron, Mickmacks and Anies,[1] the last of which are a nation of Iroquois and allies of the English. They were taken prisoners in the last war.

The Hurons are some of the same Indians as those who live at Lorette, and have received the Christian religion. They are a tall, robust people, well shaped, and of a copper color. They have short black hair which is shaved on the forehead from one ear to the other. None of them wear hats or caps. Some have earrings, others not. Many of them have the face painted all over with cinnabar; others have only strokes of it on the forehead and near the ears; and

[1] Probably Oneidas.—F.

some paint their hair with the same material. Red is the color they chiefly use in painting themselves, but I have also seen some who had daubed their face with black. Many of them have figures on the face and on the whole body, which are stained into the skin, so as to be indelible. The manner of making them shall be described later. These figures are commonly black; some have a snake painted on each cheek, some have several crosses, some an arrow, others the sun, or anything else their imagination leads them to. They have such figures likewise on the breast, thighs and other parts of the body; but some have no figures at all. They wear a shirt which is either white or blue striped and a shaggy piece of cloth, which is either blue or white, with a blue or red stripe below. This they always carry over their shoulders, or let it hang down, in which case they wrap it round their middle. Round their neck they have a string of violet wampum, with some white wampum between them. These wampum are small, of the figure of oblong pearls, and made of the shells which the English call clams (*Venus mercenaria* L.) I shall make a more particular mention of them later. At the end of the wampum strings, many of the Indians wear a large French silver coin with the king's effigy on their breasts. Others have a large shell on the breast, of a fine white color, which they value very highly; others again have no ornament at all round the neck. They all have their breasts uncovered. In front hangs their tobacco pouch made of the skin of an animal with the hairy side turned outwards. Their shoes are made of skins, and bear a great resemblance to the heel-less shoes which the women in Finland use. Instead of stockings they wrap the legs in pieces of blue cloth, as I have seen the Russian peasants do.

The Mickmacks are dressed like the Hurons, but distinguish themselves by their long straight hair of jet-black. Almost all the Indians have black straight hair; however, I have met with a few, whose hair was quite curly. But it is to be observed that it is difficult to judge the true complexion of the Canada Indians, their blood being mixed with the European, either by the adopted prisoners of both sexes or by the Frenchmen who travel in the country and often contribute their share towards the increase of the Indian families, to which the women, it is said, have no serious objection. The Mickmacks are commonly not so tall as the Hurons. I have not seen any Indians whose hair is as long and straight as theirs.

Their language is different from that of the Hurons; therefore there is an interpreter here for them on purpose.

The Anies are the third kind of Indians which came hither. Fifty of them went to war, being allies of the English, in order to plunder in the neighborhood of Montreal. But the French, being informed of their scheme, laid an ambush, and killed with the first discharge of their guns forty-four of them so that only the four who were here to-day saved their lives, and two others who were ill at this time. They are as tall as the Hurons, whose language they speak. The Hurons seem to have a longer and the Anies a rounder face. The Anies have something cruel in their looks; but their dress is the same as that of the other Indians. They wear an oblong piece of white tin in the hair which lies on the neck. One of those I saw had taken a flower of the rose mallow, out of a garden where it was in full blossom at this time, and put it in the hair at the top of his head. Each of the Indians has a tobacco pipe of gray limestone which is blackened afterwards and has a stem of wood. There were no Indian women present at this interview. As soon as the governor came in and was seated in order to speak with them the Mickmacks sat down on the ground, like Laplanders, but the other Indians took chairs.

There is no printing press in Canada, nor has there formerly been any: but all books are brought from France and all the orders made in the country are written, which [as I have shown] extends even to the paper currency. They pretend that the press is not yet introduced here because it might be the means of propagating libels against the government and religion. But the true reason seems to lie in the poverty of the country, as no printer could make a sufficient number of books for his subsistence; and another reason may be that France now has the profit arising from the exportation of books hither.

Food. The meals here are in many respects different from those in the English provinces. This depends upon the difference of custom, taste, and religion, between the two nations. French Canadians eat three meals a day, *viz.* breakfast, dinner and supper. They breakfast commonly between seven and eight, for the French here rise very early, and the governor-general can be seen at seven o'clock, the time when he has his levee. Some of the men dip a piece of bread in brandy and eat it; others take a dram of brandy and eat a piece of bread after it. Chocolate is likewise very common for breakfast, and many of the ladies drink coffee. Some eat no breakfast at all. I have never seen tea used here, perhaps because they can get coffee and chocolate from the French provinces in America, in the southern part, but must get tea from China. They consider it is not worth their while to send the money out of their country for it. I never saw them have bread and butter for breakfast. Dinner is exactly at noon. People of quality have a great many dishes and the rest follow their example, when they invite strangers. The loaves are oval and baked of wheat flour. For each person they put a plate, napkin, spoon and fork. (In the English colonies a napkin is seldom or never used). Sometimes they also provide knives, but they are generally omitted, all the ladies and gentlemen being provided with their own knives. The spoons and forks are of silver, and the plates of Delft ware. The meal begins with a soup with a good deal of bread in it. Then follow fresh meats of various kinds, boiled and roasted, poultry, or game, fricassees, ragouts, etc. of several sorts, together with different kinds of salads. They commonly drink red claret at dinner, either mixed with water or clear; and spruce beer is likewise much in use. The ladies drink water and sometimes wine. Each one has his own glass and can drink as much as he wishes, for the bottles are put on the table. Butter is seldom served, and if it is, it is chiefly for the guest present who likes it. But it is so fresh that one has to salt it at the table. The salt is white and finely powdered, though now and then a gray salt is used. After the main course is finished the table is always cleared. Finally the fruit and sweetmeats are served, which are of many different kinds, *viz.* walnuts from France or Canada, either ripe or pickled; almonds; raisins; hazel-nuts; several kinds of berries which are ripe in the summer season, such as currants, red and black, and cranberries which are preserved in treacle; many preserves in sugar, as strawberries, raspberries, blackberries, and mossberries. Cheese is likewise a part of the dessert, and so is milk, which they drink last of all, with sugar. Friday and Saturday, the "lean" days, they eat no meat according to the Roman Catholic rites; but they well know how to guard against hunger. On those days they boil all sorts of vegetables like peas, beans and cabbage, and fruit, fish, eggs, and milk are prepared in various ways. They cut cucumbers into slices and eat them with cream, which is a very good dish. Sometimes they put whole cucumbers on the table and everybody that likes them takes one, peels and slices it, and dips the slices into salt, eating them like radishes. Melons abound here and are always eaten without sugar. In brief, they live here just as well on Fridays and Saturdays, and I who am not a particular lover of meats would willingly have had all the days so-called lean days. There is always salt and pepper on the table. They never put any sugar into wine or brandy, and upon the whole they and the English do not use half so much sugar as we do in Sweden, though both nations have large sugar plantations in their West Indian possessions. They say no grace before or after their meals, but only cross themselves, a custom which is likewise omitted by some. Immediately after dinner they drink coffee without cream. Supper is commonly at seven o'clock, or between seven and eight at night, and the dishes the same as at dinner. Pudding is not seen here and neither is punch, the favorite drink of the Englishmen, though the Canadians know what it is.

August the 23rd

The "Poor Man's Horse". In many places hereabouts they use their dogs to fetch water out of the river. I saw two great dogs to-day attached to a little cart, one before the other. They had neat harnesses, like horses, and bits in their mouths. In the cart was a barrel. The dogs were directed by a boy, who ran behind the cart, and as soon as they came to the river they jumped in of their own accord. When the barrel was filled, the dogs drew their burden up the hill again to their house. I frequently saw dogs employed in this manner, during my stay at Quebec. Sometimes they put but one dog before the watercart, which was made small on purpose. The dogs were not very large, hardly of the size of our common farmer's dogs. The boys that attended them had great whips with which they urged them on occasionally. I have seen them fetch not only water but also wood and other things. In winter it is customary in Canada, for travellers to put dogs before little sledges, made on purpose to hold their clothes, provisions, etc. Poor people commonly employ them on their winter journeys, and go on foot themselves.

Almost all the wood, which the poorer people in this country fetch out of the woods in winter, is drawn by dogs, which have therefore got the name of "the poor man's horse". They commonly place a pair of dogs before each load of wood. I have likewise seen some neat little sledges for ladies to ride in, in winter. They are drawn by a span of dogs, and go faster on a good road than one would think. A middle-sized dog is sufficient to draw a single person when the roads are good. I have been told by old people that horses were very scarce here in their youth and almost all the transportation was then effected by dogs. Several Frenchmen who have been among the Esquimaux on Terra Labrador have assured me that they not only make use of dogs for drawing drays, with their provisions and other necessaries, but are likewise employed for pulling the men themselves in little sledges.

August the 25th

The high hills to the west of the town abound with springs. These hills consist of the black slate, before mentioned, and are pretty steep, so that it is difficult to get to the top. Their perpendicular height is about twenty or four and twenty yards. Their summits are destitute of trees and covered with a thin crust of earth, lying on the lime slate, and are tilled or used for pastures. It seems inconceivable therefore that these naked hills can have so many running springs, which in some places gush out like torrents. Have these hills the quality of attracting the water out of the air in the day time or at night? Or are the lime slates more apt to do it than others?

Domestic Animals. All the horses in Canada are strong, well-built, swift, as tall as the horses of our cavalry, and of a breed imported from France. The inhabitants have the custom of docking the tails of their horses, which is a rather severe treatment, as they cannot defend themselves against the numerous swarms of gnats, gadflies, and horseflies. They put the horses in tandem before their carts, which has probably occasioned the docking of their tails, as the horses would hurt the eyes of those behind them, by moving their tails to and fro. The governor-general and a few of the chief people in town have coaches; the rest make use of open two-wheeled carts. It is a general complaint that the country people are beginning to keep too many horses, by which means the cows are kept short of food in winter.

The cows have likewise been imported from France, and are of the size of our common Swedish cows. Everybody agreed that the cattle, which are born of the original French breed, never grow to the same size as the parent stock. This they ascribe to the cold winters, during which they are obliged to put their cattle into stables and give them but little food. Almost all the cows have horns; a few, however, I have seen without them. A cow without horns would be reckoned an unheard-of curiosity in Pennsylvania. Is not this to be attributed to the cold? The cows give as much milk here as in France. The beef and veal at Quebec is reckoned fatter and more palatable than at Montreal. Some look upon the salty pastures below Quebec as the cause of this difference. But this does not seem sufficient, for most of the cattle which are sold at Quebec have no meadows with arrow-headed grass (*Triglochin*) on which they graze. In Canada the oxen draw with the horns, but in the English colonies they draw with their withers, as horses do. The cows vary in color; however, most of them are either red or black.

Every countryman commonly keeps a few sheep which supply him with as much wool as he needs to clothe himself with. The better sort of clothes are brought from France. The sheep degenerate here after they are imported from France, their wool becoming coarser, and their progeny is still poorer. The want of food in winter is said to cause this degeneration.

I have not seen any goats in Canada, and I have been assured that there are none. I have seen but very few in the English colonies, and only in their towns where they are kept on account of some sick people who drink the milk by the advice of their physicians.

The harrows are triangular; two of the sides are six feet and the third four feet long. The teeth and every other part of the harrows are of wood. The teeth are about five inches long and about as far apart.

The view of the country about a quarter of a Swedish mile north of Quebec on the west side of the St. Lawrence River is very fine. The country is very steep towards the river, and grows higher as you go further from the water. In many places it is naturally divided into terraces. From the heights one can see a great distance; Quebec appears very flat to the south and the St. Lawrence River is to the east, on which vessels sail up and down. To the west are high mountains where the slope to the river begins. All the country is cultivated and laid out in grain fields, meadows, and pastures; most of the fields are sown with wheat, many with white oats, and some with peas. Several fine houses and farms are interspersed all over the country and none are ever together. The dwelling-house is commonly built of black slate and generally whitewashed on the outside. Many rivulets and brooks flow down from the heights below the mountains, the latter consisting entirely of the black slate that shatters into pieces when exposed to the air. On the slate lies earth two or three feet in depth. The soil in the grain fields is always mixed with little pieces of the slate. All the rivulets cut their beds deep into the ground so that their shores are commonly of limestone. A dark gray variety is sometimes found among the strata, which, when broken, smells like stinkstone.

Shipbuilding. They were now building several ships below Quebec, at the king's command. However, before my departure an order arrived from France prohibiting the further building of ships of war, except those which were already on the stocks; because they have found that the ships built of American oak did not last so long as those of the European variety. Near Quebec is found very little oak, and what grows there is not fit for use, being very small; therefore they are obliged to fetch their oak timber from those parts of Canada which border upon New England. But all the North American oaks have the quality of lasting longer and withstanding rot better the further north they grow, and *vice versa.* The timber from the confines of New England is brought in floats or rafts on the rivers near those parts and near St. Pierre which empty into the great St. Lawrence River. Some oak is likewise brought from the country between Montreal and Fort St. Frédéric, or Fort Champlain; but it is not held so good as the first, and the place it comes from is further distant.

August the 26th

To-day they showed some green earth which had been brought to the general, Marquis de la Galissonnière, from the upper parts of

Canada. It was a clay which cohered very fast together, and was of a green color throughout like verdi-gris.[1]

All the brooks in Canada contain *crawfish,* the same kind as ours. The French are fond of eating them and say they have vastly decreased in number since they began to catch them.

The common people in the country seem to be very poor. They have the necessaries of life but little else. They are content with meals of dry bread and water, bringing all other provisions, such as butter, cheese, meat, poultry, eggs, etc. to town, in order to get money for them, for which they buy clothes and brandy for themselves and finery for their women. Notwithstanding their poverty, they are always cheerful and in high spirits.

August the 29th

Down the St. Lawrence. By the desire of the governor-general, Marquis de la Jonquière, and of Marquis de la Galissonnière, I set out with some French gentlemen, to visit the so-called silver mine, or the lead mine, near the Bay of St. Paul. I was glad to undertake this journey, as it gave me an opportunity of seeing a much greater part of the country than I should otherwise have done. This morning therefore we set out on our tour in a boat and went down the St. Lawrence River.

The harvest was now at hand, and I saw all the people at work in the grain fields. They had begun to reap wheat and oats a week ago.

The view near Quebec is very beautiful from the river. The town lies very high and all the churches and other buildings appear very conspicuous. The ships in the river below ornament the landscape on that side. The powder magazine, which stands at the summit of the mountain, on which the town is built, towers above all the other buildings.

The country we passed by afforded a no less charming sight. The St. Lawrence River flows nearly from south to north here; on both sides of it are cultivated fields, but more on the west than on the east side. The hills on both shores are steep and high. A number of fine elevations separated from each other, large fields which

[1] It was probably impregnated with particles of copper ore.—F.

looked quite white with the grain that covered them, and excellent woods of deciduous trees, made the country round us look very pleasant. Now and then we saw a stone church, and in several places brooks fell from the hills into the river. Where the brooks are large enough the people have erected sawmills and grist mills.

After rowing for the space of a French mile and a half we came to the Isle of Orleans, which is a large island, near seven French miles and a half long, and almost two of those miles broad, in the widest part. It lies in the middle of the St. Lawrence River, is very high, and has steep and very woody shores. There are some places without trees, which have farmhouses below, quite close to the shore. The Isle itself is well cultivated and nothing but fine houses of stone, large grain fields, meadows, pastures, woods of hardwood trees, and some churches built of stone, are to be seen on it.

We passed into that branch of the river which flows on the west side of the Isle of Orleans, it being the shortest. It is reckoned about a quarter of a French mile broad, but ships cannot take this route on account of the sandbars which lie here near the projecting points of land, the shallowness of the water and the rocks and stone at the bottom. The shores on both sides still kept the same appearance as before. On the west side, or on the mainland, the hills near the river consist throughout of black slate, and the houses of the peasants are made of this kind of stone, whitewashed on the outside. Some few houses are of a different kind of stone. The chain of high, large mountains which is on the west side of the river, and runs nearly from south to north, gradually comes nearer to the river; for at Quebec they are nearly two French miles distant from the shore, but nine French miles lower down the river they are close to the shore. These mountains are generally covered with woods, but in some places the woods have been destroyed by accidental fires. About seven and a half French miles from Quebec, on the west side of the river, is a church, called St. Anne, close to the shore. This church is remarkable because the ships from France and other parts, as soon as they have gotten far enough up the St. Lawrence River to get sight of it, give a general discharge of their artillery as a sign of joy that they have passed all danger in the river, and have escaped all the sandbars in it.

The water has a pale red color and was very dirty in those parts of the river which we saw to-day, though it was everywhere computed above six fathoms deep. Somewhat below St. Anne, on the west side of the St. Lawrence River, another river, called la Grande Rivière, or the Great River, empties into it. Its water flows with such violence as to make its way almost into the middle of the branch of the St. Lawrence River which runs between the shore and the Isle of Orleans.

About two o'clock in the afternoon the tide began to flow up the river, and the wind being likewise against us we could not proceed any farther till the tide began to ebb. We therefore took up our night lodgings at a great farm belonging to the priests in Quebec, near which is a fine church called St. Joachim, after a voyage of about eight French miles. We were exceedingly well received here. The king has given all the country round about this place to the Seminary, or the priests at Quebec, who have leased it to farmers. They have built houses on it. Here are two priests, and a number of young boys, whom they instruct in reading, writing, and Latin. Most of these boys are designed for the priesthood. Directly opposite this farm to the east is the northernmost point, or the extremity of the Isle of Orleans.

All the gardens in Canada abound with red currant bushes, which were at first brought over from Europe. Everywhere they were red with berries.

The wild grapevines (*Vitis labrusca & vulpina*) grow quite plentifully in the woods. In all other parts of Canada they plant them in the gardens, near arbors and summer houses. The latter are made entirely of laths, over which the vines climb with their tendrils, and cover them entirely with their foliage so as to shelter them entirely from the heat of the sun. They are very refreshing and cool in summer. The wheat was never cut here with a scythe but with a sickle which was larger than usual. These sickles were filed full of grooves on the lower side. The grooves were so deep that the edge too was full of them. There were two kinds of haystacks here on the meadows, one the ordinary cone-shaped, the other like our contrivances for drying peas. I saw no hay sheds out in the fields in this part of Canada.

The strong contrary winds obliged us to lie all night at St. Joachim.

AUGUST THE 30TH

This morning we continued our journey in spite of the wind, which was very violent against us. The water in the river begins to get a brackish taste when the tide is highest, somewhat below St. Joachim, and the further one goes down, the more the saline taste increases of course. At first the western shore of the river has fine but low grain fields, but soon after the high mountains run close to the riverside. Before they come to the river the hilly shores consist of black slate, but as soon as the high mountains appear on the river side the slate disappears. There the stone, of which the high mountains consist, is a chalky rock, mixed with mica and quartz (*Saxum micaceo quarzoso-calcarium*). The mica is black; the quartz partly violet, and partly gray. All the four constituent parts are so well mixed together that they can not be easily separated by an instrument, though plainly distinguishable with the eye. During our journey today the breadth of the river was generally three French miles. They showed me the channels where the ships are obliged to go, which seemed to be very difficult, as the vessels are obliged to bear away from either shore, as occasion requires, or as the rocks and sandbars in the river oblige them to do. We were often obliged on this journey to experience a kind of optical illusion with respect to distances. Often when a high mountain seemed but a short distance away we found that it was several miles distant. There was the same effect upon the land as upon the sea in judging the distance to an island. [This was probably due to the clear atmosphere.—Ed.]

For the distance of five French miles we had a very dangerous passage to go through, for the whole western shore, along which we rowed, consisted of very high and steep mountains, where we could not have found a single place to land with safety for the distance of five miles, in case a high wind had arisen. There are indeed two or three openings, or holes in the mountains, into which one could have drawn the boat in the greatest danger. But they are so narrow that in case the boat could not find them in a hurry, it would inevitably be dashed against the rocks. These high mountains are either quite bare, or covered with some small firs, standing far apart. In some places there are great clefts on the mountain slopes, in which trees grow very close together and are taller than on the other parts of the mountain; so that those places looked like hedges planted on the solid rock. A little while afterward we passed a small church and some farms round it. The place is called Petite Rivière, and they say its inhabitants are very poor, which seems very probable. They have no more land to cultivate than what lies between the mountains and the river, which in the widest part is not above three musketshots, and in most parts but one broad. About seventeen French miles from Quebec the water is so salty in the river that no one can drink it, our rowers therefore provided themselves with a kettle full of fresh spring water this morning. About five o'clock in the evening we arrived at St. Paul's Bay and took our lodgings with the priests, who have a fine large house here and who entertained us very hospitably.

Bay St. Paul is a small parish, about thirteen French miles below Quebec, lying some distance from the shore of a bay formed by the river, on a low plain. It is surrounded by high mountains on every side, one large gap excepted which is over near the river. All the farms are some distance from each other. The church is reckoned one of the most ancient in Canada, which seems to be confirmed by its bad architecture and want of ornaments, for the walls are formed of timber, erected perpendicularly about two feet from each other, supporting the roof. Between these pieces of timber, they have made the walls of the church of black slate. The roof is flat. The church has no steeple, but a bell fixed above the roof, in the open air. Almost all the country in this neighborhood belongs to the priests, who have leased it to the farmers. The inhabitants live chiefly upon agriculture and the making of tar, which is sold at Quebec.

Since this country is low and situated upon a bay of the river, it may be conjectured that this flat ground was formerly part of the bottom of the river, and was formed either by a decrease of water in the river or by an increase of earth, which was carried upon it from the land by the brooks or thrown on it by storms. A great part of the plants which are to be found here are likewise marine, such as glasswort, sea milkwort, and seaside peas (*Salicornia, Glaux, Pisum maritimum*). But when I asked the inhabitants whether they found shells in the ground by digging for wells, they always answered in the negative. I received the same answer from those who live in the low fields directly north of Quebec, and all agreed that they never found anything by digging except different kinds of earth and sand.

It is to be noted that there is generally a different wind in the bay from that in the river, which arises from the high mountains covered with tall woods that surround it on every side but one. For example, when the wind comes from the river, it strikes against one of the mountains at the entrance of the bay, it is reflected, and consequently takes a direction quite different from what it had before.

I found sand of three kinds upon the shore: one is a clear coarse sand, consisting of angulated grains of quartz, and is very common on the shore; another is a fine black sand, which I have likewise found in abundance on the shores of Lake Champlain [1] and which is common all over Canada. Almost every grain of it is attracted by the magnet. Besides this, there is a garnet-colored sand [2] which is likewise very fine. This may owe its origin to the garnet-colored grains of sand which are to be found in all the stones and mountains here near the shore. The sand may have arisen from the crumbled pieces of some stones, or the stones may have been composed of it. I have found both this and the black sand on the shores in several parts of this journey, but the black sand was always the most plentiful.

August the 31st

All the high hills in the neighborhood sent up a smoke this morning as from a charcoal kiln.

Gnats are innumerable here, and as soon as a person comes out of doors they immediately attack him; and they are still worse in the woods. They are exactly the same gnats as our common Swedish ones, being only somewhat smaller than the North American kind. Near Fort St. Jean, I have also seen gnats which were the same as ours, but they were somewhat bigger, almost of the size of our craneflies (*Tipula hortorum* L.). Those which abound here are immeasurably blood-thirsty. However, I comforted myself with the thought that the time of their disappearance was near at hand.

---

[1] See page 385.

[2] See page 386.

En Route to the Mines. This afternoon we went still lower down the St. Lawrence River, to a place where, we were told, there were silver or lead mines. Somewhat below Bay St. Paul we passed a neck of land which consists entirely of a gray, rather compact limestone, lying in tipping and almost perpendicular strata. It seems to be merely a variety of the slate. The strata incline to the southeast and basset out to the southwest. The thickness of each is from ten to fifteen inches. When the stone is broken it has a strong smell, like stinkstone. We kept, as before, to the western shore of the river, which consists of nothing but steep mountains and rocks. The river is not more than three French miles broad here. Now and then we could see stripes in the rock of a fine white, loose, semi-opaque spar. In some places of the river are boulders as big as houses, which have rolled down from the mountains in spring. The places they formerly occupied are plainly to be seen.

In several places they have eel traps in the river, like those I have before described.[1]

Algonquin Words. By way of amusement I wrote down a few Algonquin words which I learned from a Jesuit who has been a long time among the Algonquins. They call water, *nypi*; *mukuman*, knife; the head, *ustigon*; the heart, *uthä*; the body, *wihas*; the foot, *ushita*; a little boat, *ush*; a ship, *nabikoän*; fire, *skute*; hay, *maskusu*; the hare, *whabus*; (they have a verb which expresses the action of hunting hare, derived from the noun); the marten, *whabistania*; the elk, *musu*[2] (but so that the final u is hardly pronounced); the reindeer, *atticku*; the mouse, *manitulsis*; beaver, *amisku*. The Jesuit who told me those particulars, likewise informed me that he had great reason to believe that if any Indians here owed their origin to Tartary, he thought the Algonquins certainly did; for their language is universally spoken in that part of North America which lies far to the west of Canada, towards Asia.

---

[1] See page 423.

[2] The famous "moose deer" is accordingly nothing but an elk; for no one can deny the derivation of moose deer from *moosu*. Considering especially that before the Iroquois or the Five Nations grew to that power, which they at present have all over North America, the Algonquins were then the leading nation among the Indians and their language was of course then a most universal language over the greater part of North America; and though they have been very nearly destroyed by the Iroquois, their language is still more universal in Canada than any of the rest.—F.

It is said to be a very rich language in vocabulary; as, for example, the verb "to go upon the ice", is entirely different in the Algonquin from "to go upon dry land", "to go upon the mountains", etc.

Late last night we arrived at *Terre d'Eboulement,* which is twenty-two French miles from Quebec, and the last cultivated place on the western shore of the St. Lawrence River. The country lower down is said to be so mountainous that nobody can live in it, there not being a single spot of ground which can be tilled. A little church belonging to this place stands on the shore near the water.

No walnut trees grow near this village, nor are there any varieties of them further north of this place. At Bay St. Paul, there are two or three walnut trees of that species which the English call butternut trees; but they are looked upon as great rarities, and there are no others in the neighborhood. Oaks of any kind will not grow near this place, either lower down or further north.

Wheat is the kind of grain which is sown in the greatest quantities here. The soil is pretty fertile, and they have sometimes gotten twenty-four or twenty-six bushels from one, though the harvest is generally ten or twelvefold. The bread here is whiter than anywhere else in Canada. The Canadians sow plenty of oats, and it succeeds better than wheat. They sow likewise a great quantity of peas, which yield a larger crop than any cereal and there are examples of them producing a hundredfold.

Here are but few birds, and those that pass the summer here migrate in autumn, so that there are no other birds than snowbirds, red partridges and ravens in winter. Even crows do not venture to expose themselves to the rigors of winter but take flight in autumn.

The bullfrogs live in the pools of this neighborhood. Fireflies are likewise to be found here.

Instead of candles they use lamps in country places, in which they burn the train-oil of porpoises, a common oil here. Where they have none of it, they use instead the train-oil of seals.

September the 1st

There was a woman with child in this village who was now in the fifty-ninth year of her age. She had not had her catamenia for eighteen years. In the year 1748 she got the small pox and now she was very large. She said she was very well and could feel the motions of the foetus. She looked healthy and her husband was living. Since her case was such an uncommon one she was brought to the royal physician, M. Gauthier, so that he might thoroughly observe her condition. He had accompanied us on this journey.

At half an hour after seven this morning we went down the river. The country near Terre d'Eboulement is high and consists of hills of a loose earth, which lies in three or four terraces above each other, and are all well cultivated and mostly turned into grain fields, though there are likewise meadows and pastures.

The great earthquake which happened in Canada, in February, 1663, and which is mentioned by Charlevoix,[1] had done considerable damage to this place. Many hills had tumbled down, and a great part of the grain fields on the lowest hills had been destroyed. They showed me several little islands which arose in the river on this occasion.

There are pieces of black limeslate scattered on those hills which consist of earth. For the space of eight French miles along the side of the river there is not a piece of slate to be seen; but instead of it there are high gray mountains, consisting of a rock which contains a purple and a crystalline quartz mixed with limestone and black mica. The face of these mountains go into the water. We now began to see the slate again.

The river is here computed to be about four French miles broad.

On the sides of the river, about two French miles inland, there are such terraces of earth as at Terre d'Eboulement; but soon after they are succeeded by high disagreeable-looking mountains.

Several brooks flow with a great noise into the river here over the steep shores. These are sometimes several yards high and consist either of earth or of rock.

A Mineral Spring. One of these brooks which flows over a hill of limestone contains a mineral water. It has a strong smell of sulphur, is very clear, and does not change its color when mixed with gall-apples. If it is poured into a silver cup, it looks as if the cup were gilt; and the water leaves a sediment of a crimson color at the bottom. The stones and pieces of wood which lie in the water are covered with a mud which is pale gray at the top and black at the bottom of the stone. This mud has not much pungency, but

---

[1] See his *Histoire de la Nouvelle France,* tom. II. p. m. 125.—F.

tastes like oil of tobacco. My hands had a sulphurous smell all day because I had handled some of the wet stones.

The slate now abounds again near the level of the water. It lies in strata, which are placed almost perpendicularly near each other, inclining a little towards W. S. W. Each stratum is between ten and fifteen inches thick. Most of them are split into thin slabs where they are exposed to the air, but in the inside, whither neither sun, air nor water can penetrate, they are close and compact. Some of these stones are not quite black but have a grayish cast.

About noon we arrived at *Cap aux Oyes*, or Geese Cape, which has probably gotten its name from the number of wild geese which the French found near it on their first arrival in Canada. At present we saw neither geese nor any kind of birds here, a single raven excepted. Here we were to examine the renowned metallic veins in the mountain; but found nothing more than small veins of a fine white spar, containing a few specks of lead ore. Cap aux Oyes is computed to be from twenty-two to twenty-five French miles distant from Quebec. I was much pleased in finding that most of the plants are the same as those which grow in Sweden; some of which are:

The sand reed (*Arundo arenaria* L.) which grows in abundance in the sand and prevents its being blown about by the wind.

The sea lyme grass (*Elymus arenarius* L.) likewise abounds on the shores. Both it and the preceding plant are called *Seigle de mer* (sea rye) by the French. I have been assured that these plants grow in great plenty in Newfoundland and on other North American shores, the places covered with them looking, at a distance, like grain fields, which might explain the passage in our northern accounts of the excellent Vinland,[1] which mentions that they had found whole fields of wheat growing wild.

The seaside plantain (*Plantago maritima* L.) is found very frequently on the shore. The French boil its leaves in a broth on their sea voyages or eat them as salad. It may likewise be pickled like samphire.

The bear berries (*Arbutus uva ursi* L.) grow in great abundance

[1] *Vinland det goda,* or the good wine-land, is the name which the old Scandinavian navigators gave to America, which they discovered long before Columbus. See Torfæi *Historia Vinlandiæ antiquae S. partis Americæ Septentrionalis,* Hafniæ 1715, 4to. and Mr. George Westmann's A. M. Dissertation on that Subject. Åbo, 1747.—F.
Westmann's thesis was not printed until 1757—Ed. see p. 776, item 35.

here. The Indians, French, English and Dutch, in those parts of North America which I have seen, call them *Sagáckhomi*, and mix the leaves with tobacco for their use. Even the children use only the Indian name for these berries.

Gale, or sweet willow (*Myrica gale* L.) is likewise abundant here. The French call it *laurier,* and some *poivrier.* They put the leaves into their broth to give it a pleasant taste.

The sea rocket (*Bunias cakile* L.) is, likewise, not uncommon. Its root is pounded, mixed with flour, and eaten here when there is a scarcity of bread.

The sorb tree, or mountain ash, the cranberry bush (*lingon*), the juniper tree, the seaside peas, the *Linnæa,* and many other Swedish plants, are likewise to be found here.

We returned to Bay St. Paul to-day. A gray seal swam behind the boat for some time, but was not near enough to be shot.

September the 2nd

Lead Ore. This morning we went to see the silver or lead veins. They lie a little on the south side of the mills, belonging to the priests. The mountain in which the veins lie has the same constituent parts as the other high gray rocks in this place, *viz.* a rock composed of a whitish or pale gray limestone, a purple or almost garnet-colored quartz, and black mica. The limestone is found in greater quantities here than the other parts, and it is so fine that the grain is hardly visible. It effervesces very strongly with *aqua fortis* (nitric acid). The purple or garnet-colored quartz is next in quantity; it lies scattered in exceedingly small grains, and strikes fire when struck with steel. The little black particles of mica follow next; and last of all, the transparent crystalline specks of quartz. There are some small grains of spar in the limestone. All the different kinds of stone are very well mixed together, except that the mica now and then forms little veins and lines. The stone is very hard, but when exposed to sunshine and the open air it changes so much as to look quite rotten, and becomes brittle; in that case its constituent particles grow quite indistinguishable. The mountain is full of perpendicular clefts in which the veins of lead ore run from E. S. E. to W. N. W. It seems the mountain had formerly gotten cracks here which afterwards filled up with a kind of stone in which the lead ore was generated. That stone which contains the lead ore is a soft, white and often semidiaphanous spar, which is very easily malleable. In it there are sometimes stripes of a snowy white limestone and almost always veins of a green kind of stone like quartz. This spar has many cracks, and divides into such pieces as quartz, but is much softer, never strikes fire with steel, does not effervesce with acids, and is not smooth to the touch. It appears to be a variety of Wallerius's[1] vitrescent spar.[2] There are sometimes small pieces of a grayish quartz in this spar, which emit strong sparks of fire when struck with steel. In these kinds of stone the lead ore is lodged. It commonly lies in little lumps of the size of peas, but sometimes in specks of an inch square, or bigger. The ore is very clear, and lies in little cubes.[3] It is generally very poor, a few places excepted. The veins of soft spar and other kinds of stone are very narrow, and commonly from ten to fifteen inches broad. In a few places they are twenty inches broad, and in one single place twenty-two and a half. The brook which intersects the mountain towards the mills, runs down so deep into the mountain that the distance from the edge to the bottom of the brook is nearly twelve yards. Here I examined the veins and found that they always kept the same breadth, not increasing near the bottom of the brook, and likewise that they were no richer below than at the top. Hence it may easily be concluded that it is not worth while sinking mines here. Of these veins there are three or four in this neighborhood at some distance from each other, but all of the same quality. The veins are almost perpendicular, sometimes deviating a little. When pieces of the green stone before mentioned lie in the running water, a great deal of the adherent white spar and limestone is consumed; but the green stone remains untouched. That part of the veins which is turned towards the air and rain has mouldered a great part of the spar and limestone, but the green stone has resisted their attacks. Sometimes deep holes are found in these veins which are filled with mountain crystals. The greatest quantity of lead or silver ore is to be found next to the rock. There

---

[1] Johan Gottschalk Wallerius (1709-1785), Swedish geologist. The Swedish edition of his *Mineralogia* had appeared in 1747.

[2] See Wallerius's *Mineralogy,* Germ. ed. p. 87. Forst. *Introd. to Mineralogy,* p. 13.—F.

[3] It is a cubic lead ore, or galena. Forster's *Introd. to Min.* p. 51.—F.

are now and then little grains of pyrites in the spar, which have a fine gold color. The green stone when pulverized and put on a red-hot shovel burns with a blue flame. Some say they can then observe a sulphurous smell, which I could never perceive, though my sense of smell is perfect. When this green stone has grown quite red-hot, it loses its green color and acquires a whitish one, but will not effervesce with *aqua fortis*.

The sulphurous springs (if I may so call them) are at the foot of the mountain which contains the silver or lead ore. Several springs join here and form a little brook. The water in those brooks is covered with a white scum and leaves a white, mealy matter on the trees and other bodies in its way; this matter has a strong sulphurous smell. Trees covered with this mealy matter, when dried and set on fire, burn with a blue flame and emit a smell of sulphur. The water does not change by being mixed with gall-apples, nor does it change blue paper into a different color. It makes no good lather with soap. Silver is tarnished and turns black if kept in this water for a little while. The blade of a knife was turned quite black after it had lain about three hours in it. It has a disagreeable smell which, they say, it spreads still more in rainy weather. A number of grasshoppers had fallen into it at present. The inhabitants used this water as a remedy against the itch.

Silver Ore. In the afternoon we went to see another vein which had been spoken of as silver ore. It lay about a quarter of a mile to the northeast of St. Paul's Bay, near a point of land called Cap au Corbeau, close to the shore of the St. Lawrence River. The mountain in which these veins lay consisted of a pale red vitrescent spar, a black mica, a pale limestone, purple or garnet-colored grains of quartz, and some transparent quartz. Sometimes the reddish vitrescent spar is the most abundant and lies in long stripes of small hard grains. Sometimes the fine black mica abounds more than the remaining constituent parts; and these two last kinds of stone generally run in alternate stripes. The white limestone, which consists of almost invisible particles, is mixed in among them. The garnet-colored quartz grains appear here and there, and sometimes form whole stripes. They are as big as pin's heads, round, shining, and strike fire with steel. All these stones are very hard, and the mountains near the sea are composed entirely of them. They sometimes lie in almost perpendicular strata of ten or fifteen inches thickness.

The strata, however, point with their upper ends to the northwest, and go upwards from the river, as if the water, which is close to the southeast side of the muontains, had forced the strata to lean toward that side. These mountains contain very narrow veins of a white and sometimes of a greenish, fine semidiaphanous, soft spar, which crumbles easily into grains. In this spar they very frequently find specks, which look like a calamine blend.[1] Now and then, but very seldom, there is a grain of lead ore. The mountains near the shore consist sometimes of a black fine-grained hornstone and a ferruginous limestone. The hornstone in that case is always present in three or four times as great a quantity as the limestone.

In this neighborhood there is likewise a *sulphurous spring*, having exactly the same qualities as that which I have before described. The broad-leaved reed mace (*Typha latifolia* L.) grows right in the spring, and succeeds extremely well. A mountain ash stood near it whose berries were of a pale yellow faded color, whereas on all other mountain ashes they have a deep red color.

Great quantities of tar are made at Bay St. Paul. We now passed near a place in which they burn tar during summer. It is exactly the same as ours in Österbotten, Finland, only somewhat smaller in size, though I have been told that there are sometimes very great manufactures of it here. The tar is made solely from *Pin rouge* or red pine. All other firs, of which there are several kinds here, are not fit for this purpose, because they do not give tar enough to repay the trouble of making it. People use only the roots which are full of resin, and which they dig out of the ground with about two yards of the trunk, just above the root, laying aside all the rest. They have not yet learned the art of drawing the resin from one side of the tree by peeling off the bark; at least they never use this method. The tar barrels are only about half the size of ours. Such a barrel holds forty-six pots,[2] and sells at present for twenty-five francs at Quebec. The tar is considered quite good.

The sand on the shore of the St. Lawrence River consists in some places of a kind of pearl sand. The grains are of quartz, small and translucent. In some places it consists of little particles of mica;

---

[1] Forster's *Introd. to Mineralogy,* p. 50. *Zincum sterilum* Linn. *Syst. Nat.* III, p. 126. Ed. XII.—F.

[2] A pot equals 1.81 quarts or 2 liters, French measure.

and there are likewise spots covered with the garnet-colored sand which I have before described and which abounds in Canada.

September the 4th

The mountains hereabouts were covered with a very thick fog today, resembling the smoke of a charcoal kiln. Many of these mountains are very high. During my stay in Canada, I asked many people who have travelled much in North America, whether they ever met with mountains so high that the snow never melted on them, to which they always answered in the negative. They say that the snow sometimes stays on the highest, *viz.* on some of those between Canada and the English colonies during a great part of the summer, but that it melts as soon as the great heat begins.

Flax. Iron Ore. Every countryman sows as much flax as he wants for his own use. They had already harvested it some time ago, and spread it on the fields, meadows, and pastures, in order to bleach it. It was very short this year in Canada. They find iron ore in several places hereabouts. Almost a Swedish mile from Bay St. Paul up in the country there is a whole mountain full of iron ore. The country round it is covered with a thick forest, and has many rivulets of different sizes which seem to make the erection of iron works very easy here. But since the government has suffered very much by the ironworks at Trois Rivières, nobody ventures to propose anything further in that way.

September the 5th

Early this morning we set out on our return to Quebec. We continued our journey at noon, notwithstanding the heavy rain and thunder we got afterwards. At that time we were just at Petite Rivière, with the tide beginning to ebb, and it was impossible for us to go against it; therefore we lay by here and went on shore.

Petite Rivière is a little village on the western side of the river St. Lawrence and lies on a little creek, from which it takes its name. The houses are built of stone and are dispersed over the country. Here is likewise a fine little church of stone. To the west of the village are some very high mountains, which causes the sun to set three or four hours sooner here than in other places. The St. Lawrence River annually cuts off a piece of land on the east side of the village so that the inhabitants fear they will in a short time lose all the land they possess here, which at most is but a musket-shot broad. All the houses here are full of children.

Schist. The slate schists on the hills are of two kinds. One is black, which I have often mentioned and on which the town of Quebec is built. The other is generally black but sometimes dark gray and seems to be a species of the former. It is called *Pierre à chaux* here. It is chiefly distinguished from the former by being cut very easily, giving a very white lime when burnt, and not easily mouldering into thin slabs in the air. The walls of the houses here are made entirely of this slate, and likewise the chimneys, those places excepted which are exposed to the greatest fire, where they place pieces of gray rock mixed with a deal of mica. The mountains near Petite Rivière consist merely of gray rock which is just the same as that which I described near the lead mines of Bay St. Paul. The foot of these mountains is composed of one of the lime-slates. A great number of the Canada mountains of gray rock rest on a calcareous schist, in the same manner as the gray rocks of Västergötland in Sweden.

September the 6th

Eels and porpoises are caught here at a certain season of the year, *viz.* at the end of September and during the whole month of October. The eels come up the river at that time and are caught in the manner I have before described. They are followed by the porpoises which feed upon them. The greater the quantity of eels, the greater the number of porpoises, which are caught in the following manner. When the tide ebbs in the river, the porpoises commonly go down along the sides of the river, catching the eels which they find there. The inhabitants of this place therefore stick little branches with leaves into the river, in a curved line or arch, the ends of which face the shore, but stand at some distance from it, leaving a passage. The branches stand about two feet distant from each other. When the porpoises come amongst them and perceive the rustling the water makes with the leaves, they dare not venture to proceed, fearing a snare or trap and endeavor to return.

Meanwhile the water has receded so much that in going back they light upon one of the ends of the arch, whose moving leaves frighten them again. In this confusion they swim backwards and forwards till all the water has ebbed off, and they lie on the bottom where the inhabitants kill them. They furnish a great quantity of train-oil.

Near the shore is a gray clay, full of ferruginous cracks, and pierced by worms. The holes are small, perpendicular, and big enough to admit a medium-sized pin. Their sides are likewise ferruginous and half petrified; and where the clay has been washed away by the water, the rest looks like ochre-colored stumps of tobacco pipe tubes.

At noon we left Petite Rivière, which lies in a little bay, and proceeded to St. Joachim. Between these two places the western shore of the St. Lawrence River consists of prominent mountains, between which there are several small bays. It has been found by long experience that there is always a wind on these mountains. And when the wind is pretty strong at the last mentioned place it is not advisable to go to Quebec in a boat, the wind and waves in that case being very strong near these mountains. We had at present an opportunity of experiencing it. In the creeks between the mountains, the water was calm, but on our coming near one of the points formed by the high mountains the waves increased and the wind was so strong that two people were forced to take care of the helm, and the mast broke. The waves are likewise greatly increased by the strong current near those points or capes.

September the 7th

A little before noon, we continued our voyage from St. Joachim.

Tree fungi are used very frequently instead of tinder. Those which are taken from the sugar maple are reckoned the best; those of the red maple are next in quality; and next to them, those of the sugar birch. For want of these they make use of those which grow on the aspen tree or tremble.

There are no other evergreen trees in this part of Canada than the thuya, the yew, and some of the fir. The thuya is esteemed for resisting decay much longer than any other wood; and next in value to it is the pine, called *pérusse* here.

They make *cheese* in several places here. That of the Isle of Orleans is, however, reckoned the best. This kind is small, thin, and round, and four such cheeses weigh about a French pound. Twelve of them sell for thirty sols. A pound of salt butter costs ten sols at Quebec, and of fresh butter, fifteen sols. Formerly they could get a pound of butter for four sols here.

The grain fields slope towards the river; they are allowed to lie fallow and to be sown alternately. The sown ones looked yellow at a distance and the fallow ones green. The weeds are left on the latter all summer for the cattle to feed upon.

The ash tree furnishes the best hoops for barrels here; and for want of it, they take the thuya, little birch trees, wild cherry trees and others.

Black Lime Schist. The hills near the river, on the western side opposite the Isle of Orleans, are very high and pretty steep. They consist for the most part of black schist. There are likewise some spots which consist of a rock which at first sight looks like a sandstone, and is composed of gray quartz, a reddish limestone, a little gray limestone, and some pale gray grains of sand. These parts of the stone are small and pretty equally mixed with each other. The stone looks red, with a grayish cast, and is very hard. It lies in strata, one above another. The thickness of each stratum is about five inches. It is remarkable that there are both elevated and hollow impressions of pectinites on the surface where one likewise meets with the petrified shells themselves; but on breaking the stone it does not even contain the least vestige of an impression or petrified shell. All the fossils are small, about the length and breadth of an inch. The particles of quartz in the stone strike fire with steel, and the particles of limestone effervesce strongly with *aqua fortis.* The upper and lower surfaces of the strata consist of limestone and the inner parts of quartz. Large quantities of this stone are quarried in order to build houses, pave floors, and make stair-cases of it. Great quantities of it are sent to Quebec. It is to be noted that there are petrefactions in this stone, but never any in the black calcareous schist.

The women dye their woolen yarn yellow with seeds of gale (*Myrica gale* L.), which is called *poivrier* here, and grows abundantly in wet places.

This evening M. Gauthier and I went to see the waterfall at Montmorency. The country near the river is high and level and laid out in meadows. Above them the high and steep hills begin, which are covered with a crust of earth and turned into grain fields. In some very steep places and near the rivulets, the hills consist of mere black lime schist, which is often crumbled into small pieces, like earth. All the fields below the hills are full of such pieces of schist. When some of the larger pieces are broken, they smell like stinkstone. In some more elevated places, the earth consists of a pale red color; and the schists are likewise reddish.

The *waterfall near Montmorency* is one of the highest I ever saw. It is in a river whose breadth is not very great and falls over the steep side of a hill, consisting entirely of black lime schist. The fall is now at the head of the bay. Both sides of the bay consist merely of the same schist, which is very much cracked and eroded, and though sloping one could hardly walk up it. The hill of schist under the waterfall is quite perpendicular, and one cannot look without astonishment at the quantity of water going over it. The rain of the preceding days had increased the water in the river to such an extent that it was terrifying to see such an amount of water hurling itself over the precipice. The breadth of the fall is not above ten or twelve yards. Its perpendicular height M. Gauthier and I guessed to be between a hundred and ten and a hundred and twenty feet; and on our return to Quebec, we found our guess confirmed by several gentlemen who had actually measured the fall and found it to be nearly as we had estimated. The people who live in the neighborhood exaggerate in their accounts of it, absolutely declaring that it is three hundred feet high. Father Charlevoix[1] is too sparing in giving it only forty feet in height.[2] At the bottom of the fall, there is always a thick fog of vapors, spreading about the water, being resolved into them by its violent fall. This fog occasions almost perpetual rain here, which is more or less heavy, in proportion to its distance from the fall. M. Gauthier and myself, together with the man who showed us the way, were willing to come nearer to the falling water in order to examine more accurately how it came down from such a height and how the stone

[1] See his *Histoire de la Nouv. France,* tom. V. p. m. 100.—F.

[2] Actual height of falls is 250 feet.

behind the water looked. But, when we were about twelve yards away from the fall, a sudden gust of wind blew a thick spray upon us, which in less than a minute had wet us thoroughly as if we had walked for half an hour in a heavy shower. We therefore hurried as fast as we could and were glad to get away. The noise of the fall is sometimes heard at Quebec which is two French miles to the south; and this is a sign of a northeast wind. At other times it can be well heard in the villages a good way to the north, and it is then reckoned an undoubted sign of a southwest wind or of rain. The black lime schist on the sides of the fall lies in dipping and almost perpendicular strata. In these lime slate strata, are the following kinds of stone.

Fibrous gypsum. This lies in very thin leaves between the cracks of the schist. Its color is a snowy white. I have found it in several parts of Canada in the same black limestone.

Pierre à Calumet. This is the French name of a stone disposed in strata between the lime schist and of which they make almost all the tobacco pipe heads in the country. The thickness of the strata varies. I have seen pieces nearly fifteen inches thick, but they are commonly between four and five inches. When the stone is long exposed to the open air or heat of the sun it becomes a yellowish color, but on the inside it is gray. It is of such compactness that its particles are not distinguishable by the naked eye. It is pretty soft, and will bear cutting with a knife. From this quality the people likewise judge the suitability of the stone for tobacco heads; for the hard pieces of it are not so fit for use as the softer ones. I have seen some of these stones split into thin leaves on the outside, where they were exposed to the sun. All the tobacco pipe heads, which the common people in Canada use, are made of this stone, and ornamented in different ways. A great part of the gentry likewise use them, especially when they are on a journey.

The Indians have employed this stone for the same purposes for several ages past, and have taught it to the Europeans. The heads of the tobacco pipes are naturally of a pale gray color, but they are blackened while they are quite new to make them look better. People cover the head all over with grease and hold it over a burning candle or any other fire, by which means it gets a good black color, which is increased by frequent use. The tubes of the pipes are always made of wood[1] and a brass wire holds them to the head.

There is no coal near this fall, or in the steep hills close to it. However, the people in the neighboring village showed me a piece of coal, which, they said, had been found on one of the hills near the fall.

We arrived at Quebec very late at night.

SEPTEMBER THE 8TH

Intermittent fevers of any kind are very rare at Quebec, M. Gauthier affirms. On the contrary, they are very common near Fort St. Frédéric and near Fort Detroit, which is a French colony, and between Lake Erie and Lake Huron, in forty-three degrees north latitude. Most of the natives told me something which I have mentioned before concerning both Englishmen and Frenchmen, namely, that native Europeans live longer than native Canadians. They also assured me that the second generation born here, and still more the third, does not reach the age of the preceding one, though one reason was ascribed to the dangerous journeys of the men for furs among the savages. It was said to be very rare to see a centenarian in Canada, though here and there an octogenarian was to be found.

Some of the people of quality make use of ice cellars to keep beer cool during the summer and to preserve fresh meat in the great heat. These cellars are commonly built of stone, under the house. The walls of them are covered with boards, because the ice is more easily consumed by stones. In winter they fill them with snow, which is beat down with the feet and covered with water. They then open the cellar holes and the door to admit the cold.—It is customary in summer to put a piece of ice into the water or wine which is to be drunk.

All the salt which is used here is imported from France. They had also made good salt in Canada from sea water; but since

[1] All over Poland, Russia, Turkey, and Tartary, they smoke pipes made of a kind of stone-marle, to which they fix long wooden tubes; for which latter purpose, they commonly employ the young shoots of the various kinds of Spiræa, which have a kind of pith easily to be thrust out. The stone-marle is called generally sea-scum, being pretty soft; and by the Tartars, in Crimea, it is called *keffekil*. And as it cuts so easily, various figures are curiously carved in it, when it is worked into pipe-heads, which often are mounted with silver.—F.

This is apparently the modern meerschaum pipe.

France keeps the salt trade entirely to itself, they had not continued to make it.

The Esquimaux are a particular kind of American savage, who live only near the water and never far in the country, on Terra Labrador, between the most outward point of the mouth of the St. Lawrence River and Hudson Bay. I have never had an opportunity of seeing one of them. I have spoken with many Frenchmen who have seen them and had them on board their own vessels. I shall here give a brief history of them, according to their unanimous accounts.

The Esquimaux are entirely different from the Indians of North America in regard to their complexion and their language. They are almost as white as Europeans, and have little eyes: the men, also, have beards. The Indians, on the contrary, are copper-colored, and the men have no beards. The Esquimaux language is said to contain some European words.[1] Their houses are either caverns or clefts in the mountains or huts of turf above ground. They never sow or plant vegetables, living chiefly on various kinds of whales, on seals (*Phoca vitulina* L.) and walruses (*Trichechus rosmarus* L.). Sometimes they likewise catch land animals, on which they feed. They eat most of their meat raw. Their drink is water and people have seen them drinking the sea water which was like brine.

Their shoes, stockings, breeches, and jackets are made of seal skins well prepared, and sewed together with sinews of whales, which may be twisted like threads and are very tough. Their clothes, the hairy side of which is turned outwards, are sewed together so well that they can walk up to their shoulders in water without wetting their underclothes. Under their upper clothes they wear shirts and waistcoats made of sealskins, prepared so well as to be quite soft. I saw one of their women's dresses: a cap, a waistcoat and skirt, made all of one piece of sealskin well prepared, soft to the touch, and the hair on the outside. There was a long train,

[1] The Moravian brethren in Greenland, coming once over with some Greenlanders to Terra Labrador, the Esquimaux ran away at their appearance; but they ordered one of their Greenlanders to call them back in his language. The Esquimaux, hearing his voice, and understanding the language, immediately stopped, came back, and were glad to find a countryman, and wherever, they went, among the other Esquimaux, they gave out, that one of their brethren was returned. This proves the Esquimaux to be of a tribe different from any European nation, as the Greenland language has no similarity with any language in Europe.—F.

about six inches wide, on the back of the skirt. In front it scarcely reached to the middle of the thigh, but under it the women wore breeches and boots, all of one piece. The Esquimaux women are said to be handsomer than any women of the American Indians, and their husbands are accordingly more jealous.

A Kayak. I have likewise seen an Esquimaux boat. The outside of it consisted entirely of skins, the hair of which had been taken off, so that they felt as smooth as vellum. The boat was near six feet long, but very narrow, and very sharp-pointed at the extremities. On the inside of such a boat, they place two or three thin boards, which give a kind of form to the boat. It is covered with skins at the top, excepting near one end, where there is a hole big enough for a single person to sit and row in and keep his thighs and legs under the cover. The shape of the hole resembles a semicircle, the base or diameter of which is turned towards the larger end of the boat. The hole is surrounded by wood, on which a soft skin is fastened, with straps in its upper part. When the Esquimau makes use of his boat he puts his legs and thighs under the deck, sits down at the bottom of the boat, draws the skin before mentioned round his body, and fastens it well with the straps; the waves may then beat over his boat with considerable violence, and not a single drop comes into it; the clothes of the Esquimaux keep him dry. He has an oar in his hand, which has a blade or paddle at each end; it serves him for rowing and keeping the boat in equilibrium during a storm. The paddles of the oar are very narrow. The boat will hold but a single person. Esquimaux have often been found safe in their boats many miles from land, in violent storms, where ships found it difficult to save themselves. Their boats float on the waves like bladders, and they row them with an incredible velocity. I am told they have boats of different shapes. They have likewise larger boats of wood, covered with skins, in which several people may sit and in which their women commonly go to sea.

Arms. Bows and arrows, javelins and harpoons, constitute their arms. With the last they kill whales and other large marine animals. The points of their arrows and harpoons are sometimes made of iron, sometimes of bone, and sometimes of the teeth of the walrus. Their quivers are made of seals' skins. The needles with which they sew their clothes are likewise made of iron or of bone. All their iron they get by some means or other from the Europeans.

They sometimes go on board the European ships in order to exchange some of their goods for knives and other iron. But it is not advisable for Europeans to go on shore, unless they be numerous, for the Esquimaux are false and treacherous and cannot suffer strangers amongst them. If they find themselves too weak, they run away at the approach of strangers; but if they think they are an over-match for them, they kill all that come in their way, without leaving a single one alive. The Europeans, therefore, do not venture to let a greater number of Esquimaux come on board their ships than they can easily master. If they are ship-wrecked on the Esquimaux coasts, they may as well be drowned in the sea as come safe to the shore: this many Europeans have experienced. The European boats and ships which the Esquimaux get into their power are immediately cut to pieces and robbed of all their nails and other iron which they work into knives, needles, arrow-heads, etc. They make use of fire for no other purposes than the working of iron and the preparing of the skins of animals. Their meat is eaten raw. When they come on board a European ship and are offered some of the sailors' food, they will never taste of it till they have seen some Europeans eat it. Though nothing pleases other savage nations so much as brandy, yet many Frenchmen have assured me that they never could prevail on the Esquimaux to take a dram of it. Their mistrust of other nations is the cause of it; for they undoubtedly imagine that they are going to poison them, or do them some harm; and I am not certain whether they do not judge right. They have no earrings, and do not paint the face like the American Indians. For many centuries past they have had dogs whose ears are erect and never hang down. They use them for hunting and instead of horses in winter for drawing their goods on the ice. They themselves sometimes ride in sledges drawn by these dogs. They have no other domestic animal. There are, indeed, plenty of reindeer in their country; but it is not known that either the Esquimaux or any of the Indians in America have ever tamed them. The French in Canada, who are in a manner the neighbors of the Esquimaux, have taken a deal of pains to carry on some kind of trade with them and to endeavor to engage them in a more friendly intercourse with other nations. For that purpose they took some

Esquimaux children, taught them to read, and educated them in the best manner possible. The intention of the French was to send these children back to the Esquimaux, that they might inform them of the kind treatment the French had given them and thereby incline them to conceive a better opinion of the French. But unhappily all the children died of the small pox and the scheme was dropped. Many persons in Canada doubted whether the scheme would have succeeded, though the children had survived. For they say there was formerly an Esquimau taken by the French and brought to Canada, where he stayed a good while and was treated with great civility. He learnt French pretty well, and seemed to relish the French way of living. When he was sent back to his countrymen, he was not able to make the least impression on them in favor of the French, but was killed by his nearest relations as half a Frenchman and a foreigner. This inhuman proceeding of the Esquimaux against all strangers is the reason why none of the Indians of North America ever give quarter to the Esquimaux if they meet them, but kill them on the spot, though they frequently pardon their other enemies, and incorporate the prisoners with their nation.

For the use of those who are fond of comparing the languages of several nations, I have here inserted a few Esquimaux words, communicated to me by the Jesuit Saint Pié. One, *kombuc*; two, *tigal*; three, *ké*; four, *missilagat*; water, *willalokto*; rain, *killaluck*; heaven, *taktuck*, or *nabugakshe*; the sun, *shikonak*, or *sakaknuk*; the moon, *takock*; an egg, *mannejuk*; the boat, *kayack*; the oar, *pacotick*; an arrow, *katso*; the head, *niakock*; the ear, *tchiu*; the eye, *killik*, or *shik*; the hair, *nutshad*; a tooth, *ukak*; the foot, *itikat*. Some think that the Esquimaux are nearly of the same origin as the Greenlanders, or Skralingers,[1] and pretend that there is a great affinity in the language.[2]

[1] The early Scandinavians who visited America in the tenth and eleventh centuries called the native North Americans *skrælingar* or *skrælings;* Dan. *skrælling,* a wretch. Whether they were Indians or Esquimaux is not absolutely certain, but to-day the opinion seems to favor the Indian theory. The real meaning of the term *skræling* is also unknown, but may be connected with an Old Norse word for "screaming" or "making a noise" (mod. Sw. *skrälla* or *skråla*), as the savage Indians were wont to do when going into battle, and as the natives did who encountered the Northmen. To-day the vast majority of the Greenlanders are Esquimaux. Kalm evidently here thought of the natives met by his ancient ancestors as Esquimaux.

[2] The above account of the Esquimaux may be compared with Henry Ellis's *Ac-*

Plum trees of different sorts, brought over from France, succeed very well here. The present year they did not begin to flower till this month. Some of them looked very well, and I am told the winter does not hurt them.

September the 11th

The Marquis de la Galissonnière is one of the three noblemen, who, above all others, have gained high esteem with the French admiralty in the last war. The three are the Marquis de la Galissonnière, de la Jonquière, and de l'Etenduere.[1] The first of these was now above fifty years of age, of a low stature, and somewhat humpbacked, but of a very agreeable appearance. He had been here for some time as governor-general, and was soon going back to France. I have already mentioned something concerning this nobleman; but when I think of his many great qualities, I can never give him a sufficient encomium. He has a surprising knowledge in all branches of science, and especially in natural history, in which he is so well versed that when he began to speak with me about it I imagined I saw our great Linné under a new form. When he spoke of the use of natural history, of the method of learning, and employing it to raise the state of a country, I was astonished to see him take his reasons from politics, as well as natural philosophy, mathematics and other sciences. I own that my conversation with this nobleman was very instructive to me; and I always drew a deal of useful knowledge from it. He told me several ways of employing natural history to the purposes of politics, the science of government, and to make a country powerful in order to weaken its envious neighbors. Never has natural history had a greater promoter in this country; and it is very doubtful whether it will ever have his equal here. As soon as he got the place of governor-general, he began to take those measures for getting information in natural

count of a Voyage to Hudson's Bay, by the Dobbs Galley and California, etc. and *The Account of a Voyage for the Discovery of a North West Passage by Hudson's Straights*, by the Clerk of the California. Two vols. 8vo. And lastly, with Crantz's *History of Greenland*. two vols. 8vo.—F.

[1] This should be l'Estanduere. See Marchand, *op. cit.* 2, 183, note. He distinguished himself in a naval battle with the English, under vice-admiral Hawke, the following year.

history which I have mentioned before. When he saw people who had for some time been in a settled place of the country, especially in the more remote parts, or had travelled in those parts, he always questioned them about the trees, plants, earths, stones, ores, animals, etc. of the place. He likewise inquired what use the inhabitants made of these things; in what state their husbandry was; what lakes, rivers, and passages there were; and a number of other particulars. Those who seemed to have clearer notions than the rest were obliged to give him circumstantial descriptions of what they had seen. He himself wrote down all the accounts he received; and by this great application, so uncommon among persons of his rank, he soon acquired a knowledge of the most distant parts of America. The priests, commandants of forts, and of several distant places were often surprised by his questions, and wondered at his knowledge, when they came to Quebec to pay their visits to him; for he often told them that near such a mountain or on such a shore, etc. where they often went hunting, there were some particular plants, trees, soils, ores, etc. for he had gotten a knowledge of those things before. Hence it happened, that some of the inhabitants believed he had a preternatural knowledge of things, as he was able to mention all the curiosities of places, sometimes nearly two hundred Swedish miles from Quebec, though he had never been there himself, and though the others, on the other hand, had lived there for years. A person who did not know this gentleman well enough would have considered him dry and only moderately pleasant, in social relations, especially for one who had not penetrated into the sciences. But the more one became acquainted with him the better his good qualities appeared and the greater became the cause for respecting a person who was characterized by everything big. Never was there a better statesman than he; and nobody could take better measures or choose more proper means for improving a country and increasing its welfare. Canada was hardly acquainted with the treasure it possessed in the person of this nobleman, when it lost him again. The king wanted his services at home and could not leave him so far off. He was going to France with a collection of natural curiosities, and a quantity of young trees and plants, in boxes full of earth. I cannot describe all the favors he showed me. It was greater than I could have expected in my own fatherland. I do not know whether the natives or the sciences will miss him most, because he was the tenderest of fathers for both, and for the latter the biggest patron and promoter that any place has been able to show. Happy the country that has such a chief! There it is not necessary to lament about egoistic and imaginary obstacles for promoting deeds of public welfare. Such a chief gives encouragement to all things that benefit a fatherland.

Black lime schist (*or slate*) has been repeatedly mentioned during the course of my journey. I will here give a more minute description of it. The mountain on which Quebec is built, and the hills along the St. Lawrence River consist of it for several miles on both sides of Quebec. About a yard from the surface this stone is quite compact and without any cracks, so that one cannot perceive that it is slate, its grain being imperceptible. It lies in strata, which vary from three or four inches to twenty inches thickness or more. In the mountains on which Quebec is built, the strata do not lie horizontal, but dip so as to be nearly perpendicular, the upper ends pointing northwest and the lower ones southeast. For this reason the corners of these strata always stand out at the surface into the streets and cut the shoes in pieces. I have likewise seen some strata inclining to the north but nearly as perpendicular as the former. Horizontal strata, or nearly such, have been found too. The strata are divided by narrow cracks, which are commonly filled with fibrous white gypsum that can sometimes be pried loose with a knife, if the layer of stratum of slate above it is broken in pieces. And in that case it has the appearance of a thin white leaf. The larger cracks are almost filled up with transparent quartz crystals of different sizes. One part of the mountain contains vast quantities of these crystals, from which that point of the mountain which lies to the S. S. E. of the palace has got the name of Pointe de Diamante, or Diamond Point. The small cracks which divide the stone run generally at right angles. The distances between them are not always equal. The outside of the stratum, or that which is turned towards the other stratum, is frequently covered with a fine, black, shining membrane which looks like a kind of pyrous hornstone. In it there is sometimes a yellow pyrites in the form of small grains. I never found fossils or any other kinds of stone in it besides those I have just mentioned. The whole mountain on which Quebec is situated, consists entirely of lime schist from top to bottom. When this stone is broken, or scraped with a knife, it gives a strong smell like stinkstone. That part of the mountain which is exposed to the open air, has crumbled into small pieces, which have lost their black color and turned a pale red instead. Almost all the public and private buildings at Quebec are built of this schist, and likewise the walls round the town, and round the monasteries and gardens. It is easily quarried, and cut to the size desired. But it has the property of splitting into thin slabs, parallel to the surface of the stratum whence they are taken, after lying during one or more years in the air, and exposed to the sun. However, this quality does no damage to the walls in which they are placed; for the stones are laid on purpose in such a position that the cracks always run horizontally, and upper stones press so much upon the lower ones that they can only get cracks and split on the outside without going further inwards. The slabs always grow thinner as the houses grow older.

The Quebec Climate. In order to give my readers some idea of the climate of Quebec and of the different changes of heat and cold at the several seasons of the year, I will here insert some particulars extracted from the meteorological observations of the royal physician, M. Gauthier. He gave me a copy of those which he had made from October, 1744, to the end of September, 1746. The thermometrical observations I shall omit, because I do not believe them accurate. M. Gauthier made use of de la Hire's thermometer, in which the degrees of cold cannot be exactly determined since all the quicksilver in it is compressed into the globe at the bottom before the intense cold sets in. The observations are made throughout the year, between seven and eight in the morning and two and three in the afternoon, but he has seldom made any after noon. His thermometer was likewise inaccurate, for it was placed in a bad location, sometimes in an open window, and sometimes in the sun. —The new style is used in dates.

The year 1745

January. The 29th of this month the St. Lawrence River was covered over with ice near Quebec. In the observations of other years it is found that the river is sometimes covered with ice in the beginning of January or the beginning of December.

February. Nothing remarkable happened during the course of this month.

March. They say this has been the mildest winter they ever felt; even the eldest persons could not remember one so mild. The snow was only two feet deep, and the ice in the river opposite Quebec had the same thickness. On the twenty-first there was a thunder-storm, a soldier was struck and hurt very much. On the 19th and 20th they began to make incisions into the sugar maple and to prepare sugar from its sap.

April. During this month they continued to extract the sap of the sugar maple for making sugar. On the 7th the gardeners began to make hotbeds. On the 20th the ice in the river broke loose near Quebec and went down. This rarely happens so soon, for the St. Lawrence River is sometimes covered with ice opposite Quebec on the 10th of May. On the 22nd and 23rd, there fell a quantity of snow. On the 25th they began to sow near St. Joachim. The same day they saw some swallows. The 29th they sowed grain all over the country. Ever since the 23rd the river had been clear at Quebec.

May. The third of this month the cold was so great in the morning that the Celsius or Swedish [centigrade] thermometer was four degrees below the freezing point; however, it did not hurt the grain. On the 16th all the spring sowing was done. On the 5th the sanguinaria, narcissus, and violet began to bloom. The 17th the wild cherry trees, raspberry bushes, apple trees and lime trees, began to put out their leaves. The strawberries were in flower about that time. On the 29th the wild cherry trees were in blossom. On the 26th, part of the French apple trees and plum trees bloomed.

June. By the 5th of this month all the trees had leaves. The apple trees were in full flower. Ripe strawberries were to be had on the 22nd. Here it is noted that the weather was very fine for the growth of vegetables.

July. The grain began to form into ears on the 12th, and had ears everywhere on the 21st. (It is to be observed that they sow nothing but spring seed here.) Soon after it began to flower. Hay making began the 22nd. All this month the weather was excellent.

August. On the 12th there were ripe pears and melons at Montreal. On the 20th the wheat was ripe round Montreal, and the harvest began there. On the 22nd the harvest began at Quebec. On

the 30th and 31st, there was a very light hoarfrost on the ground.
September. The harvest of all kinds of grain ended on the 24th and 25th. Melons, watermelons, cucumbers, and fine plums were very plentiful during the course of this month. Apples and pears were likewise ripe, which is not always the case. On the last days of this month they began to plow the land. The following is one of the observations of this month: "The old people in this country say that the grain was formerly seldom ripe till the 15th or 16th of September, and sometimes on the 12th, but no sooner. They likewise assert that it was never perfectly ripe. But since the woods have been sufficiently cleared the beams of the sun have had more room to operate and the grain ripens sooner than before." It is further remarked that the hot summers are always very fruitful in Canada, but that in most years only one-tenth of the grain ever arrives at perfect maturity.
October. During this month the fields were plowed, and the weather was very fine all the time. There was a little frost for several nights, and on the 28th it snowed. Towards the end of this month the trees began to shed their leaves.
November. They continued to plow till the 10th of this month, when the trees had shed all their leaves. Till the 18th the cattle went out of doors, a few days excepted when bad weather had kept them inside. On the 16th there was some thunder and lightning. There was not yet any ice in the St. Lawrence on the 24th.
December. During this month it is observed that the autumn has been much milder than usual. On the 1st a ship could still set sail for France; but on the 15th the St. Lawrence was covered with ice on the sides, though open in the middle. In the Charles River the ice was thick enough for horses with heavy loads to pass over it. On the 26th the ice in the St. Lawrence River was washed away by a heavy rain; but on the 28th, part of that river was again covered with ice.

The next observations show that this winter, too, has been one of the mildest. I now resume the account of my journey.

This evening I left Quebec with a fair wind. The governor-general of Canada, the Marquis de la Jonquière, ordered one of the king's boats and seven men to bring me to Montreal. The middle of the boat was covered with blue cloth under which we were secured from the rain. This whole journey I made at the expense of the French king.[1] We went three French miles to-day.

September the 12th

We continued our journey all day.

The small kind of maize, which ripens in three months' time, was ripe about this date and harvested and hung up to dry.

The weather about this time was like the beginning of our August, old style. Therefore, it seems, autumn commences a whole month later in Canada than in the central part of Sweden.

Kitchen Gardens. Near each farm there is a kitchen garden in which onions are most abundant, because the French farmers eat their dinners of them with bread, on Fridays and Saturdays, or fasting days. However, I cannot say the French are strict observers of fasting, for several of my rowers ate meat to-day, though it was Friday. The common people in Canada may be smelled when one passes by them on account of their frequent use of onions. Pumpkins also are abundant in the farmers' gardens. They prepare them in several ways, but the most common is to cut them through the middle, and place each half on the hearth, open side towards the fire, till it is roasted. The pulp is then cut out of the peel and eaten. Better class people put sugar on it. Carrots, lettuce, Turkish beans, cucumbers, and currant shrubs, are planted in every farmer's little kitchen garden.

Tobacco. Every farmer plants a quantity of tobacco near his house, in proportion to the size of his family. It is necessary that one should plant tobacco, because it is so universally smoked by the common people. Boys of ten or twelve years of age, as well as the old people, run about with a pipe in their mouth. Persons of the better class do not refuse either to smoke a pipe now and then. In the northern parts of Canada they generally smoke pure tobacco; but further north and around Montreal, they take the inner bark of the red Cornelian cherry (*Cornus sanguinea* L.), crush it, and mix it with the tobacco, to make it weaker. People of both sexes,

[1] According to Jos. E. Roy in *Voyage de Kalm en Canada* (1900) Kalm's total expenses in Canada amounted to 2,182 livres, all paid by the French Government.

and of all ranks, use snuff very much. Almost all the tobacco which is consumed here is the product of the country, and some people prefer it even to Virginian tobacco: but those who pretend to be connoisseurs reckon the last kind better than the other.

Manners and Customs. Though many nations imitate the French customs, I observed, on the contrary, that the French in Canada in many respects follow the customs of the Indians, with whom they have constant relations. They use the tobacco pipes, shoes, garters, and girdles of the Indians. They follow the Indian way of waging war exactly; they mix the same things with tobacco; they make use of the Indian bark boats and row them in the Indian way; they wrap a square piece of cloth round their feet, instead of stockings, and have adopted many other Indian fashions. When one comes into the house of a Canadian peasant or farmer, he gets up, takes his hat off to the stranger, invites him to sit down, puts his hat on and sits down again. The gentlemen and ladies, as well as the poorest peasants and their wives, are called Monsieur and Madame. The peasants, and especially their wives, wear shoes which consist of a piece of wood hollowed out, and are made almost as slippers. Their boys and the old peasants themselves wear their hair behind in a queue, and most of them wear red woolen caps at home and sometimes on their journeys.

Food. The farmers prepare most of their food from milk. Butter is seldom seen, and what they have is made of sour cream, and therefore not so good as English butter. A good deal of this butter has a slight taste of tallow. Congealed sour milk is found everywhere, in stone vessels. Many of the French are very fond of milk, which they eat chiefly on fast days. However, they have not so many methods of preparing it as we have in Sweden. The common way is to boil it, and put bits of wheat bread and a good deal of sugar into it. The French here eat nearly as much meat as the English on those days when their religion allows it. For excepting the soup, the salads and the dessert, all their other dishes consist of meat variously prepared.

At night we slept at a farmhouse near the river called Petite Rivière, which entered here into the St. Lawrence River. This place was reckoned sixteen French miles from Quebec, and ten from Trois Rivières. The tide was still strong here. Here is the last place where the hills along the river consist of black lime schist; further on they are composed merely of earth.

Fireflies flew about the woods at night, though not in great numbers; the French call them *Mouches à feu.*

The houses in this neighborhood are all made of wood. The rooms are pretty large. The inner roof rests on two, three, or four large, thick spars, according to the size of the room. The chinks are filled with clay instead of moss. The windows are made entirely of paper. The chimney is erected in the middle of the room; that part of the room which is opposite the fire is the kitchen; that which is behind the chimney serves the people for sleeping and entertaining strangers. Sometimes there is an iron stove behind the chimney.

September the 13th

Near *Champlain*, which is a place about five French miles from Trois Rivières, the steep hills near the river consist of a yellow and sometimes ochre-colored sandy earth, in which a number of small springs arise. The water in them is generally filled with yellow ochre, which is a sign that these dry sandy fields contain a great quantity of the same iron ore which is dug at Trois Rivières. It is not conceivable whence that number of small rivulets arise, the ground above being flat, and exceedingly dry in summer. The lands near the river are cultivated for about an English mile into the country, but behind them there are thick forests and low grounds. The woods, which collect a quantity of moisture and prevent the evaporation of the water, force it to make its way under ground to the river. The shores of the latter are here covered with a great deal of black iron sand.

Towards evening we arrived at Trois Rivières, where we stayed no longer than was necessary to deliver the letters which we had brought with us from Quebec. After that we went a French mile further north before we took up our night's lodging.

Three Old People. This afternoon we saw three remarkable old people. One was an old Jesuit, called father Joseph Aubery, who had been a missionary to the converted Indians of St. François. This summer he ended the fiftieth year of his mission. He therefore returned to Quebec to renew his vows there, and he seemed to be healthy and in good spirits. The other two people were our landlord and his wife; he was above eighty years of age, and she was not much younger. They had now been fifty-one years married. The year before, at the end of their fiftieth year of their marriage, they had gone to church together and offered up thanks to God Almighty for the great grace he had given them. They were yet quite well, content, merry and talkative. The old man said that he was at Quebec when the English besieged it, in the yaar 1690, and that the bishop went up and down the streets, dressed in his pontifical robes, with a sword in his hand, in order to encourage the soldiers.

The old man said that he thought the winters were formerly much colder than they are now. There fell likewise a greater quantity of snow when he was young. He could remember the time when pumpkins, cucumbers, etc. were killed by the frost about mid-summer, and he assured me that the summers were warmer now than they used to be. About thirty and some odd years ago there had been such a severe winter in Canada that the frost killed many birds; but the old man could not remember the particular year. Everybody admitted that the summers in 1748 and 1749 had been warmer in Canada than they had been in many years.

The soil is considered quite fertile here, and wheat yields a nine or tenfold harvest. But when this old man was a boy and the country was new and rich everywhere, they could get a twenty or four-and-twentyfold crop. They sow but little rye here, nor do they sow much barley, except for the use of the cattle. They complain, however, that when they have a bad crop, they are obliged to bake bread of barley in place of wheat.

September the 14th

This morning we got up early and pursued our journey. After we had gone about two French miles we reached Lake St. Pierre, which we crossed. Many plants which are common in our Swedish lakes swim on the top of this water. This lake is said to be covered every winter with such strong ice that a hundred loaded horses could go over it together with safety.

A crawfish, or river lobster, somewhat like a crab but quite minute, about two geometrical lines long and broad in proportion, was frequently drawn up by us with the aquatic weeds. Its color was a pale greenish white.

The cordate pontederia (*Pontederia cordata* L.) grows plentifully on the sides of a long and narrow canal or waterway, in the places frequented by our water lilies (*Nymphæa*). A great number of hogs wade far into this kind of strait and sometimes duck the greatest part of their bodies under water in order to get at the roots which they are very fond of.

As soon as we had passed over Lake St. Pierre, the face of the country was entirely changed, and became as agreeable as could be wished. The isles, and the land on both sides of us, looked like the prettiest pleasure gardens; and this continued till we neared Montreal.

Near every farm on the riverside there were some boats, hollowed out of the trunks of single trees but commonly neat and well made, having the proper shape of boats. In only one single place did I see a boat made of the bark of trees.

September the 15th

We continued our journey early this morning. On account of the force of the stream which came down against us, we were sometimes obliged to let the rowers go on shore and draw the boat up. At four o'clock in the evening we arrived at Montreal; and our voyage was reckoned a happy one, because the violence of the river flowing against us all the way, and the changeableness of the winds, commonly protract it to two weeks.

September the 19th

Grapes. Several people here in town have gotten the French grapevines and planted them in their gardens. They have two kinds of grapes, one of a pale green, or almost white; the other, of a reddish brown. From the white ones they say white wine is made, and from the red ones, red wine. The cold in winter obliges them to put dung round the roots of the vines, without which they would be killed by the frost. The grapes are beginning to ripen now; the white ones ripen a little sooner than the red ones. They make no wine of them here, because it is not worth while; but they are served for dessert like other berries. They say these grapes do not grow so big here as in France.

Watermelons (*Cucurbita citrullus* L.) are cultivated in great plenty in the English and French American colonies, and there is hardly a peasant here who has not a field planted with them. They are cultivated chiefly in the neighborhood of the town, and they are very rare in the north part of Canada. The Indians plant great quantities of watermelons at present, but whether they have done it of old is not easily determined, for an old Oneida Indian (of the six Iroquois nations) assured me that the Red Men did not know watermelons before the Europeans came into the country and showed them to the Indians. The French, on the other hand, asserted that the Illinois Indians had abundance of this fruit when the French first came to them, and that they declared they had planted them since time immemorial. However, I do not remember having read that the Europeans, who first came to North America, mention the watermelons, in speaking of the dishes of the Indians at that time. How great the summer heat is in those parts of America which I have passed through can easily be conceived, when one considers that in all those places they never sow watermelons in hot beds but in the open fields in spring, without so much as covering them, and they ripen in the season. Here are two species of them, *viz.* one with a red pulp, and one with a white one. The first is more common to the southward, with the Illinois Indians, and in the English colonies; the last is more abundant in Canada. The seeds are sown in the spring, after the cold has entirely left, in a good rich ground, at considerable distance from each other, because their stalks spread far and require much room if they are to be very productive. They were now ripe at Montreal, but in the English colonies they ripen in July and August. They usually require less time to ripen than the common melons. Those in the English colonies are usually sweeter and more agreeable than the Canadian ones. Does the greater heat contribute anything towards making them more palatable? Those in the province of New York are, however, reckoned the best. They contain a large percentage of water, and are cut into slices when eaten. They are always consumed raw, fresh.

The watermelons are very juicy; and the juice is mixed with the cooling pulp, which is very refreshing in the hot summer season. Nobody in Canada, Albany, or any other part of New York, could produce an example that the eating of watermelons in great quantities had hurt anybody; and there are examples of sick persons eating them without any danger. Further to the south, the frequent use of them, it is thought, brings on intermitting fevers and other bad distempers, especially in such people as are less used to them. Many Frenchmen assured me that when people born in Canada came to the Illinois Indians and ate several times of the watermelons there, they immediately got a fever; and therefore the Illinois advised the French not to eat of a fruit so dangerous to them. They themselves are subject to attack by fevers, if they cool their stomachs too often with watermelons. In Canada they keep them in a room which is a little heated, which means they keep fresh two months after they are ripe; but care must be taken that the frost does not spoil them. In the English plantations they keep them fresh in dry cellars during a part of the winter. They assured me that they keep better when they are carefully broken off from the stalk, and afterwards singed with a redhot iron in the place where the stalk was attached. In this manner they may be eaten at Christmas, and after. In Pennsylvania, where they have a dry sandy earth, they dig a hole in the ground, put the watermelons carefully into it with their stalks, by which means they keep very fresh during a great part of the winter. Few people, however, take this trouble with the watermelons; because they being very cooling, and the winter being very cold too, it seems to be less necessary to keep them for eating in that season, which is already very cold. They are of the opinion in these parts that cucumbers are more refreshing than watermelons. The latter are very strongly diuretic. The Iroquois call them *onóheserakáhti*.

Pumpkins of several kinds, oblong, round, flat or compressed, crook-necked, small, etc. are planted in all the English and French colonies. In Canada they fill the chief part of the farmers' kitchen gardens, though the onions are a close second. Each farmer in the English plantations has a large field planted with pumpkins, and the Germans, Swedes, Dutch and other Europeans settled in their colonies plant them. They constitute a considerable part of the Indian food; however, the natives plant more squashes than common pumpkins. They declare that they had the latter long before the Europeans discovered America, which seems to be confirmed by the accounts of the first Europeans that came into these parts, who mentioned pumpkins as common food among the Indians. The French here call them *citrouilles*, and the English in the colonies pumpkins. They are planted in spring, when they have nothing to fear from the frost, in an enclosed field, and in a good, rich soil. They are likewise frequently put into old hot beds. In Canada they ripen towards the beginning of September, but further south they are ripe at the end of July. As soon as the cold weather commences they remove all the pumpkins that remain on the stalk, whether ripe or not, and spread them on the floor in a part of the house, where the unripe ones grow perfectly ripe if they are not laid one upon the other. This is done round Montreal in the middle of September; but in Pennsylvania I have seen some in the fields on the nineteenth of October. They keep fresh for several months and even throughout the winter, if they be well secured in dry cellars (for in damp ones they rot very soon) where the cold cannot enter, or, which is still better, in dry rooms which are heated now and then to prevent the cold from damaging the fruit.

Pumpkins are prepared for eating in various ways. The Indians boil them whole, or roast them in ashes and eat them, or sell them thus prepared in the town; and they have, indeed, a very fine flavor when roasted. The French and English slice them and put the slices before the fire to roast; when they are done they generally put sugar on the pulp. Another way of roasting them is to cut them through the middle, take out all the seeds, put the halves together again, and roast them in an oven. When they are quite done, some butter is put in while they are warm, which being imbibed into the pulp renders it very palatable. The settlers often boil pumpkins in water, and afterwards eat them either alone or with meat. Some make a thin kind of pottage of them, by boiling them in water and afterwards macerating the pulp. This is again boiled with a little of the water, and a good deal of milk, and stirred about while it is boiling. Sometimes the pulp is kneaded into a dough with maize or other flour; of this they make pancakes. Some make puddings and tarts of pumpkins. The Indians, in order to preserve the pumpkins for a very long time, cut them in long slices which they fasten or twist together and dry either in the sun or by the fire in a room. When they are thus dried, they will keep for years, and when boiled they taste very well. The Indians prepare them thus at home and on their journeys and from them the Europeans have adopted this method. Sometimes they do not take the time to boil the pumpkin, but eat it dry with dried beef or other meat; and I own they are eatable in that state, and very welcome to a hungry stomach. At Montreal they sometimes preserve them in the following manner: they cut a pumpkin in four pieces, peel them, and take the seeds out. The pulp is put into a pot with boiling water, in which it must boil from four to six minutes. It is then put into a strainer and left in it till the next day that the water may run off. Then it is mixed with cloves, cinnamon, and lemon peel and preserved in syrup, the quantity of the latter being the same as that of the pulp. After this operation it is boiled together till all the syrup is absorbed and the white color of the pulp is lost.

September the 20th

The crop this year in Canada was reckoned the finest they had ever had. In the province of New York, on the contrary, the crop was very poor. The autumn was very fine this year in Canada.

September the 22nd

Indian Trade. The French in Canada carry on a great trade with the Indians; and though it was formerly the only trade of this extensive country, its inhabitants were considerably enriched by it. At present they have besides the Indian goods, several other articles which are exported. The Indians in this neighborhood who go hunting in winter like the other Indian nations, commonly bring their furs and skins to sell in the neighboring French towns; however, this is not sufficient. The Red Men who live at a greater distance never come to Canada at all; and lest they should bring their goods to the English, or the English go to them, the French are obliged to undertake journeys and purchase the Indian goods in the country of the natives. This trade is carried on chiefly at Montreal, and a great number of young and old men every year undertake long and troublesome voyages for that purpose, carrying with them such goods as they know the Indians like and want. It is not necessary to take money on such a journey, as the Indians do not value it; and indeed I think the French, who go on these journeys, scarcely ever take a sol or penny with them.

Goods Sold to the Natives. I will now enumerate the chief goods which the French carry with them for this trade, and which have a good sale among the Indians:

1.[1] *Muskets, powder, shot,* and *balls.* The Europeans have taught the Indians in their neighborhood the use of firearms, and so they have laid aside their bows and arrows, which were formerly their only arms, and use muskets. If the Europeans should now refuse to supply the natives with muskets, they would starve to death, as almost all their food consists of the flesh of the animals which they hunt; or they would be irritated to such a degree as to attack the colonists. The savages have hitherto never tried to make muskets or similar firearms, and their great indolence does not even allow them to mend those muskets which they have. They leave this entirely to the settlers. When the Europeans came into North America they were very careful not to give the Indians any firearms. But in the wars between the French and English, each party gave their Indian allies firearms in order to weaken the force of the enemy. The French lay the blame upon the Dutch settlers in Albany, saying that the latter began in 1642 to give their Indians firearms, and taught the use of them in order to weaken the French. The inhabitants of Albany, on the contrary, assert that the French first introduced this custom, as they would have been too weak to resist the combined force of the Dutch and English in the colonies. Be this as it may, it is certain that the Indians buy muskets from the white men, and know at present better how to make use of them than some of their teachers. It is likewise certain that the colonists gain considerably by their trade in muskets and ammunition.

2, a. *Pieces of white cloth*, or of a coarse uncut material. The Indians constantly wear such cloth, wrapping it round their bodies. Sometimes they hang it over their shoulders; in warm weather they fasten the pieces round the middle; and in cold weather they put them over the head. Both their men and women wear these pieces of cloth, which have commonly several blue or red stripes on the edge.

[1] The goods are not numbered in the original.

b. *Blue or red cloth.* Of this the Indian women make their skirts, which reach only to their knees. They generally choose the blue color.

c. *Shirts and shifts of linen.* As soon as an Indian, either man or woman, has put on a shirt, he (or she) never washes it or strips it off till it is entirely worn out.

d. *Pieces of cloth,* which they wrap round their legs instead of stockings, like the Russians.

3. *Hatchets, knives, scissors, needles,* and *flint.* These articles are now common among the Indians. They all get these tools from the Europeans, and consider the hatchets and knives much better than those which they formerly made of stone and bone. The stone hatchets of the ancient Indians are very rare in Canada.

4. *Kettles of copper or brass,* sometimes tinned on the inside. In these the Indians now boil all their meat, and they produce a very large demand for this ware. They formerly made use of earthen or wooden pots, into which they poured water, or whatever else they wanted to boil, and threw in red hot stones to make it boil. They do not want iron boilers because they cannot be easily carried on their continual journeys, and would not bear such falls and knocks as their kettles are subject to.

5. *Earrings* of different sizes, commonly of brass, and sometimes of tin. They are worn by both men and women, though the use of them is not general.

6. *Cinnabar.* With this they paint their face, shirt and several parts of the body. They formerly made use of a reddish earth, which is to be found in the country; but, as the Europeans brought them vermilion, they thought nothing was comparable to it in color. Many persons told me that they had heard their fathers mention that the first Frenchmen who came over here got a heap of furs from the Indians for three times as much cinnabar as would lie on the tip of a knife.

7. *Verdigris,* to paint their faces green. For the black color they make use of the soot off the bottom of their kettles, and daub the whole face with it.

8. *Looking glasses.* The Indians like these very much and use them chiefly when they wish to paint themselves. The men constantly carry their looking glasses with them on all their journeys;

but the women do not. The men, upon the whole, are more fond of dressing than the women.

9. *Burning glasses*. These are excellent utensils in the opinion of the Indians because they serve to light the pipe without any trouble, which pleases an indolent Indian very much.

10. *Tobacco* is bought by the northern Indians, in whose country it will not grow. The southern Indians always plant as much of it as they want for their own consumption. Tobacco has a great sale among the northern Indians, and it has been observed that the further they live to the northward, the more tobacco they smoke.

11. *Wampum*, or as it is here called, *porcelaine*. It is made of a particular kind of shell and turned into little short cylindrical beads, and serves the Indians for money and ornament.[1]

12. *Glass beads*, of a small size, white or other colors. The Indian women know how to fasten them in their ribbons, bags and clothes.

13. *Brass* and *steel wire*, for several kinds of work.

14. *Brandy*, which the Indians value above all other goods that can be brought them; nor have they anything, though ever so dear to them, which they would not give away for this liquor. But on account of the many irregularities which are caused by the use of brandy, the sale of it has been prohibited under severe penalties; however, they do not always pay implicit obedience to this order.

These are the chief goods which the French carry to the Indians and they do a good business among them.

Furs Bought from the Natives. The goods which they bring back from the Indians consist almost entirely of furs. The French take them in exchange for their goods, together with the necessary food provisions which they may want on the return journey from the Indians. The furs are of two kinds; the best are the northern ones, and the poorer sort those from the south. In the northern parts of America there are chiefly the following skins of animals: bears, beavers, elks (*originacs*),[2] reindeer (*cariboux*), wolf-lynzes (*Loups cerviers*), and martens. They sometimes get marten skins from the south, but they are red and good for little. *Pichou du nord* is perhaps the animal which the English, near Hudson Bay, call

[1] An imitation wampum was made of porcelain, and sold to the Indians, hence the Canadian name *porcelaine*, presumably.

[2] Canadian-Basque form for *originaux* or *orignaux*.

the wolverene. To the northern furs belong that of the bear, which is rare, and of the fox, which is not very frequent, and generally black; and several other skins.

The skins of the southern parts are taken chiefly from: wild cattle, stags, roebucks, otters, *pichoux du sud*, of which P. Charlevoix makes mention,[1] and are probably a species of cat-lynx, or perhaps a kind of panther; foxes of various kinds, raccoons, cat-lynxes, and several others.

It is inconceivable what hardships the people in Canada must undergo on their hunting journeys. Sometimes they must carry their goods a great way by land. Frequently they are abused by the Indians, and sometimes they are killed by them. They often suffer hunger, thirst, heat, and cold, and are bitten by gnats, and exposed to the bites of poisonous snakes and other dangerous animals and insects. These [hunting expeditions] destroy a great part of the youth in Canada, and prevent the people from growing old. By this means, however, they become such brave soldiers, and so inured to fatigue that none of them fears danger or hardships. Many of them settle among the Indians far from Canada, marry Indian women, and never come back again.

The prices of the skins in Canada, in the year 1749, were communicated to me by M. de Couagne, a merchant at Montreal, with whom I lodged. They were as follows:

Great and middle sized bear skins, cost five livres; skins of young bears, 50 sols; lynxes, 25 sols; *pichoux du sud*, 35 sols; foxes from the southern parts, 35 sols; otters, 5 livres; raccoons, 5 livres; martens, 45 sols; wolf-lynxes (*Loups cerviers*), 4 livres; wolves, 40 sols; carcajous,[2] an animal which I do not know, 5 livres; skins of the visons, a kind of marten [the American mink], which live in the water, 25 sols; raw skins of elks (*originacs verts*), 10 livres; stags

[1] There seems to be no clear distinction in contemporaneous sources between the *pichou du nord* and the *pichou du sud*. The word is originally of Indian origin. Pierre F. X. de Charlevoix in his *Histoire et Description Générale de la Nouvelle France* (III, 407, in the three-volume edition) "uses the plural *Pichoux* for two species of wild cats: the one with a short tail, which is the common American wild cat; and the other, a larger animal that goes by the name of *cougar* or *puma* (*Felis concolor* L.). The latter is called also *catamount, mountain lion* and *American lion*." See William A. Read, *Louisiana-French* (1931), pp. 101-102.

[2] The wolverene. The term is somtimes applied to the Canadian lynx, cougar, or American badger.

(*cerfs verts*), 8 livres; bad skins of elks and stags (*originacs et cerfs passés*), 3 livres; skins of roebucks, 25, or 30 sols; red foxes, 3 livres; beavers, 3 livres.

I will now insert a list of all the different kinds of skins, which are to be gotten in Canada, and which are sent from there to Europe. I obtained it from one of the greatest merchants in Montreal. They are as follows:

Prepared roebuck skins, *chevreuils passés.*
Unprepared ditto, *chevreuils verts.*
Tanned ditto, *chevreuils tanés.*
Bears, *ours.*
Young bears, *oursons.*
Otters, *loutres.*
Pécans, [Woodshock, [or fisher] a species of Canadian marten (Marchand)].
Cats, *chats.*
Wolves, *loup de bois.*
Lynxes, *loups cerviers.*
North pichoux, *pichoux du nord.*
South pichoux, *pichoux du sud.*
Red foxes, *renards rouges.*
Cross foxes, *renards croisés.*
Black foxes, *renards noirs.*
Gray foxes, *renards argentes.*
Southern or Virginian foxes, *renards du sud ou de Virginie.*
White foxes, from Tadoussac, *renards blancs de Tadoussac.*
Martens, *martres.*
Visons, or *foutreaux.*
Black squirrels, *écureuils noirs.*
Raw stags skins, *cerfs verts.*
Prepared ditto, *cerfs passés.*
Raw elk skins, *originacs verts.*
Prepared ditto, *originacs passés.*
Reindeer skins, *cariboux.*
Raw hind skins, *biches vertes.*
Prepared ditto, *biches passées.*
Carcajoux. [Wolverene or Labrador badger].
Muskrats, *rats musqués.*

Fat winter beavers, *castors gras d'hiver.*
Ditto summer beavers, *castors gras d'été.*
Dry winter beavers, *castors secs d'hiver.*
Ditto summer beavers, *castors secs d'été.*
Old winter beavers, *castors vieux d'hiver.*
Ditto summer beavers, *castors vieux d'été.*

Native Copper. To-day I got a piece of native copper from Lake Superior. They find it there almost pure, so that it does not need melting over again, but is immediately fit for working. Father Charlevoix [1] speaks of it in his *History of New France.* One of the Jesuits at Montreal who had been at the place where this metal is native told me that it is generally found near the mouths of rivers and that there are pieces of pure copper too heavy for a single man to lift up. The Indians there say they formerly found a piece about seven feet long and nearly four feet thick, all pure copper. As it is always found in the ground near the mouths of rivers, it is probable that the ice or water carried it down from a mountain; but, notwithstanding the careful search that has been made, no place has been found where the metal lies in any great quantity but only in loose pieces.

Lead Ore. The head or superior of the priests of Montreal gave me a piece of lead ore to-day. He said it was taken from a place only a few French miles from Montreal, and it consisted of compact, shining cubes of lead ore. I was told by several persons here that further south in the country there is a place where they find a great quantity of this lead ore in the ground. The Indians nearby melt it and make balls and shot of it. I got some pieces of it consisting of a shining lead ore with narrow stripes through it and of a white hard earth or clay which effervesces with *aqua fortis.*

I likewise received some reddish brown earth to-day, found near the Lac des Deux Montagnes, or Lake of Two Mountains, a few French miles from Montreal. It may be easily crumbed into dust between the fingers. It is very heavy, and more so than the earth of that kind generally is. Outwardly it has a kind of glossy appearance, and when it is handled by the fingers for a time it looks as if it were covered with silver. It is, therefore, probably a kind of lead earth or an earth mixed with iron mica.

---

[1] See his *Hist. de la Nouv. Fr.,* Tom. VI., p. 415.—F.

The Women of Canada. The ladies in Canada are generally of two kinds: those who come over from France, and those who are natives. The former possess the politeness peculiar to the French nation; the latter may be divided into those of Quebec and Montreal. The first of these are equal to the French ladies in good breeding, having the advantage of frequently conversing with the French gentlemen and ladies, who come every summer with the king's ships, and stay several weeks at Quebec, but seldom go to Montreal. The ladies of this last place are accused by the French of being contaminated by the pride and conceit of the Indians, and of being much wanting in French good breeding. What I have mentioned above[1] about their dressing their head too profusely is the case with all the ladies throughout Canada. Their hair is always curled, even when they are at home in a dirty jacket and a short coarse skirt that does not reach to the middle of their legs. On Sundays and visiting days they dress so gayly that one is almost induced to think their parents in origin and social position to be among the best in the realm. The Frenchmen, who consider things in their true light, complain very much that a great number of the ladies in Canada have gotten into the pernicious custom of taking too much care of their dress, and squandering all their fortune and more upon it, instead of sparing something for future times. They are no less attentive to having the newest fashion; the best and most expensive dresses are discarded and cut to pieces; and they smile inwardly when their sisters are not dressed according to their fancy. But what they get as new fashions are often old and discarded in France by the time they are adopted in Canada, for the ships come but once every year from abroad, and the people in Canada consider that as the new fashion for the whole year which the people on board brought with them or which they imposed upon them as new.

The ladies in Canada, and especially at Montreal, are very ready to laugh at any blunders strangers make in speaking; but they are very excusable. People laugh at what appears uncommon and ridiculous.[2] In Canada nobody ever hears the French language spoken by any one but Frenchmen, for strangers seldom come

---

[1] See diary for July 25, 1749.

[2] Apparently the Canadian ladies laughed at Kalm's French.

there, and the Indians are naturally too proud to learn French, and compel the French to learn their language. Therefore it naturally follows that the sensitive Canadian ladies cannot hear anything uncommon without laughing at it. One of the first questions they put to a stranger is whether he is married; the next, how he likes the ladies in the country, and whether he thinks them handsomer than those of his own country; and the third, whether he will take one home with him. There are some differences between the ladies of Quebec and those of Montreal; those of the latter place seemed to be generally handsomer than those of the former. The women seemed to me to be somewhat too free at Quebec, and of a more becoming modesty at Montreal. The ladies at Quebec, especially the unmarried ones, are not very industrious. A girl of eighteen is reckoned very poorly off if she cannot enumerate at least twenty lovers. These young ladies, especially those of a higher rank, get up at seven, and dress till nine, drinking their coffee at the same time. When they are dressed they place themselves near a window that opens into the street, take up some needlework and sew a stitch now and then; but turn their eyes into the street most of the time. When a young fellow comes in, whether they are acquainted with him or not, they immediately lay aside their work, sit down by him, and begin to chat, laugh, joke, and invent "double-entendres" and make their tongues go like a lark's wings; this is considered *avoir beaucoup d'esprit*. In this manner they frequently pass the whole day, leaving their mothers to do all the work in the house. In Montreal the girls are not quite so flighty, and more industrious. It is not uncommon to find them with the maid in the kitchen. They are always at their needle work or doing some necessary business in the house. They are likewise cheerful and content; and nobody can say that they lack either wit or charms. Their fault is that they think too well of themselves. However, the daughters of all ranks, without exception, go to market, buy watermelons, pumpkins, and other food and carry it home themselves. They rise as soon and go to bed as late as any of the people in the house. I have been assured that in general their fortunes are not great and are rendered still more scarce by the number of children and the small revenues in a house. The girls at Montreal are very much displeased that those at Quebec get husbands sooner than they. The reason for this is that many young gentlemen who come over from France with the ships are captured by the ladies at Quebec, and marry them; but as these gentlemen seldom go up to Montreal the girls there are not often so happy as those of the former place.

September the 23rd

Sault au Recollet. This morning I went to Sault au Recollet, a place three French miles north of Montreal to describe the plants and minerals there and chiefly to collect seeds of various plants. Near the town there are farms on both sides of the road; but as one advances the country grows woody and varies in regard to height. It is generally very rough and there are pieces both of ordinary granite and a kind of gray limestone. The roads are bad and almost impassable for carriages. A little before I arrived at Sault au Recollet the woods came to an end, and the country was either cultivated or turned into meadows and pastures. Otherwise there was nothing especially pleasant on this journey. The parts visited could not be compared with those around Montreal.

Lime Kilns. About a French mile from the town are two lime kilns on the road. They are built in the ground of a gray infusible limestone, on the outside, and of pieces of granite rock nearest the fire. The height of the kiln from top to bottom is eighteen feet.

The limestone which they burn here is of two kinds. One is quite black and so compact that its constituent particles cannot be distinguished, some dispersed grains of white and pale gray spar excepted. Now and then there are thin cracks in it filled with a white small-grained spar.

I have never seen any fossils in this stone, though I looked very carefully for them. This stone is common on the Isle of Montreal, about ten or twenty inches below the upper soil. It lies in strata of five or ten inches thickness. This stone is said to give the best lime; for, though it is not so white as that of the following gray limestone, it makes better mortar, and almost turns into stone, growing harder and more compact every day. In repairing a house made partly of this mortar, it has happened that stones of the house crumbled sooner than the mortar itself.

The other kind is gray and sometimes a dark gray limestone,

consisting of a compact calcareous stone, mixed with grains of spar of the same color. When broken, it has a strong smell of stinkstone. It is full of petrified striated shells or pectinites. The greatest part of these petrifactions are, however, only impressions of the hollow side of the shells. Now and then I found also petrified pieces of the shell itself, though I could never find the same shells in their natural state on the shores; and it seems inconceivable how such a quantity of impressions could come together, as I shall presently mention.

I have had great pieces of this limestone, consisting of little else than pectinites lying close to one another. This limestone is found on several parts of the isle where it lies in horizontal strata of the thickness of five or ten inches. This stone yields a great quantity of white lime, but it is not so good as the former, because it grows damp in wet weather.

Fir wood is reckoned the best for the lime kilns and the thuya wood next to it. The wood of the sugar maple and other trees of a similar nature are not fit for it, because they leave a great quantity of coals.

Gray pieces of granite are to be seen in the woods and fields hereabouts.

The leaves of several trees and plants began now to get a pale hue; especially those of the red maple, the smooth sumach (*Rhus glabrum* L.), the *Polygonum sagittatum* L., and several of the ferns.

A great cross is erected on the road, and the boy who accompanied me told me that a person was buried there who had wrought great miracles. Those who went by touched their caps when they passed the cross.

At noon I arrived at Sault au Recollet, which is a little place situated on a branch of the St. Lawrence River that flows with a violent current between the Isle of Montreal and the Island of Jesus. It has gotten its name from an accident which happened to a Recollet friar called Nicolas Viel, in the year 1625. He went into a boat with a converted Indian and some native Hurons in order to go to Quebec; but on going over this place in the river, the boat upset, and both the friar and his proselyte were drowned. The Indians (who have been suspected of occasioning the upsetting of the boat) swam to the shore, saved what they could of the friar's effects and kept them.

The country hereabouts is full of stones, and settlers have but lately begun to cultivate it, for all the old people could remember the places covered with tall woods, which are now turned into grain fields, meadows, and pastures. The priests say that this place was formerly inhabited by some converted Hurons. These Indians lived on a high mountain, at a little distance from Montreal, when the French first arrived here, and the latter persuaded them to sell that land. They did so, and settled here at Sault au Recollet, and the church which still remains here was built for them, and they have attended divine service in it for many years. As the French began to increase on the Isle of Montreal, they wished to have it entirely to themselves, and persuaded the Indians again to sell them this spot and go to another. The French have since prevailed upon the Indians (whom they did not like to have with them because of their drunkenness and rambling idle life) to leave this place again and go to settle at the Lac des Deux Montagnes, where they are at present and have a fine church of stone. Their church at Sault au Recollet is of wood, looks very old and dilapidated, though its inside is tolerably good, and is used by the Frenchmen in this place. They have already brought a quantity of stones hither, and intend building a new church very soon. The botanical observations which I made during these days, I shall reserve for another publication.

Though there had been no rain for several days, the moisture in the air is so great that as I spread some papers on the ground this afternoon, in a shady place, intending to put the seeds I collected into them, they were so wet in a few minutes that they were useless. The whole sky was very clear and bright, and the heat as intolerable as in the middle of July.

Husbandry. One half of the grain fields are left fallow, alternately. The fallow grounds are never plowed in summer, so the cattle can feed upon the weeds that grow on them. All the seed used here is spring seed, as I have before observed. Some plow the fallow grounds late in autumn, others defer that business till spring; but the first way is said to give a much better crop. Wheat, barley, rye, and oats are harrowed, but peas are plowed into the ground. Farmers sow commonly about the fifteenth of April, and begin with the peas. Among the many kinds of peas which are to be gotten here, they prefer the green ones to all others for sowing. Peas require a high, dry, poor soil, mixed with coarse sand. (They did not know what it meant to stake the peas.) The harvest time commences about the end and sometimes in the middle of August. Wheat returns are generally fifteenfold and sometimes twentyfold; oats from fifteenfold to thirtyfold. The crop of peas is sometimes fortyfold, but at other times only tenfold, for it varies very much. The plow and harrow are the only implements of husbandry they have and those are not of the best sort. The manure is spread out in spring. The soil consists of a gray stony earth, mixed with clay and sand. They sow no more barley than is necessary for the cattle, for they make no malt here. They sow a good deal of oats, but merely for the horses and other cattle. Nobody knows here how to make use of the leaves of deciduous trees as a food for cattle, though the forests are furnished with no other than trees of that kind, and though the people are commonly forced to feed their cattle at home during five months of the year. No hobbled cattle were seen, and no ditches, unless there was enough water actually to drown the fields.

I have already repeatedly mentioned that almost all the wheat which is sown in Canada is spring wheat, that is, such as is sown in spring. Near Quebec it sometimes happens, when the summer is less warm, or the spring later than usual, that a great part of the wheat does not ripen entirely before the cold commences. I have been assured that some people who live on the Isle of Jesus sow wheat in autumn, which is better, finer, and gives a more plentiful crop than the spring wheat; but it does not ripen more than a week before the other wheat.

September the 25th

In several places hereabouts, they enclose the fields with a stone fence instead of wooden pales. The large amount of stones which are to be gotten here renders the labor very trifling.

Here is an abundance of beech trees in the woods, and they now have ripe seeds. The people in Canada collect them in autumn, dry them and keep them till winter, when they eat them instead of walnuts and hazel nuts; and I am told they taste very good.

There is a salt spring, as the priest of this place informed me, seven French miles from here, near the river d'Assomption; they made a fine white salt from it during the war. The water is said to be very briny.

Fruit and Nut Trees. Some kinds of fruit trees succeed very well near Montreal, and I had here an opportunity of seeing some very fine pears and apples of various sorts. Near Quebec the pear trees will not grow because the winter is too severe for them, and sometimes they are killed by the frost in the neighborhood of Montreal. Plum trees of several sorts which were first brought over from France, succeed very well and withstand the rigors of winter. Three varieties of American walnut trees grow in the woods, but the walnut trees brought over from France died almost every year down to the very root, bringing forth new shoots in the spring. Peach trees cannot well thrive in this climate; a few bear the cold, but for greater safety they are obliged to put straw round them. Chestnut trees, mulberry trees, and the like, have never yet been planted in Canada.

Land Owned by Clergy and Noblemen. The whole cultivated part of Canada has been given away by the king to the clergy and some noblemen; but all the uncultivated parts belong to him, as likewise the place on which Quebec and Trois Rivières are built. The ground on which the town of Montreal is built, together with the whole isle of that name, belongs to the priests of the order of St. Sulpicius, who live at Montreal. They have given the land in tenure to farmers and others who are willing to settle on it, so much that they have no more upon their hands at present. The first settlers paid a trifling rent for their land; for frequently the whole lease for a piece of ground, three arpens broad and thirty long, consisted of a couple of chickens; and some pay twenty, thirty, or forty sols for a piece of land the same size. But those who came later had to pay near two éçus (crowns) for such a piece of land, and thus the land rent became very unequal throughout the country. The revenues of the Bishop of Canada do not arise from any landed property. The churches are built at the expense of the congregations. The inhabitants of Canada do not yet pay any taxes to the king; and he has no other revenues from it than those which arise from the custom house.

A Mill. The priests of Montreal have a mill here where they take the fourth part of all that is ground. However, the miller receives a third part for his share. In other places he gets half of it. The priests sometimes lease the mill for a certain sum. Besides them, nobody is allowed to erect a mill on the Isle of Montreal, they having reserved that right to themselves. In the agreement drawn up between the priests and the inhabitants of the isle, the latter are obliged to get all their grain ground in the mills of the former. The mill is built of stone with three waterwheels and three pairs of stones. I noticed first that the wheels and axles were made of white oak; secondly that the cogs in the wheel and other parts were made either of the sugar maple or of *Bois dur* (*Carpinus ostrya*), because that was considered the hardest wood there; third, that the millstones had come from France and consisted of a conglomerate and quartz grains, both of the size of hazelnuts and ordinary sand, all bound together by white limestone. The stones were said to be sufficiently hard. The kernels were shaken down from the funnel in the manner described before.

They make a good deal of sugar in Canada of the juice running out of the incisions in the sugar maple, the red maple, and the sugar birch; but that of the first tree is most commonly used. The way of preparing it has been more minutely described by me in the *Memoirs of the Royal Swedish Academy of Sciences,* 1751.[1]

September the 26th

Autumn. Early this morning I returned to Montreal. Everything began now to look like autumn. The leaves of the trees were faded or reddish, and most of the plants had lost their flowers. Those which still preserved them were the following;[2]

Several sorts of asters, both blue and white.
Golden rods of various kinds.
Common milfoil.

---

[1] See Bibliography, item 16.

[2] *Asteres. Solidagines. Achillea millefolium. Prunella vulgaris. Carduus crispus. Oenothera biennis. Rudbeckia triloba. Viola Canadensis. Gentiana Saponaria.*

Common self-heal.
The crisped thistle.
The biennial oenothera.
The rough-leaved sun flower, with trifoliated leaves.
The Canada violet.
A species of gentian.

Wild grapevines are abundant in the woods hereabouts, climbing up very high trees.

Indian Food. I have made inquiry among the French, who travel far into the country, concerning the food of the Indians. Those who live far north I am told cannot plant anything on account of the great cold. They have, therefore, no bread, and do not live on vegetables; meat and fish are their only food, and chiefly the flesh of beavers, bears, reindeer, elks, hares, and several kinds of birds. Those Indians who live far southward, eat the following things. Of vegetables they plant corn, wild kidney beans (*Phafeoli*) of several kinds, pumpkins of different sorts, squashes, a kind of gourd, watermelons and melons (*Cucumis melo* L.). All these plants have been cultivated by the Indians long before the arrival of the Europeans. They likewise eat various fruits which grow in their woods. Fish and meat constitute a very large part of their food. And they like chiefly the flesh of wild cattle, roe-bucks, stags, bears, beavers and some other quadrupeds. Among their dainty dishes they reckon the water taregrass (*Zizania aquatica* L.), which the French call *folle avoine*, and which grows plentifully in their lakes, in stagnant waters, and sometimes in rivers which flow slowly. They gather its seeds in October, and prepare them in different ways, and chiefly as groats, which taste almost as good as rice. They make also many a delicious meal of the several kinds of walnuts, chestnuts, mulberries, acimine (*Annona muricata* L.), chinquapins (*Fagus pumila* L.), hazel nuts, peaches, wild prunes, grapes, whortleberries of several sorts, various kinds of medlars, blackberries and other fruit and roots. But the species of grain so common in what is called the Old World were entirely unknown here before the arrival of the Europeans; nor do the Indians at present ever attempt to cultivate them, though they see the use which the settlers make of the culture of them, and though they are fond of eating the dishes which are prepared from them.

SEPTEMBER THE 27TH

Beavers. Beavers are abundant all over North America and they are one of the chief articles of trade in Canada. The Indians live upon their flesh during a great part of the year. It is certain that these animals multiply very fast; but it is also true that vast numbers of them are annually killed and that the Indians are obliged at present to undertake distant journeys in order to catch or shoot them. Their decreasing in number is very easily accounted for, because the Indians, before the arrival of the Europeans, only caught as many as they found necessary to clothe themselves with, there being then no trade with the skins. At present a number of ships go annually to Europe, laden chiefly with beavers' skins; the English and French endeavor to outdo each other by paying the Indians well for them, and this encourages the latter to extirpate these animals. All the people in Canada told me that when they were young all the rivers in the neighborhood of Montreal, the St. Lawrence River not excepted, were full of beavers and their dams; but at present they are so far destroyed that one is obliged to go several miles up the country before one can meet one. I have already remarked above that the beaver skins from the north are better than those from the south.

The Beaver a "Fish". Beaver meat is eaten not only by the Indians but likewise by the Europeans, and especially by the French, on their fasting days; for his Holiness the Pope has, like many of the old zoologists, classified the beaver among the fishes, since he spends most of his time in water. The meat is reckoned best if the beaver has lived upon vegetables, such as the aspen and the beaver tree (*Magnolia glauca* L.); but when he has eaten fish, it does not taste so well. To-day I tasted this meat boiled for the first time; and though everybody present besides myself thought it a delicious dish, yet I could not agree with them. I think it is eatable, but has nothing delicious about it. It looks black when boiled and has a peculiar taste. In order to prepare it well it must be boiled from morning till noon, that it may lose the strange taste which it has. The tail is likewise eaten, after it has been boiled in the same manner and roasted afterwards; but it consists of fat only, though they would not call it so, and cannot be swallowed by one who is not used to eating it.

Much has already been written concerning the dams or houses of the beavers; it is therefore unnecessary to repeat it. Sometimes, though but seldom, they catch beavers with white hair. In American cities one can now get as fine beaver hats made as one ever could in France or England.

The Fasting of Catholics. In connection with the eating of beavers the fasting of the Catholics appeared to me a bit strange. Those who first inaugurated the fast days did it undoubtedly with good and holy intentions to keep the people from eating too much meat, which is injurious to health, fattens the body too much, and makes it inadaptable for many things. But it seemed to be quite enough for them during the ordinary fast days to abstain from meat. If they could afford it they lived everywhere sumptuously and fed their body just as on the other days of the week; for they then had more courses prepared of eggs, of all kinds of fish, prepared with oils and fats, all kinds of milk dishes, and many especially sweet and good tasting fruits with a quantity of wine. So that for the most part wherever you ate on a fast day the table was better provided with varieties of food than on any other days, and still they called it fasting, *jours maigres* they named them.

Wine is almost the only liquor which people above the common class drink. They make a kind of spruce beer of the top of the white fir[1] which they drink in summer; but the use of it is not general and it is seldom drunk by people of quality. Great sums go annually out of the country for wine, as they have no grapevines here of which they could make a liquor that is fit to be drunk. The common people drink water, for it is not yet customary here to brew beer of malt; and there are no orchards large enough to supply the people with apples for making cider. Some of the people of rank who possess large orchards, sometimes out of curiosity get a small quantity of cider made. The people of quality here, who have been accustomed from their youth to drink nothing but wine, are greatly at a loss in time of war, when all the ships which bring wine are intercepted by the English privateers. Towards the end of the last war, they gave two hundred and fifty francs, and even one hundred éçus, for a *barrique,* or hogshead, of wine.

---

[1] *Epinette blanche.* The way of brewing this beer is described at large in the *Memoirs of the Royal Swedish Academy of Sciences* for the year 1751, p. 190.—F.

Prices of Commodities. The present price of several things I have been told by some of the most prominent merchants here is as follows: an average horse cost forty francs and upwards; a good horse is valued at a hundred francs or more. A cow is now sold for fifty francs; but people can remember the time when they were sold for ten éçus. A sheep costs five or six livres at present; but last year, when everything was dear, it cost eight or ten francs. A hog one year old, and of two hundred or a hundred and fifty pounds weight, is sold at fifteen francs. M. Couagne, the merchant, told me that he had seen a hog of four hundred pounds weight among the Indians. A chicken is sold for ten or twelve sols, and a turkey for twenty sols. A minot[1] of wheat sold for an éçu last year; but at present it costs forty sols. Corn is always the same price as wheat because there is little of it here, and it is all used by those who trade with the Indians. A minot of oats costs sometimes from fifteen to twenty sols; but of late years it has been sold for twenty-six or thirty sols. Peas have always the same price as wheat. A pound of butter costs commonly about eight or ten sols; but last year it rose up to sixteen sols. A dozen eggs used to cost but three sols; however, now they are sold for five. They make no cheese at Montreal; nor is there any to be had, except what is gotten from abroad. A watermelon generally costs five or six sols, but if of a large size, from fifteen to twenty.

There are as yet no manufactures established in Canada, probably because France will not lose the advantage of selling off its own goods here. However, both the inhabitants of Canada and the Indians are very badly off for want of them in times of war.

Marriages. Those persons who wish to be married must have the consent of their parents. However, the judge may give them leave to marry if the parents oppose their union without any valid reason. Likewise if the man be thirty years of age, and the woman twenty-six, they may marry without waiting for their parents' consent. All they have to do is to go to the priest, who reads the banns three Sundays in succession in church, after which the ceremony may take place in the church in the presence of as few or as many people as they desire. Priests do not like to perform marriages in the homes.

---

[1] A French measure, about the same as two bushels in England.—F.

September the 29th

This afternoon I went out of town to the southwest part of the island in order to view the country and the husbandry of the people, and to collect several seeds. Just before the town are some fine fields, which were formerly cultivated but which now serve as pastures. To the northwest appears the high mountain which lies west of Montreal. It is very fertile and covered with fields and gardens from the bottom to the summit. On the southeast side is the St. Lawrence River, which is very broad here; and on its sides are extensive grain fields and meadows, and fine houses of stone which look white at a distance. At a great distance southeast appear the two high mountains near Fort Chambly, and some others near Lake Champlain, raising their tops above the woods. All the fields hereabouts are filled with stones of different sizes, and among them there is now and then some black limestone. About a French mile from the town the highroad goes along the river which is on the left; and on the right all the country is cultivated and inhabited. The farmhouses are three, four or five arpens distant from each other. The hills near the river are generally high and pretty steep; they consist of earth, and the fields below them are filled with pieces of granite and black limestone. About two French miles from Montreal the river runs very rapidly and is full of rocks; in some places there are waves. However, those who go by boats into the southern parts of Canada are obliged to work through such places.

Right outside of the city were a couple of windmills; they were built like others I have seen in this land, except that instead of having thin boards for wings they had linen, which was removed after a grinding.

Most of the farmhouses in this neighborhood are of stone, partly of the black limestone, and partly of other stones in the neighborhood. The roof is made of shingles or of straw. The gable is always very high and steep. Other buildings, such as barns and stables, are of wood.

Wild geese and ducks began now to migrate in great flocks to the southern countries.

October the 2nd

The two preceding days, and today, I employed chiefly in collecting seeds.

Last night's frost caused a great alteration in several trees. Walnut trees of all sorts were now shedding their leaves very fast. The flowers of a kind of nettle (*Urtica divaricata* L.) were entirely killed by the frost. The leaves of the American lime tree were likewise damaged. In the kitchen gardens the leaves of the pumpkins were all killed. However, the beech, oak, and birch, did not seem to have suffered at all. The fields were all covered with a hoarfrost. The ice in the pools of water was a geometrical line and a half in thickness.

The biennial oenothera (*Oenothera biennis* L.) grows in abundance on open woody hills, and fallow fields. An old Frenchman, who accompanied me as I was collecting its seeds, could not sufficiently praise its property of healing wounds. The leaves of the plant must be crushed and then laid on the wound.

The *Soeurs de Congregation* is an organization of religious women different from nuns. They do not live in a convent, but have houses both in the town and country. They go where they please, and are even allowed to marry if an opportunity offers; but this, I am told, happens very seldom. In many places in the country there are two or more of them: they have their house commonly near a church and generally the parsonage is on the other side of the church. Their business is to instruct young girls in the Christian religion, to teach them reading, writing, needlework and other feminine accomplishments. People of fortune board their daughters with them for some time. They have their boarding, lodging, beds, instruction, and whatever else they want, on very reasonable terms. The home where the whole community of these ladies live, and from which they are sent out into the country, is at Montreal. A lady that wants to join them must pay a large sum of money toward the common expenses, and some people believe it to be four thousand livres. If a person be once received, she is sure of a subsistence during her lifetime.

La Chine is a pretty village, three French miles to the southwest of Montreal, but on the same isle, close to the St. Lawrence River.

The farmhouses lie along the river side, about four or five arpens from each other. Here is a fine church of stone, with a small steeple; and the whole place has a very agreeable location. Its name is said to have had the following origin. When the unfortunate M. Salée was here, who was afterwards murdered by his own countrymen further up in the country, he was very intent upon discovering a shorter road to China by means of the St. Lawrence River. He talked of nothing at that time but this new short way to China. But as his project of undertaking the journey in order to make this discovery was stopped by an accident which happened to him here, and he did not at that time come any nearer China, this place got its name, as it were, by way of a joke.

This evening I returned to Montreal.

October the 5th

Government. The governor-general at Quebec, is, as I have already mentioned before, the chief magistrate in Canada. Next to him is the intendant at Quebec; then follows the governor of Trois Rivières. The intendant has the greatest power next to the governor-general; he pays all the money of the government, and is president of the board of finances and of the court of justice in this country. He is, however, under the governor-general; for if he refuses to do anything to which he seems obligated by his office, the governor-general can give him orders to do it, which he must obey. He is allowed, however, to appeal to the government in France. In each of the capital towns, the governor is the highest person, then the lieutenant-general, next to him a major, and after him the captains. The governor-general gives the first orders in all matters of consequence. When he comes to Trois Rivières and Montreal, the power of the governor ceases, because he always commands wherever he is. The governor-general commonly goes to Montreal once every year, and usually in winter, and during his absence from Quebec, the lieutenant-general commands there. When the governor-general dies, or goes to France before a new one has come in his stead, the governor of Montreal goes to Quebec to take command in the interim, leaving the major in command at Montreal.

Trade. One or two of the king's ships are annually sent from

France to Canada, carrying recruits to supply the places of those soldiers who have either died in service or have gotten leave to settle in the country and turn farmers, or to return to France. Almost every year France sends a hundred or a hundred and fifty people over in this manner. With these people it likewise sends over a great number of persons who have been found guilty of smuggling in France. They were formerly condemned to the galleys, but at present they send them to the colonies, where they are free as soon as they arrive, and may choose what manner of life they please, but are never allowed to go back to France without the king's special license. The king's ships likewise bring a great quantity of merchandise which the king has bought to be distributed among the Indians on certain occasions. The inhabitants pay very little to the king. In the year 1748 a beginning was made however, by laying a duty of three per cent on all the French goods imported by the merchants of Canada. A regulation was also made at the time that for all furs and skins exported to France from here one should pay a certain duty; but for what is carried to the colonies one pays nothing. The merchants of all parts of France and its colonies are allowed to send ships with goods to this place, and similarly the Quebec merchants are at liberty to send their goods to any place in France and its colonies. But the merchants at Quebec have but few ships, because the sailor's wages are very high. The towns in France which trade chiefly with Canada are Rochelle and Bourdeaux; next to them are Marseilles, Nantes, Hâvre-de-Grace, St. Malo, and others. The king's ships which bring goods to this country come either from Brest or from Rochefort. The merchants at Quebec send flour, wheat, peas, wooden utensils, etc. in their own bottoms, to the French possessions in the West Indies. The walls round Montreal were built in 1738 at the king's expense, on condition that the inhabitants should, little by little, pay off the cost to the king. The town at present pays annually 6000 livres for them to the government, of which 2000 are given by the seminary of priests. At Quebec the walls have likewise been built at the king's expense, but he did not redemand the expense of the inhabitants, because they had already the duty upon goods to pay, as above mentioned. The beaver trade belongs solely to the Indian Company in France, and nobody is allowed to carry it on here except the people appointed by that company. Every other fur trade is open to everybody. There are several places among the Indians far in the country where the French have stores of their goods; and these places they call *les postes*. The king has no other fortresses in Canada than Quebec, Fort Chambly, Fort St. Jean, Fort St. Frédéric, or Crown Point, Montreal, Frontenac, and Niagara. All other places belong to private persons. The king keeps the Niagara trade all to himself. Everyone who intends to trade with the Indians must have a license from the governor-general, for which he must pay a sum proportionate to the advantages for trade. A merchant who sends out a boat laden with all sorts of goods, and four or five persons with it, is obliged to give five or six hundred livres for the permission; and there are places for which they give a thousand livres. Sometimes one cannot buy the license to go to a certain trading place because the governor-general has granted or intends to grant it to some acquaintance or relation of his. The money arising from the granting of licenses belongs to the governor-general; but it is customary to give half of it to the poor: whether this is always strictly observed or not I shall not pretend to determine.

The Catholic Church Service. No other religion was tolerated here except the Catholic. It was said by all those who had been in France that people of both sexes in Canada were more devout than they were in France; nowhere could they go to church more regularly than here. Most of the service was in Latin. It seemed as if the whole service was too much of an external *opus operatum*. Most of it consisted in the reading of prayers with a rapidity which made it impossible to understand them, even for those who understood Latin. I could only get a word now and then and never a whole sentence, so that the common man could certainly get nothing of it nor derive any benefit from it. Even the best Latin scholar could not possibly keep his thoughts together and pray fervently at such break-neck speed. In fact, this must have been impossible for the priests themselves. The sermon was in French, and all quotations from the Scriptures were first given in the Latin Vulgate and then translated into French. Even the clericals, however, had difficulty in speaking Latin, since the words they needed did not appear in their missals. It was customary both in the city and country, both upon rising and retiring, to kneel in prayer; but whether this was in Latin or French I did not wish to be inquisitive enough to ask.

Although I paid particular attention to the matter, I never saw a Bible in any house, either in French or Latin, except at the residences of the clergy. But I saw a few French and Latin prayer books, though most of them were prayers to the Holy Virgin rather than to Almighty God.[1]

---

[1] This passage is omitted in Forster. Marchand condenses a portion of it from the Dutch translation, but in doing so makes Kalm declare that he had never seen a Bible in the hands of any priest or monk.

SUPPLEMENTARY DIARY

[Here begins the recently discovered continuation of Kalm's Swedish journal of his American travels, which was published at Helsingfors, Finland, in 1929 by Fredr. Elfving. Since this part consists primarily of brief and more or less unpolished notes and accounts, unedited by the author himself, a certain freedom of form has necessarily been adopted in the following translation of it to make it a little more readable and more in conformity with the letter and spirit of the preceding Englished portion. However, in order to preserve a certain local color, Kalm's spelling of foreign words has been retained, wherever practical. The meteorological observations at the head of each daily entry have been assigned to the appendix, as were those of the original edited and published by Kalm].

October 7, 1749

In Montreal

Obstacles to the continuation of my journey, also the cause why I was not allowed to pass through Forts Frontenac and Niagara, but was forced to return by the same route as I had come, that is, through Fort St. Frédéric.—I was now almost ready to set out from here and was occupied yesterday with putting my seeds into small bags or cornets in order that I might to-day go from here up the river to Fort Frontenac; but just as I was about to start I received the Governor-general Marquis la Jonquière's letter, which frustrated this plan. On my departure from Quebec I received from him a passport to go through Fort St. Frédéric (now Crown Point), since I considered it impossible to get anyone in Oswego to take me from there to Albany. On my arrival in Montreal I was permitted not only to talk with the English who were bringing home the French prisoners (the latter of whom said that I could easily go through Oswego), but I also talked with the commander of Fort Frontenac, who informed me that on the plain was found an abundance of Indian rice (*Fol. avoine*), also red cedar and herbs, the medicinal value of which he praised most highly. Besides, he knew them well and promised to show them to me when I came there; thereupon he proceeded ahead to the place. As I was not able to get a tenth of the Indian rice in the region about Montreal which I ought to have had, and as red cedar and the other plants were not to be found here, I was compelled to write again to the governor-general and ask permission of him to plan my route through Fort Frontenac and Niagara to Albany. I gave as reasons that I had been sent by her Royal Highness [the Queen of Sweden]; that she had ordered me unconditionally to procure a generous supply of seeds of the Indian rice and other useful herbs and plants; that I was to take a route which neither I (nor any other botanist) had ever travelled before; and that consequently I hoped to discover much that was new and useful, at the same time fulfilling the request and hopes of the Swedish Academy of Sciences, not to mention other reasons. Monsieur Longueuil, governor of Montreal, placed no difficulties in the way of granting this request. But I considered it necessary that the governor-general should be consulted about the matter. Just as I was flattering myself the most over the great discoveries I was about to make on this journey and what [specimens] I should be able to collect, I received the following letter:

A Quebec le 26 S:bre 1749

Je suis bien faché, Monsieur, de ne pouvoir pas changer l'arrangement que j'ai pris pour votre voyage, et d'être obligé de vous refuser la permission que vous me demandez de passer par les forts Frontenac et Niagara. En faisant votre route par le Fort S:t Frédéric, vous aurez tout l'agréement que pouvez desirer, j'ai donné des ordres exprés pour cela, et je vous souhaite beaucoup de plaisir et de satisfaction dans votre voyage.

J'ai l'honneur d'être parfaitement, Monsieur, votre très humble et très obéissant serviteur

La Jonquière.

[Quebec, September 26, 1749]

[I regret very much, sir, not to be able to change the arrangements which I have made for your journey, and to be compelled to refuse you the permission for traveling via Forts Frontenac and

Niagara. In taking your route via Fort St. Frédéric you will have all the comforts you desire—I have given definite orders for that—and I wish you much pleasure and satisfaction on your trip.

I have the honor, etc.

La Jonquière.[1]]

As soon as I received this letter I went at once to the governor in the city, M. Longueuil, and advanced many reasons why he should allow me to make the journey as I had planned it for myself and showed him how important it was for me. But he gave as a reply that he could not deviate a hair's breadth from the orders he had from Governor-general la Jonquière, which he showed to me, and wherein the words were even more rigid, namely: "Ne permettez pas le sieur Kalm de passer par Fort Frontenac et Niagara etc." I saw then that it was impossible for me to do otherwise than take the path of least resistance.[2] Yet in order that I might still fulfill my duty to the Royal Academy of Sciences I sought once more to prevail upon M. La Longueuil of this city to give me permission to take a route which would be far more productive for me. For that purpose I prepared the following memorial in French, as well as I was able, and had been busy with it the greater part of the day:

Monsieur.

M'excusez, Monsieur, si je suis forcé de Vous incommoder; m'excusez aussi, Monsieur, si je ne puis pas m'expliquer si bien dans la langue française, comme je bien souhaiterais; mais il suffit pour moi, si Vous, Monsieur, pouvez comprendre ce que je vais de dire.

J'avais l'honneur de recevoir hiere, quand j'étois sur le point de partir d'ici, la lettre très-gracieuse, que Monsieur le Gouverneur General Marquis la Jonquière m'a fait l'honneur de m'écrire, et dans cela la reponse sur ce que j'avois l'honneur de demander de lui, d'avoir la permission de retourner d'ici par Fort Frontenac à Nouvelle Angleterre; Monsieur Marquis la Jonquière m'écrit qu'il ne peut pas m'accorder cela, mais que je suis obligé de prendre la route par Fort S:t Frédéric.

---

[1] The part in brackets is a translation of the French.

[2] The reason for denying Kalm's request was technically a military one. Kalm had just come from the English colonies and was to return to them. No chances were taken with anyone. Kalm, though otherwise treated with the greatest hospitality, was undoubtedly closely watched, especially when in the neighborhood of a fortification.

Comme d'un côté le temps ne me permet pas ou d'aller moimême à Quebec pour demander de nouveau cette permission, ou d'envoyer un exprès pour cela, parceque je serais obligé de perdre le demi d'une semaine et davantage, qui beaucoup à perdre pour moi dans . . . tems d'année quand plusieurs grains sont meurs et prêts de tomber; et comme je d'un autre côté ay raison de croire, que Monsieur le Gouverneur General peut être, n'a bien compris le sujet de mon voyage, car je ne puis douter, que si M:r la Jonquière l'a bien compris, c'est tout impossible, qu'il a pû me refuser une demande, que je n'ay pas fait pour mes plaisirs, mais pour suivre les ordres et l'instruction, que par les ordres du Prince et Princesse hereditaires de la Suede l'Academie royale des sciences m'a donné; comça, Monsieur, je suis obligé à peu de mots de Vous dire le sujet de mon voyage, et en même tems a Vous très humblement, que vous me voulez accorder la permission de passer par Fort Frontenac.

Si tôt, Monsieur, que la Suede avoit cette joye inexprimable de voir la Princesse hereditaire, la soeur du Roi de Prusse, arriver à Stockholm, le premier soin de cette grande Princesse etoit de suivre les traces et l'example de son glorieux Pere pour rendre un royaume florissant et en état de pouvoir faire une veritable assistance aux ses allies contre leurs enemies; c'étoit pour cela, qu'elle parloit avec les senateurs du Royaume de Suede, qui étoient membres de l'Academie royale des Sciences de Suede, et depuis aussi avec tous les autres membres du dite Academie, de penser sur tous les moyens d'une affaire si importante.

L'Academie ne manqua pas de donner pour reponse à cette grande Princesse, qu'entre autres moyens pour reussir dans une proposition si utile et avantageuse pour la Suede, ce sera très important, si on envoyera qu'elqu'un de membres de la même Academie à l'Amerique Septentrionale; on savoit, que dans Canada et dans Nouvelle Angleterre ils se trouveroient plusieurs arbres, grains, herbes, froment et . . . , qui n'etoient pas dans l'Europe, et cependant, ou étoient très bonnes pour manger, ou pour la teinture, ou pour autres usages; on sçavoit aussi, que le froid étoit si dur dans l'Amerique Septentrionale comme dans la Suede, et par conséquent, que toutes les arbres et les herbes, qui peuvent croître dans Amerique Septentrionale, peuvent avec la même facilité croître et être plantés dans la Suede, sans être tué ou détruits par le froid, comme ces plantes et arbres, qu'on a fait introduire de France et d'autre pays chaudes d'Europe; si on pouvoit avoir des grains de ces arbres et herbes, surtout de ceux, de lesquelles on sçavoit quelque utilité, c'étoit un grand moyen pour rendre la Suede encore plus florissante.

Si tôt que la Princesse a reçu cette reponse, Elle venoit Ellemême dans l'Academie des Sciences, et ordonnoit, que la même Academie choisiraient un de ses membres pour cet voyage le plûtôt que pouvoit se faire: Elle même voulait aller chez le Roy pour lui prier de donner ordres à ses ambassadeurs à Paris et à Londres de procurer pour celui, que l'Academie des Sc. jugeroit habile d'entreprendre cet voyage, tout l'agreement, toute soureté, toute liberté de voyager par tout dans Amerique Septentrionale òu il voudra, sans être empêché de suivre l'instruction que lui sera donné pour satisfaire aux plaisirs de grande Princesse si utiles pour la Suede. Le Roy donna tout à l'heure ces ordres avec plaisir, et on étoit persuadé, que le Roy de France n'aimeroit quelque chose tant que satisfaire entierement les desirs d'une Princesse, pour laquelle il avoit toujours eu une telle estime. Elle pressa Elle-même Monsieur Lanmarie l'Ambassadeur de France d'aussi écrire à sa cour pour cela; peu après j'étois choisi pour entreprendre cet voyage, et je recevois les ordres de l'Academie de me preparer pour la même. La Princesse même aussi bien que Monsieur Lanmarie m'assuroient, que je pouvois être très assuré de cela, qu'ici dans Canada j'avois liberté de voyager par tout comme j'étois dans ma patrie, tout dans la même maniere, qu'on a permis dans Suede aux messieurs Academiciens de Paris de fair par tout; de Nouvelle Angleterre Elle n'oseroit dire le même, parcque la cour de Suede et Celui d'Angleterre n'étoient dans une telle alliance comme celui de France et Suede. La Princesse me faisoit donner une instruction pour mon voyage, de laquelle, Monsieur, voici quelques articles, et jugez de la, si c'est permis à moi de suivre mes plaisirs.

Article 2. Il faut, que vous voyager dans ces endroits d'Amerique Septentrionale, qui pour le froid ont le plus grand rapport avec la Suede (N) sur tout dans Canada, parceque le froid là est si grand, comme dans la Suede et les peu des herbes, que nous avons ici dans la Suede de Canada peuvent resister au froid si beaucoup, comme ceux de la Suede même.—Article 3. Quand vous trouvez quelque arbre, ou quelque plante, qui est connue pour quelque grand utilité, ou pour manager, ou pour la teinture, ou pour un excellent bois, etc. en prenez les grains toujours le plus au nord que vous pouvez trouver cette arbre ou cette plante.

Artic. 4. Les grains que nous specialement demandons exprés, que vous, quand vous retournez, aurez en grande quantité, sont les grains de l'arbre Meurier, Chataigne, Noix de toutes sortes, Bled d'Inde, Fol. Avoine, Myrtus de quoi on fait les chandelles, Cedar rouge et blanc, toutes les plantes que les sauvages mangent, Sassafras, Erable de quoi on fait le sucre, Chinkapins, Pommes de Terres, Taho, Taki, Raisins sauvages, etc.—Artic. 5. Nous avons de vous cette confiance, que vous comme un sçavant pouvez trouver telles plantes, aussi, de quoi l'utilité sera aussi grande pour votre patrie, comme les predits.

Selon ces ordres, Monsieur, et selon cette instruction je suis parti de Suede. A Londres l'Ambassadeur de Suede ne pouvoit pas recevoir quelque passeport pour moi du Roy d'Angleterre pour quelque méfiances qui alors étoient.

[Sir:[1]

Pardon me, Sir, if I am obliged to bother you; excuse me also Sir, if I am not able to express myself as well in the French language as I should wish, but it will suffice if you, Sir, are able to understand what I am going to say.

I had the honor of receiving yesterday, when I was on the point of leaving here, the very gracious letter which M. the Governor-general Marquis la Jonquière had done me the honor of writing, and in it was the reply to the matter which I asked of him, namely, to have permission to return from here by way of Ft. Frontenac to New England. M. la Jonquière wrote me that he was not able to grant that, but that I must take the route via Ft. Frédéric.

Since, on the one hand, time does not permit me to go myself to Quebec to ask that permission again, or to send a messenger for it, because I should be obliged to lose half a week or more, which is too much for me to lose at this time of the year when all the seeds are ripe and ready to fall; and since I, on the other hand, have reason to believe that Monsieur the Governor-general, has not, perhaps, understood very well the purpose of my trip, because I

[1] The following in brackets is a translation of the French letter.

do not doubt that if Monsieur la Jonquière had understood it, it would have been quite impossible to refuse a request which I make not for my own pleasure but in order to follow the instructions which have been given me at the command of the Hereditary Prince and Princess by the Swedish Royal Academy of Sciences. Therefore, Monsieur, I am obliged to say a few words to you on the subject of my voyage and at the same time beg you to grant me permission to pass Ft. Frontenac.

As soon, Monsieur, as Sweden had that inexpressible joy of seeing the Hereditary Princess, sister of the King of Prussia, arrive in Stockholm, the first duty of that grand Princess was to follow the footsteps and example of her glorious father in making a flourishing Kingdom and a state able to give real assistance to her allies against her enemies. It was for that reason that she spoke with the senators of the Kingdom of Sweden who were members of the Royal Academy of Sciences, and later with all the other members of the Academy also [about a proposed scientific journey], thinking of all the means to carry out a plan so important.

The Academy did not hesitate to offer in response to that grand Princess other suggestions for succeeding in a proposition so useful and so advantageous to Sweden. It would be very important if they sent a member of that same Academy to North America. It was known that in Canada and New England there were many trees, grains, herbs, cereals, etc., which were not grown in Europe, and which nevertheless were very good for food, dyes or other uses. One knew also that the cold in [some parts of] North America was as severe as in Sweden, and that, as a result, all the trees and plants which could grow in North America could with the same success be planted and grown in Sweden, without being killed or destroyed by the frost, as the plants and trees had been which had been introduced from France and other warm countries of Europe. If they were able to get some seeds of American trees and plants, and especially those which they knew to be useful, this would be a potent means of making Sweden still more prosperous.

As soon as the Princess received that response, she came herself to the Academy of Sciences and commanded it to choose one of its members for that voyage, as soon as he could undertake it. She even wished to go to the King's palace to beg him to give orders to the ambassadors in Paris and London to procure for him whatever the Academy of Sciences considered necessary to undertake the voyage, such as permission and freedom to travel everywhere in North America that he wished, and without being hindered in following the instructions that would be given him, to the satisfaction of the Princess and the benefit of Sweden.

The King at once gave these orders and it was believed that the King of France desired nothing so much as the entire satisfaction of the Princess, for whom he had always had a great esteem.

She herself urged Monsieur Lanmarie, the Ambassador of France, also, to write to his court about this matter. A little afterward I was chosen to undertake that voyage, and I received orders from the Academy of Sciences to prepare myself for it. The Princess herself as well as M. Lanmarie informed me that I could be assured in this matter; that here in Canada I would have the liberty to travel everywhere, as if I were in my own country, just as they permitted gentlemen from the Academy of Paris to go about everywhere in Sweden. The Princess gave me instructions for my voyage; here are some of the articles from them, Monsieur; judge from them whether I am to be permitted to follow out my desires.

Article 2. It is necessary that you travel in those parts of North America which have the reputation in Sweden of being the coldest in all Canada, because the cold there is as great as that in Sweden, and the few plants which we have in Sweden from Canada are able to resist the cold as well as those native of Sweden.

Article 3. Whenever you find a tree or plant which is known for some particular use, as a food or dye, or for its excellent wood, etc., take the seeds always from the farthest north that you are able to find that tree or plant.

Article 4. The seeds that we wish you to bring in great quantities are seeds of the meurier [mulberry] tree, the chestnut, nuts of all kinds, Indian corn [maize], Indian rice, myrtle from which candles are made, bayberry, red and white cedar, all the plants which the Indians eat, sassafras, maples, from which they make sugar, dwarf chestnuts (chinkapins), potatoes, taho (*Peltandia Virginica*), taki (*Oroutium aquaticum*) wild grapes, etc.

Article 5. We have confidence in you that as a scientist you will be able to find plants whose usefulness to the fatherland will be as great as is predicted.

According to these orders, Monsieur, and conforming to these instructions I left Sweden. In London the Swedish Ambassador was unable to procure for me a passport from the King of England on account of some lack of confidence.]

Birch-bark is said to be quite scarce in Canada and birch-bark canoes daily more expensive.

Birch-bark Canoes. All the strips and ribs in them are made of white cedar (*Thuya*); the space between the latter varying in breadth between that of a palm and the width of three digits. The strips are placed so close to one another that one cannot see the birch-bark between them. All seams are held together by spruce roots or ropes made of the same material split. In all seams the birch-bark has been turned in double. The seams are made like a tailor's cross-stitch. In place of pitch they use melted resin on the outside seams. If there is a small hole in the birch-bark, resin is melted over it. The inner side of the bark or that nearest the tree always becomes the outer side of the boat. The whole canoe consists ordinarily of six pieces of birch-bark only, of which two are located underneath and two on either side. The bark strip directly underneath is sometimes so long that it covers three fourths of the canoe's length. I have not yet seen a boat whose bottom consisted of one piece only. Birch-bark canoes are dangerous to navigate, because if the sail is forced down in stormy weather, it may splinter the bottom of the boat. If one knocks against a sharp or rough stone, a large piece of the bottom of the canoe may be ripped out. It is therefore evident that these boats are continually subjected to adventures and must often be repaired. On that account no one should set out in them without bringing resin and even birch-bark along, though the latter can generally be procured wherever one goes. Likewise it is possible to procure the spruce roots nearly everywhere, and, lacking these, pine roots are said to be equally serviceable.

[After this the Diary reproduces a "Description of some Esquimaux words," taken from Mr. [Arthur] Dobbs' *An Account of the Countries Adjoining to Hudson Bay*, 1744. This list, which consists of 152 words and phrases, has not been reprinted].

October the 8th

Monsieur Picquet, a missionary,[1] an odd man who has travelled much here in Canada, called on me to-day. He tried in every way to convince me that Father Charlevoix, who has described Canada,[2] was a big liar who had gone far astray from the truth.

All day I was occupied with putting my seeds into strong paper cornets, with labelling them, and with packing my other things. In the evening I wrote to Governor-general Marquis de la Jonquière, thanked him for all his favors, yet at the same time made it known how it grieved me that I was not permitted to direct my course via Fort Frontenac and Niagara, especially since it is not an easy matter to send someone hither again such a long distance from Sweden. I also wrote to Monsieur Gauthier (his name is pronounced as *Gō'-thié*).

October the 9th

From Monsieur St. Lucas [I learned the following]: Algonquin is the mother, whose daughters are

Nepissin	Saki
Outayouis	Masguta
Saulteurs	Kikaps
Lutouatauani	Tête de boule
Renards	Gens de terre, on Hudson Bay

S(i)eoux is a language by itself.

Puants is an entirely different language, guttural in sound; [those who speak it are a] brave [people]; they have quite a fondness for the French.

Têtes de boule (quite stupid) are fond of the French but do not wish to have them among themselves for any length of time, as

[1] François Picquet (1708-1781) was noted for his influence among the Indians. He fought in the Seven Years' War; was wounded at Quebec, 1759; and after the battle of the Plains of Abraham escaped to New Orleans in Indian dress, the English having put a price on his head. He died in France in great poverty.

[2] *Histoire et description générale de la Nouvelle-France,* 1-3, Paris, 1744, by Pierre François Xavier de Charlevoix (1682-1761). Pére Charlevoix was noted as teacher, explorer and historian. His name is too well known to need elaboration.

they believe them capable of magic. If they see a Frenchman's gun they think that death is imminent. As soon as a Frenchman comes to them, they bring forth all their fur products in order to make the Frenchman depart soon.

October the 10th

Length of Montreal, 723 toises (un tois-6 feet); width, where broadest, 190; where narrowest toward the north, only 90.

A rainy fall indicates a winter without snow and *vice versa.*

The current in the St. Lawrence is so strong that when one wants to go from Montreal up to Fort Frontenac, one requires from nine to ten days for the trip. But when one goes down stream it takes ordinarily only two days. Between Montreal and Fort Frontenac the distance is 60 French miles, or as far as from Montreal to Quebec. Common people ordinarily called Fort Frontenac Fort Cataracoui.

From Montreal to Prairie à Magdal is reckoned three French miles, but they must be very short ones; from Prairie to Fort S. Jean four miles, some said five; but Monsieur la Croix, who has travelled that distance more (often) than anyone else and one of his relatives who said he had measured it, said that it was not more than four lieues or French miles; but he is mistaken.

[I must] write down where I lived in Montreal, and also note with what love and kindness Monsieur de Couagne, Mademoiselle Charlotte and Mademoiselle Couagnette received me—just as if I had been a child of theirs, nothing less.

October the 11th

I had indeed intended to continue my journey to-day, but since we did not get enough horses, because the road over which we were to travel was pretty bad, I had to postpone my journey to the following day. I shall wait and see if we can get any to-morrow. In every other way Baron Longueuil, Governor of Montreal, had done all that could be expected on his part, for he had given the sergeant who accompanied me strict orders to procure horses for me from the local residents. But those who lived nearest had already put their horses under shelter [evidently at some distance away] and the others had theirs out in the large forests.

French Language. All are of the opinion that in Canada the ordinary man speaks a purer French than in any province in France, yes, that in this respect it can vie with Paris itself. Those Frenchmen who were born in Paris must in this particular commend the inhabitants of Canada. The majority of them, men as well as women, could not only read whatever had been written, but also could write fairly well. I saw women who wrote as well as the best penman could have written. For my part I was ashamed that I could not write as well. That women write well here is largely due to the fact that in this country one has to learn to write one kind of letters only, namely the Latin or French. Besides, every girl is eager to write a message to her lover without having to ask the assistance of another. One thing I noticed especially in the French language was that in the style of writing it has not the advantage of the Swedish. I found that as far as the art of writing in all languages is concerned, that the Swedish is the most natural, because we write rarely more letters than we read, and we pronounce almost all the letters we write; we do not as a rule write any superfluous letters. The opposite holds in the French as well as in the English language, in which one writes many letters which are unnecessary and which one neither reads nor pronounces, e. g. *ils parlent* is written *ls, ent* more than is necessary, as one needs only *i parl,* and all the other characters are useless, and so mostly in all other cases. As a result one-third more characters are found in a French book than are necessary or, in other words, a French book could be one-third smaller than it is, if the unnecessary characters were removed. It was a pleasure to see how the women in Canada, who had not paid much attention to the French spelling and art of writing, followed in their writing the natural method so that they rarely wrote more characters than were essential and just as a Swede would write. As an instance I asked them to write *elles parlent* without telling them how it is spelled, only that it referred to more than one person. They wrote *ell parl* as a Swede would have written it, and likewise in nearly all words which I asked them to reproduce. This I observed to be so in Quebec, Montreal and here; if there were any words which they were familiar with through reading, they would write them correctly (written this A. M. in Prairie).

October the 12th

The Journey. I had bethought myself of setting forth from Prairie during the morning before noon, but as the priest there, a rather civil and educated man, Monsieur Lignerie, together with the captain of said place, sent a message to me and urged me, if it were possible, to postpone my journey until after the mass so that the people could hear it, I found it my duty not to refuse such a just request. Monsieur Lignerie made the service shorter than was customary. After the mass at noon we set forth. I had as baggage four carts, two horses for each, in addition to the horse I rode, all of which was provided for me by the government in this country and paid for with the money of the French crown. The roads were unrivaled in wretchedness, wet and winding so that my horse sank in the mire up to his belly in most places. The weather was also rather mean and so rainy that one could scarcely lift up one's eyes. A large part of the trees had lost their leaves and the woods appeared rather barren.

The larch (*Larix*) grew in abundance on both sides of the road, in some places in sandy soil; the seeds were now ripe and I took a lot of them. The trees which stood nearest the road were not large, about twenty-four feet tall, but the persons who accompanied me said that one sometimes finds them as tall as the largest pines and proportionately broad. The tree is said to be good for lumber, likewise for fuel. They called them everywhere here the red American larch (*Epinette rouge*), quite distinct from that which bore that name in Quebec. Everyone knew that it lost its leaves during the winter and some of them had already begun to fade.

Indian Dances. In the evening of the fourth I arrived here at St. Jean. Several Indians were here who were out on a hunting expedition. When an Indian goes hunting, he does not go alone but takes his whole family with him, also his belongings; that is, his wife, children and dog. He then travels around in the forests shooting all kinds of animals. He eats the meat and preserves the skins to sell, for which he receives in return his clothes and ornaments, also his gunpowder, shot and other articles purchased from the Europeans. The Commander, Monsieur de Ganne (Gannes?), honored me by allowing some of the Indians to dance for me during the evening. The men danced first and then the women. (One of them, who was the leading dancer now stands beside me and is looking at what I am writing; he has a . . . over himself which he has daubed full of vermilion, smokes his pipe, etc.). Before they came out to dance, they painted their faces red and adorned themselves according to their custom. The women as well as the men painted their faces red. A drum was lent them which they struck regularly, one beat after the other, singing at the same time. One of them got up and began to dance; he pulled off his shirt and had on only enough to shield his nakedness, namely a blue cloth which was tied about his waistline and went from his back between his legs and up in front. Over this hung a cloth like a short apron or skirt. This cloth between the legs was of such a kind that it covered both the podices. When he danced he had an axe in his hand which he turned to and fro in the air. He indicated the time carefully with his feet. Sometimes the beats of the drum were further apart, sometimes quite close, and the Indian danced accordingly. Now and then he talked to the others who sang and beat upon the drum and they answered him. Sometimes they sang continuously, for the most part these words: Here I am, Here I am, etc. He turned now to this side, now to that, while he danced. He stood most of the time on one spot, but sometimes he hopped rapidly with both feet together over the whole floor or round about the floor. But the most amusing dance of all was the war dance which they danced when they were to go forth to war. In this they make known all the manners and motions which they use in warfare. He had an axe in his hand, danced a little while standing, then he threw himself down and began to creep about on his knees. Now he imitated with his hands the motion of paddling, as though to search out an enemy; now he looked backwards and with signs of the hand, nodding of the head, etc. he intimated that he saw the enemy and that the others should proceed slowly. Sometimes when on all fours he pressed his body near to the ground to make known that the enemy was close at hand. Again he sprang up hastily, ran away, struck with his axe and ran back to the others crying that he had now conquered. Sometimes when he crept on all fours he pressed his head hastily down upon the ground to hide himself. Now he proceeded slowly on hands and knees; he would motion with his hands to show how they remove the leaves as they go so that the enemy may not find their tracks. Sometimes when he was crawling about on all fours, he gathered together the sand on the floor, washed his hands with it, then heaped it up again and washed arms, waistline, etc. Then again when still crawling about, he turned suddenly backwards to make known that he thought the enemy had caught sight of him and that now it was time to withdraw. Sometimes while on all fours he suddenly sprang up, fell upon one of those who stood close at hand, threw this person under himself, gave him a good tug, ran away and thus intended to show that he had obtained the skull of his enemy, whereat he either sang or talked until he came to the others. While he was crawling on the floor he was rather quiet, but in the other dances he very frequently gave forth the tribal war cry, which indeed sounded horrible. Several others then danced, but all in the same manner. The women danced most simply; they all stood in a row, side by side, moved their feet forwards and back again, so that when they moved the right foot forward, they moved the left one back and then the opposite. When they had danced awhile thus, and turned their faces toward one side, they turned about, moving their feet in the same fashion as described. When they had danced for a time in this fashion they turned about in the same manner as they had danced before, and that constituted their whole dance. The hands were kept constantly hanging by their sides. When the men danced they often moved their hands back and forth. Once in awhile they all danced in a ring, the men foremost, the one after the other, and then the women often in the same manner, but they generally moved their hands back and forth at their sides, that is to say, one forward, when the other was back; that in brief was their dance.

October the 13th

The Latitude of Various Places in Canada. Monsieur La Croix had a brass compass on which was engraved the latitude of various places in Canada, namely:

Paris	49	Cataracoui	45
Kebec (Quebec)	47	F. . .	44
Outouvas (Ottawa)	47	Michilimaquina	46
Tadoussac	48	Niagara	43

Les Sauteurs	48	Le Detroit	43
Trois Rivières	46	Les Illinois	40
L. Royalle	46	Les Miamis	40
Missisagé	46	Folles Avoines	44
Chambli	45	Pontaovas	41
Montréal	45	Fort Louis	41

The common man in Canada is more civilized and clever than in any other place of the world that I have visited. On entering one of the peasant's houses, no matter where, and on beginning to talk with the men or women, one is quite amazed at the good breeding and courteous answers which are received, no matter what the question is. One can scarcely find in a city in other parts, people who treat one with such politeness both in word and deed as is true everywhere in the homes of the peasants in Canada. I travelled in various places during my stay in this country. I frequently happened to take up my abode for several days at the homes of peasants where I had never been before, and who had never heard of nor seen me and to whom I had no letters of introduction. Nevertheless they showed me wherever I came a devotion paid ordinarily only to a native or relative. Often when I offered them money they would not accept it. Frenchmen who were born in Paris said themselves that one never finds in France among country people the courtesy and good breeding which one observes everywhere in this land. I heard many native Frenchmen assert this. The women in the country were usually a little better dressed than our [Swedish] women. They always had night-gowns, and the girls curled and powdered their hair on Sundays. During the week the men went about in their homes dressed much like the Indians, namely, in stockings and shoes like theirs, with garters, and a girdle about the waist; otherwise the clothing was like that of other Frenchmen. The women in the country frequently had such shoes too, except on Sundays. Everywhere the girls were alert and quick in speech and their manner rather impulsive; but according to my judgment and as far as I could observe, they were not as lustful and wanton as foreigners generally claim the French to be. They are somewhat free of speech, but indeed I believe them sufficiently restrained.

Thuya. This fortress, St. Jean, was constructed entirely of wood.

In place of masonry work they had put up heavy logs of arbor vitae of eighteen to twenty-four feet in length and height, one log being placed quite close to the other. They had chosen this tree because no other tree had been found in the whole of Canada which withstood the rot in the ground and was as durable as this one. I counted the annual rings on the largest and found them as follows: N. B. The annual rings were tolerably plain. One log had 92 annual rings, with a diameter of 12 inches; another had 139 annual rings, with a diameter of 15 inches; and still another had 136 annual rings with a diameter of 15 inches. One with 134 annual rings had a diameter of 16 inches was the largest I saw, and another with 142 annual rings had a diameter of 16 inches.

The sugar maples grew in great abundance in the woods here. They had already to a large extent lost their leaves, but the small trees still had fresh, green leaves. It was the largest only which had seeds; the others were without them, doubtless because they stood so close together in the forests and the small trees seldom had the opportunity of being exposed to the sun. I call small those which were from 48 to 60 feet high, as the sugar maple grows to a rather great height in Canada and is among the tallest trees in the country. The seeds had to some extent already fallen from the trees, and those which still had any hardly fell over when cut down, before all the seeds had fallen off and lay on the ground. I believe that I have written before that the sugar maple has like other kinds of maple two seeds side by side, but it differs in this respect from the others, that one of these seed pods is always empty, a hollow capsule only. This is invariable. Thus, for example, when I gathered a couple gallons of seeds from this sugar maple, half of them were useless. It was curious that almost everyone with whom I talked in Canada, although they had tapped the trees and made sugar from more than a thousand sugar maples, they did not know when I asked them whether the sugar maple had any seed. Many were of the opinion that they had already fallen at midsummer. The right time for these leaves to fall is at the end of the month of September, or more correctly, when the first frost has come in the fall; because shortly thereafter the leaves fall and with them the seed from the tree. Yet the seeds often remain after the leaves have fallen. The place where this tree grows is in level and low-lying forests in a rich and fertile soil.

Poisson armé [1] was said to be common in Lake Champlain, also in the brook and river which flow out of it and by Fort St. Jean. It was likewise plentiful at Niagara; yet no use had been found for it. This fish destroys and devours all other fishes.

Indians. A great number of the natives, i. e. the confederates of the French, had already begun to dress like the French: the same kind of jacket and vest, while on journeys they wore the same red cap or hat. But one could not persuade them to use trousers, for they thought that these were a great hindrance in walking. The women were not so quick to give up the customs of their forefathers and clothe themselves according to the new styles, but stuck to the old fashions in everything. But wait! Some had . . . caps of homespun or of coarse blue broad-cloth. When the French are travelling about in this country, they are generally dressed like the natives; they wear then no trousers, but do not carry Indian weapons. Monsieur Croix related that when the Indians go out during the summer to steal a march upon their enemies, they bind green grass about their heads, creep along the ground, pressing their bodies as close as possible to the earth, and move very stealthily to the place where their enemies are or those whom they wish to surprise. The enemy then cannot see them, but he thinks that it is the green grass only which is moving, and quick as a flash the adversary is upon his throat. When they . . . dance their war dance, they often bind their heads with green grass to depict this. The natives farther south among the Illinois have another way of sur-

---

[1] Through the efforts and courtesy of Professor Albert G. Feuillerat of Yale the editor has been able to identify this fish, an identification which, quite naturally, proved impossible to the Finnish editor of this part, in 1929. Professor Feuillerat quotes: "The *poisson armé* is the appropriate name for those species of gar [or garfish] that infest the waters of Louisiana [the French name having been brought from Canada to the South by the Acadians]. The gar has long, narrow jaws full of sharp teeth, and its body is protected by hard rhombic scales. It is highly destructive of other kinds of fish and its flesh is rank and tough. The alligator gar (*Lepisosteus tristoechus,* Bloch and Schneider) attain a length of eight to ten feet. Two other fresh water species are the long-nosed gar (*Lepisosteus osseus* L.) and the short-nosed gar (*Lepisosteus osseus* Raf.).

The Choctaw Indians knew the gar as the "strong fish"— *náni kállo* or *náni kamássa.* The Indians made use of the gar's sharp teeth to scratch or bleed themselves with, and their pointed scales to arm their arrows, says William Bartram (*Travels,* 176).

From *Louisiana French* by William A. Read, Ph. D. University Studies, No. 5, Louisiana State University Press, 1931, p. 61."

prising their enemy, namely by imitating the sounds of all kinds of birds and quadrupeds, a practice which they make use of when they run about in the woods at night. They lie in wait at a place where they know the Frenchmen or their enemies are, with their rifles cocked, and imitate the sound of some bird, etc. to entice the enemy to shoot at it. When the enemy comes close, he knows nothing before the others bear down upon him. They have a way of enticing the roe bucks to them. They tie the head of a roe to the back of their own head, crawl along the ground where they know the roe deer are, make sounds like one of these animals, which immediately comes to them. But as soon as the Indian gets the animal as close to himself as he wishes, he fires his gun, which he has had cocked and ready. When the French travel with their wares among those natives who live in the southernmost regions, they have to keep careful watch during the night and be alert in daytime, since they do not know what kind of Indians they come in contact with, and since a great many of them do not wish to let slip the opportunity to kill the Frenchmen in order to get their goods. The natives are tremendously rugged. I saw them going about these days with only a shirt on and a weapon hanging over it, often without shoes [moccasins], though they had on their . . . or stockings. The men wore no trousers, the women a short, thin skirt; neither of the sexes had anything on their heads. Thus they travelled at this time through the forests on their hunting trips, both in good and bad weather. They lay in this manner during cold and rainy nights in the damp and wet forests without having any other clothes to put under or on top of themselves at night than those they wore during the day. Consequently they carried their beds with them wherever they went. When they came in to Montreal to buy anything and when they left, the women had to carry heavy loads on their backs, but the men went as gentlemen without carrying anything except their guns, their pipes and their tobacco pouches.

October the 14th

At seven o'clock in the morning we set out in the name of the Lord in a canoe which was quite small and rather heavily loaded. Three natives accompanied us in a similar boat. There was a gentle wind, but it was directly against us.

We saw American natives frequently on both sides of the river on which we were paddling. They had their quarters for the night on the shore, as it was the season when they were out hunting deer.

Almost all the trees had faded leaves; some of them were bare. The red oak and also the aspen still had fairly green leaves.

The ordinary species of birch grew thickly on the sides of the river in the lowlands. The land on both sides of the river on which we were travelling was low and level. The birches had snowwhite, smooth bark or one just like ours.

I was shown one place on the western side of the river, about a mile and a half from St. Jean, where a small woodland had started up. It was said that in Count Frontenac's time 1800 natives who had come from Crown Point had camped on the place.

Yesterday we saw large numbers of wild geese in flight.

The red oak had nearly everywhere green leaves, as had the white oak here and there.

Expenses. At Fort St. Jean I received from the stores on the King of France's account two pounds of gunpowder, eight pounds of lead and shot and a lot of knives to be given to the natives whom I might encounter in the forests, in return for game which they have. Moreover the keg was filled with brandy for the same purpose. The fresh meat which I had brought along was cooked here and prepared for the journey by the commander's cook. I took along ½ quire of paper. They also asked me to state my needs and the mere mention of them meant fulfillment. The natives who had danced received considerable brandy in return, likewise the persons who acted as guides and drivers for me, not to mention the food and other provisions for them.

Bulrushes (*Scirpus pallidus altiss.*) grew profusely everywhere on the banks of the river.

Horsetails were also plentiful in some spots on the river banks.

Milium festucoides (millet grass) I shall call a grass which I found for the first time at Prairie à Magdal[1] near the shores of the St. Lawrence River and again to-day on the shores of this river which flows by Fort St. Jean. . . .

I took seeds from this grass. Some of the seeds had already fallen.

[1] La Prairie de la Madeleine, now called Laprairie.

Dactylus foliosus . . . I shall call a kind of grass which grew in lowlands near the banks of the river. . . . The seeds were now mature and the grass itself had begun to wither. I do not know what this grass can be used for, yet I took seeds from it and preserved a specimen of the plant.

Prinos. The inkberry grew as a rule here in the woods in level and somewhat low-lying places. The branches were now full of red berries which I tasted and found rather bitter. A man who accompanied me had told me the same; he called the shrub *bois de marque*, but did not know any use for it. I found this shrub quite common near the city of Montreal and learned that the shrub, the leaves of which I preserved at La Chine under the name of Andromeda, was absolutely the same.

The Canadians are generally good marksmen. I have seldom seen any people shoot with such dexterity as they. A bird that flew so close to them that they could reach it with their bullet or shot, had difficulty in escaping with its life. There was scarcely one of them who was not a clever marksman and who did not own a rifle.

The flowering fern (*Osmunda filix baccif.* Charlevoix) grows abundantly in low places in the woods. I gathered some of its seeds.

The buttonbush (*Cephalanthus*) flourished in the lowlands. The seeds seemed to be ripe.

We stopped for the night a little south of the windmill, still on the western side of the lake. They reckoned that it was about ten leagues (thirty English miles) from this place to St. Jean.

Three native women also came in their canoe and took shelter for the night next to us. They had no man with them, yet each of them had a gun, for they had set out to shoot ducks. One was married, the other two were said to be single. They were Abenaquis Indians. The native who accompanied us during the whole journey was an Iroquois Indian. It is singular that an Abenaquis and an Iroquois rarely take lodgings together, yet they now and then intermarry. The women who had come hither had their funnel-shaped caps, trimmed on the outside with white glass beads. They also had on the French women's waists and jackets which I had never before seen natives wearing. Their evening meal consisted of corn and native Iroquois beans boiled together.

The beach pea (*Lathyrus*) flourished among the rocks on the shore; it was very luxuriant as it hugged the rocks. The seeds had mostly fallen from the pods, although the stem itself and the leaves were green. It seems advantageous to sow these among rocks on the shore and thus make them useful.

The wild bean (*Feverolles*) also grew abundantly on the shore among the rocks. Its seeds were ripe, but had not yet to any extent dropped out of the pods. Some of the leaves and stalks had already begun to wither and fade.

Here the shores were full of cobble stones, most of them of the black variety of which Fort St. Frédéric is built. Parts of bed rock which consisted entirely of this variety of stone were also seen.

The French called the linden "bois blanc" (white wood). The Indian women used its bark in place of hemp for laces with which to sew up their shoes. They were busy during the evenings sewing up their footwear with this material and I could have sworn that it was a fine hemp cord they used. They take the bark, boil it in water for a long time, pound it with a wooden club until it becomes soft, fibrous and like swingled hemp. They sat twisting them on their thighs.

They had made mats of the rushes (*Scirpus pal. altiss.*) upon which they lay at night. These mats were very good looking.

The canoes were not allowed to remain in the water or near the river banks during the night; instead everything was taken out of them, they were carried up on the shore, turned upside down and left there until they were needed. The reason for this was that if a storm should come up during the night, they might be dashed to pieces against the rocks in the river; also this precaution prevented water from entering them and thus bring about rotting.

October the 15th

At daybreak in the morning we continued our journey; the weather was a little cool. The shores were everywhere lined with black stones.

Arbor vitae grew nearly everywhere on the shore especially where crags and boulders or stones were plentiful. It preferred to grow in such places. I saw it nowhere of any great height, 24, 30 to 36 feet, seldom higher. In some spots it was rather thick. Most of the seeds were now ripe. Some of the cones had already opened and lost theirs, especially those growing in strong sunlight. The seeds of the others were about to drop. The arbor vitae grows readily in swamps and marshes.

The bed rock on the shore consisted of black limestone which was deposited in strata in the same way as that found at Fort St. Frédéric. I discovered in many places in it that petrified substance known as *Cornu ammonis,* which resembles somewhat a ram's horn because of its curves. The diameter of some was about a foot, yet they had three to four circles only. There was considerable space between the circles or twists so that they did not lie close upon one another.

The cranberry tree (*Femina*), a kind of *opulus,* flourished in some places on the shore. We consumed great quantities of the berries which were ripe. They had a pleasant, acid flavor and tasted right well. Even if we had had some other fruit here, we should not have scorned these.

Yesterday and to-day also we saw black squirrels in the woods. This squirrel is quite common about Fort St. Frédéric, but north of Montreal it is rather scarce and hard to find. They have these instead of the gray squirrel. Whenever either of these becomes aware of a human being in the woods, it begins to chatter and make considerable noise as it sits in trees, and one has difficulty in making it keep quiet. I have often been angry at it on that account.

The *brown partridges* which are described in the article on Raccoon were found in great numbers here in these forests. We shot several of them, which at this season were very fat birds. The Frenchmen called them *perdrix.*

Yesterday we shot a fat robin redbreast, also described in the account of Raccoon. He is called *Merula* by the Canadians.

The Indian dogs which had erect ears were said to be without equal in discovering wild cats and beavers. At this season the natives were busy hunting deer, but at the same time they took pains to see if they could discover any beaver dams, and if they found them they cut their mark into them. When a native comes to such a place and discovers that another has cut his mark into it before him, he does not touch it nor does he go there later to shoot the beavers, but considers it as a place that belongs to another which he is not supposed to touch.

Pike. The native who accompanied us to-day killed with his oar a pike which came close to his canoe when he was near the shore. It was one of our ordinary pikes, about two feet in length.

Bears are found in great numbers in this country, including the white bears in the Hudson Bay region. The ordinary bears found here are as a rule not as angry as ours, yet if they are harmed they seek to vent their anger upon the person who injured them and then sometimes do as great damage as ours. When one shoots a bear here, the meat is eaten and is considered almost as valuable as pork. The fat or suet is retained and melted; the oil made thereof is preserved. Not only the natives but also the French, especially on their journeys, use this oil in place of butter for stewing and preparing their food. Just recently when I was at Fort St. Jean, Madame la Croix had no oil with which to prepare the salad. She used bear oil therefore, and the salad tasted almost as good as with the usual cotton seed oil, though the flavor was a bit peculiar. When the French are travelling far up into the country their only food is corn. The latter is prepared thus: they put it in lye for one hour until the hull becomes loosened; then they wash it well so that the taste of the lye is removed. The kernels are then dried and carried along in bags on their journeys. They take these, add a little bear oil, some fat of the roe deer and hog's lard mixed, boil it and eat it. When the natives have lean and dried meat, they pour bear oil into a dish and dip strips of the meat before mentioned into it and eat it. The Indians, particularly the women, often oil their hair with it. It is said that the hair grows better because of it and that this oiling prevents the hair from becoming matted.

Polecat. The animal which the English call polecat, the French call *Bête puante* (stinking animal), also *Enfant du diable* (child of the devil), also *Pekan.* Monsieur Chaviodreuil said that he had eaten the meat of this animal a few times and found it as good as pork. But, one must be careful that the bladder is not injured and the contents mixed with the meat.

The *common juniper* (*Juniperus vulg.*) grew on a crag fifty-four miles (18 lieues) from Fort St. Jean on the side of a bay which the river made there and which was named for the roe deer. The juniper bushes were quite small and almost hugged the earth. I found no berries on them. Last summer when I was at Fort St. Frédéric the soldiers told me that juniper bushes were common in that neighborhood, although I do not remember at present whether I saw them. Old soldiers who now accompanied me and had formerly been stationed at Fort St. Frédéric said that they were found on many promontories round about that region and also believed that they were to be found at Lac St. Sacrement.

The French call those trees *Cedar rouge* which the English call red cedar and which the Swedes in New Sweden call *Röd En* (red juniper); whereas the French call arbor vitae *Cedar blanc* (white cedar), but according to the English, white cedar is a kind of cypress which grows in the swamps of New Sweden. Monsieur Lusignan, former commander at Fort St. Frédéric, assured me that there is another kind of *Cedar blanc* which grows in morasses near St. Joseph's River and according to all descriptions and comparisons was the cypress just mentioned. As far as the red juniper is concerned, I found it to-day in considerable numbers on the same point where I found the ordinary juniper. It did not grow there to any great size, the largest being about six inches in diameter. It grew in the clefts of the mountains and in other poor and barren places. Arbor vitae and the latter grew there together. Only a few had any berries. This is the place farthest north where I have found this red juniper. I saw that it liked the same habitat as the common juniper, because they grew there together. Monsieur Valsen, the commander at Fort Frontenac, told me that the red variety is unusually abundant about that fortress. The berries on the shrubs I discovered to-day were for the most part ripe. It receives its name from the fact that when it grows old, the wood inside or the pith (not the surface) becomes almost as red as blood. This is considered the most durable of all woods in the English provinces; in Canada where this shrub does not grow, they consider the arbor vitae as the most durable. The latter is, again, not found in the English provinces, so naturally I do not know which of the two is the more durable.

Gulls of the common variety flew in great numbers over the lake to-day.

The oars used in Canada were as a rule made of maple. This wood was preferred to all others. Lacking this they made the oars of Norway maple or ash. They did not propel the canoes in the same fashion as we do, turning our back in the direction we desired to go, and drawing the oars toward us, for they paddled forwards. In the same fashion they paddled the small canoes which were hollowed out of a part of a tree, but they rowed the boats in the same way as we.

The rocky ledges nearest the shore consisted entirely of the dense black limestone mentioned before. Sometimes small lighter-colored spar particles, frequently even fossil shells were discovered in them. They were arranged in strata of different thicknesses. Sometimes the stratum was three or less inches thick, again it was six, twelve or more inches. I saw in one place strata up to sixty inches in thickness. They were not horizontal, but lower toward the south and elevated toward the north. They were here and there cut off by perpendicular veins. There was petrified shell found in this rock, also the Cornu ammonis or likenesses of ram's horns. Frequently this rock crumbled into very small pieces or nearly dust; elsewhere it was as hard as granite. This made up the foundation of nearly all the islands in the lake, and if one happened upon those whose sides were steep, one had difficulty in getting up on them. They were overgrown with arbor vitae and other conifers.

The place or point where the common juniper and the red juniper or cedar grew is on that part of the map[1] which is designated as belonging to Monsieur Vincent, and just at the point where the red ink line runs out on the right of the same name, almost directly opposite Grand Isle.

In the evening we put up for the night just north of Rivière au Sable. This river is clearly indicated on the map, but has been given no name. It is on the western side of the lake, right opposite Isle Valcour. It is said to have received its name from the fact that along the shore lies a long sand bar of about six feet in height, with the lake on the eastern side and low-lying morasses and land on the western. A great many black ducks which the French called *outards* were swimming outside of this bar.

[1] The map is not reproduced in the original.

The beach pea (*Lathyrus maritimus*) grew in great quantities on the shore, mostly in the dry sand. It was green and luxuriant, but I could not discover any fruit on it or that it had had any this year.

Sweet-gale (*Myrica gale*) flourished here in low places. I took cones and catkins from this. The French called it *poivrié*, and I believe that it was said that one can dye yellow and other colors with its catkins. The leaves had begun to fade.

The buttonbush (*Cephalanthus*) grew here likewise abundantly. It was remarkable that although the greater number of them grew in the morass situated above here as their natural habitat, yet they were found in many places on the sand banks. Either they grew there when the place had been a swamp, and as the sand had been blown there they kept their earlier position and adapted themselves to the sand; or seeds had been carried thither and begun to germinate in the sand. The leaves of those growing on the sand banks had nearly all fallen, but those on the bushes in the morass were still left, though about to fall. The seeds were more nearly ripe on those growing in the sand than on those in the swamp.

Two kinds of pine (*Pinus*) flourished on the sand bars here and were fairly large.

The natives usually make a fire during the night, both summer and winter, when they camp in the woods. One would think that as a result there would be many forest fires during the summer, but I was given to understand that although at times fires do start during the summer, it seldom happens, for the natives themselves are very careful to put out the fires wherever they have made them, inasmuch as it serves their own interest. If a fire should break out and destroy the forest and the vegetation, the roe deer would flee from this region with the result that their hunting would be much less successful.

October the 16th

We continued on our journey in the morning, but as the weather was against us, with a stiff wind blowing and our small boat heavily loaded, we neither dared nor could proceed. After rowing about a league we were compelled to seek the shore and wait until the weather changed or the wind spent itself. We found a haven at the mouth of a small stream, which it was said was a tributary of the same river on whose banks we had camped at night. The entire shore was sandy; some of the sand had the same characteristics as desert sand, namely that it was swirled about by the wind. The monkey flower (*Mimulus*) grew in moist places on the banks of the river. The seeds were ripe and had partly fallen from the pods.

Drift Sand. The sand mentioned above which was found everywhere on the shore, had almost the same qualities as drift sand, that is, it was whirled about by the wind. Whenever it had come in contact with a tree and come to a standstill, it formed a mound. It consisted of small, fine, clear, round and globular quartz, among which were to be found small particles of black mica. A few unusual plants which were found nowhere else grew in this sand.

The giant reed (*Arundo*) . . . , *or* the reed which the Dutch planted on their sandy banks, grew abundantly everywhere on the ridges here. The spikelets were already dry and the seeds either ripe or scattered. The leaves were still partially green, that is, the leaves of this year's growth, but it had also on the same stem or root some dry leaves, which were last year's growth. It was interesting to see how thick and profusely it grew on the sandy mounds with one small tuft close to the other.

The southernwood (*Artemisia abrotanum*). A species of this also grew in large quantities everywhere in the dry sand. A specimen was preserved and seeds gathered. The stems of most plants were dry and the seeds had fallen. Some had not yet advanced to this stage and some were still in bloom. It grew everywhere in the sand, one plant a short distance from the other. It is said to be a biennial because those plants which had borne fruit this year had dried up entirely, but everywhere green leaves without stems were visible. These I learned were those which get stems and bear fruit next year.

The beach pea flourished in many places in the sand. Nowhere did I find any pods upon them.

A kind of small willow (*Salix*), twenty-four to thirty inches tall, with laniferous leaves was found growing here and there. A specimen was preserved.

These [mentioned] were the plants which grew in the sand; otherwise there were several others that flourished here, some on the shore and some in the low-lying country, as, for example: the *water hemlock* (*Cicuta ramis bulbiferis*).

I recall an herb which grew here and which I called *Dentarioides umbellifera*. There was one plant only. I called it "Dentarioides" because its seeds resembled those of the toothwort, though attached to the end of the branches, and "umbellifera" because it is said to be in the shape of an umbel. The seeds were labelled with the former name and the herb was preserved.

The *bulrush* (*Scirpus culmo triquetro*) grew on the shore here and there in the damp sand. The spikelet was on the upper side; it was now full grown. Seeds were gathered and a few specimens of this, a perennial, were preserved.

Of the *bog rush* (*Juncus*) I took two varieties, one which resembled somewhat the *Juncus capit. Psyllii*, but had round leaves, the other had a cluster or clusters on the sides. Both were growing in wet sand on the shore. Seeds were taken from both.

A small *St. John's wort* came forth through the wet soil, (*Hypericum parvum in humidis proveniens*). Seeds were gathered and a specimen prepared. This wort always grows in very wet places and is about four inches high, perhaps slightly taller. Leaves are close to one another on the stem, and the creeping root here and there sends forth new shoots.

A larger St. John's wort came forth through the wet soil. (*Hypericum majus in humidis proveniens*). This wort grew in the same places as the former. It was not only very rare, but also eleven times larger, and among the largest Hyperica. I gathered some seed but could not save a specimen, as the leaves fell off as soon as I touched them.

I call by the name *Phlox* (?) a plant which flourished in sandy, dry places. The seeds had nearly all disappeared. At the root were green leaves. I got a few seeds and preserved a plant.

Buttonwood (*Platanus occidentalis*). One of the men who accompanied me went up the river near which we were stopping, and when he came back he brought with him some small balls of the fruit of the buttonwood which he called the cotton-tree (*Cotonier*). This is one of the places farthest north where it is found. The seed did not seem to be ripe. This tree is not found in the vicinity of Montreal.

In the afternoon the wind came up and we continued our journey. As the wind increased and our boat began to leak badly we did not dare to sail any farther, but were compelled to seek a haven at a place opposite the Isle au Chapon, on the western side of the lake.

Description of Shore. The shore here as well as in many other places consisted entirely of rounded stones of varied composition. I do not know whether there had been more water in this lake formerly than now. I shall describe how the lake appears. Next to the water the shore is largely filled with rounded stones as just mentioned, either of the complex variety or the black limestone. The shore rises gradually up to six fathoms from the edge of the water, where there is ordinarily a rather steep bank of earth of about two to four fathoms height. The land above is usually level and free from rocks. This is how it appears at least in most places. Here and there are small steep cliffs of bed rock next to the water; in other places there are large rocks at the water's edge, then a high and steep bank of earth with round stones in it, and above it high mountains. Again there are sand banks which are next to the water, and often back of them is marshy land. Once in a while there are gentle, sloping sandy shores with a somewhat steep sand hill about thirty-six to sixty feet from the water's edge, above which the country is level. Occasionally the shore is so low and marshy that it is impossible to walk upon it without sinking down into the mud. This then, in brief, is the description of this lake's shore-line. Frequently there is a high bank next to the water, and back of it lowlands, level and large in area. On nearly all sides of the lake, a little distance away, are large and rather high mountains, nearly all overgrown with forests, and it is these mountains they hold responsible for the uncertain weather and the gales. I saw no visible indications that the water had decreased in the lake. It might have been possible, judging from the higher rocks which were three to five feet higher than the present water-line, but I am of the opinion that that condition might have obtained during the spring after a winter with much snow.

Hedysarum . . . flourished everywhere on the shore, even among the rocks, so that I wondered where it gathered any nourishment. It was entirely covered with seeds which clung to the clothes as one passed by. Many stems came from one root. It was now sending forth new shoots for next year. All this season's branches had died. It seems to be unusually well suited for sowing in stony places and on shores which are of little use to anyone. This plant makes splendid fodder for cattle. I gathered a sufficient quantity of its seeds.

Peas (*feverolles*) grew in the same spots as the preceding and may not be unsuited for fodder. They are good for planting in such places as I have mentioned above.

Tobacco Pouches. The native's way of protecting and carrying their tobacco with them was to cut it up and place it in the skin of an otter, a marten or of some other small animal. The feet, where they join the body, had been sewed up and an opening left under the chin. In other respects the skin was complete, as it had been taken from the animal, with head, legs and feet. The hairy side was on the outside. These skin pouches were adorned on the outside with red tassels, tin and brass trimmings. The Indian carried this tobacco pouch upon his arm, along with his tinder-box and pipe, wherever he went. As a rule the ordinary Frenchman made use of this same custom and the pouch of otter-skin was the most commonly used. A skin prepared thus cost thirty sous.

October the 17th

A Storm. At sunrise in the morning we continued our journey with the wind against us, though it was not yet very strong. About eight o'clock when we were almost opposite the Isle de Quatre Vents, the wind changed and became northeast, which was the best wind for us. We hoisted our sail and proceeded [on our journey], but our good fortune did not last long. The wind became steady, but it increased so that the waves began to be rather high. We found it wisest to seek the shore, and it was just in the nick of time, as before we reached land the wind had become so strong that we had to furl our sail, for fear that it might shatter our fragile boat. It looked rather bad for us to begin with, inasmuch as the shore nearest us was full of stones and so steep that we could not make a landing without endangering our lives. We were certain that even if we saved our lives, all that we had brought with us would be lost or spoiled. We were compelled, therefore, in spite of great danger, to row around a peninsula in order to get behind it into quiet waters. The waves were terribly high and the wind came in squalls, as it had done daily since we came to this lake. But God be praised! we got ashore safe and sound. May the Lord calm the winds so that we may soon get away from here! The place where we are now is in the inlet, which is formed by the promontory we rowed around and an extended cliff. From this cliff it is said to be six French miles to Fort St. Frédéric. It was 11 A. M. when we reached land.

Rocks embellished with peculiar markings were found about on the shore. I call them embellished from the fact that the water had eaten into them and had made figures just as if someone had carved a lot of foreign characters upon the whole rock. I have found here several such rocks. If one of these were carried far up into the forests to some little hillock and left there, some European, on coming to that place, would believe that it were a grave and that the symbols on the rock were a foreign and unknown script which the people who had come here in earlier times had made use of, and that it contained the life history of the one who was buried there. The rock was composed of a gray, closely grained sandstone mixed with a limestone of the same grain and color. Within were seen stripes of dark gray pyrites. Perhaps the same pyrites had covered the outside, and either the air, sun, water or rain had caused erosion and in some way had formed these characters.

We also found cockroaches in the place where we were resting. When we had made a fire by the side of an old decayed stump on the shore, a large number of these insects came creeping out of it driven out of their winter quarters by the fire and smoke. The French neither recognized them nor knew their name.

The *Shepherdia* (or *Lepargyræa Canadensis*) flourished everywhere on the shore of Lake Champlain, but I could never find any fruit on it. One of my followers called it *bois à perdrix* (a shrub for the partridge). I am not certain that he knew this shrub, yet he said that it had red berries and the partridges liked them.

The following were perennial plants, or at least they lasted for several years. They had begun to send forth new shoots at their roots: *Asclepias variegata* (*Periploca herbe à la puce*), a purplish weed; a species of *Hedysarum*; and the *Secale culmo multispiculo* [a variety of cereal grass].

A Traveler's Food. The French when making their journeys far up into the country to visit the Indians have as a rule during that long period of from one half to three years nothing else to live on than the hulled corn as described before, the fat of various animals which they mix with the corn and boil, and the game which they shoot in the forests. This is actually all the food they live on for such a long period, and still during this time they are making the most difficult and tiresome journeys of all. Yet they are in spite of it happy and merry, good-natured and healthy. It was on this account that the people of Canada considered corn a kind of grain which ought to be highly prized.

Oil Wells. A Frenchman in Montreal told me of an acquaintance of his who had travelled on the frontier of Carolina and there at the edge of a river had seen a well, out of which flowed an oil which floated on top of the water in the river.

Canadian Beverages. Canadians never used malt in brewing. The common man's drink was water. The better class, who had the means, used French wine, mostly red though sometimes white. Some made a drink from the spruce as described before. Cider was made occasionally by a person of rank just out of curiosity. Punch they thoroughly despised, and laughed at the English who made it. They said that they scarcely wished to taste it, although they had been in the English provinces several times. At mealtimes and whenever they ate, they drank frequently. Each one had his glass, which was filled. Between meals they seldom or never drank. The women drank much water, but also wine and wine mixed with water. The French drink much wine mixed with water at their meals. When one of them visited another between meals, the guest was never offered any refreshments except in the case of a close relative; the conversation furnished the only entertainment. Between the noonday and evening meals they sometimes would eat some fruit or sweetmeat, and they called that a collation.

In *Lake Champlain* I did not see anything that was new to me, neither plants nor anything else floating in it. The water was very clean, extremely clear and of excellent taste, with the exception of some places near the shore where some brook in the neighborhood emptied into the lake. Here the water tasted musty and somewhat swampy. Generally there were on both sides of this lake great stretches of level lowlands, which some day when they are populated and cultivated will make a glorious country, because the soil is so fertile. The greater part of the land about the lake has already been donated by the King to certain families of the gentry; consult my map.[1] The land about Fort St. Frédéric is said to belong to the King still, although it is to a great extent inhabited. On the eastern side of the lake are seen in the distance high mountains which separate Canada from New England.—The Abenaquis (Abnaki) Indians who wander about in these woods on the border are the Englishman's worst enemy.—They reckoned forty leagues (about 120 miles) from Fort St. Frédéric to Fort Chamblais. Fort St. Jean is four leagues this side of Fort Chamblais. The low-lying fields about the lake were said to be marshy.

Fishing. I have mentioned before various methods employed in fishing here. Now I wish to add another method to these. At Fort St. Jean I saw people fish by the light of torches in the same manner as we do. One of the men who was with me described how they sweep the bottom of the large lakes in Canada with seines, especially on Lake Superior, etc., and his account was just as if I had described the way in which we fish in the winter with a drag-net in Österbotten and Wöro [Finland].

Decrease in the Number of Animals, Birds, and Fishes. It is said that beavers and other animals, whose skins are sent to France, were formerly very numerous in the neighborhood of Montreal and the populated places in Canada. Now they have about disappeared there and it is necessary to travel far to shoot or bargain for them, and in the future it will be necessary to go still farther. Various kinds of birds have decreased in number, also the fish in some localities, though that is not yet so evident. Wolves are plentiful in this region.

Venomous animals that are a source of danger and injury were said not to exist in that part of Canada which is north of Fort Frontenac and Lake Ontario or north of Lake Champlain. But south of these places is found the rattlesnake, which is not to be trifled with. On this account Canada is considered to be very fortunate.

It was believed that the swallows migrated from here to a warmer region during the winter. The cuckoo was unknown in America.

---

[1] Whether or not Kalm prepared a special map to show the location of the various grants is not clear. It is not reproduced in the original. But see large map at end of this work.

Crickets were to be found in every house in Canada, but in the English provinces I have not yet seen them.

Fish of Peculiar Properties. Monsieur Valsen, Commander at Fort Frontenac, told me that a fish is caught in Lake Ontario which as yet has no name. If one eats this fish all the hairs on the body loosen so that one becomes bald. There are said to be several examples of this. Therefore the natives never eat this fish.

No *bees* are to be found in Canada, as they are said to perish during the winter. They cannot thrive there.

The *pearl oyster* has not yet been found in this country.

Oysters are found at the mouth of the St. Lawrence River, but seldom come the long distance to Quebec. The traveller occasionally brings to Montreal preserved oysters from New York. They are rather hard, but as they are scarce, one accepts them with the same appetite as one eats and accepts lobster in Stockholm, in spite of the fact that it is often half-decayed when it is received, and an inhabitant of Bohuslän [1] would never touch it.

The socalled tea bushes (*Chiogenes hispidula*) or creeping snowberries were found everywhere in the woods. My companions called them *grains de perdrix* because the partridges liked to eat their seeds.

A load of wood was said to cost 100 sous (4 shillings) at this time in Montreal. Formerly it was not priced so high. Monsieur Chavodreuil told me that when he first came to Montreal from France in 1716, he found in the city not more than three houses of stone; even some of the churches were built of wood. At present most of the houses are of stone and likewise all churches.

How the Natives Paint Designs on Their Bodies. I have related before that the Indians paint various designs on their bodies and that these are put on in such a way that they remain as long as the natives live. On their faces they paint figures of snakes, etc. Several of the French, especially the common people, who travel frequently about the country in order to buy skins, have in fun followed the example of the natives. However, they never paint their faces as the natives do, but another part of the body, as their chest, back, thighs and especially their legs. The designs they paint are

[1] A Swedish province on the southwestern coast of Sweden, where fresh lobster is plentiful.

made up of stripes, or they represent the sun, our Crucified Saviour, or something else which their fancy may dictate. As a rule the natives who are masters of the art adorn the Frenchmen. The color most used is black, and I do not recall seeing any other. Men who accompanied me, told me that they also use red paint and that black and red are the only colors used. The red dye comes from cinnabar which they here call vermilion. The black dye is made as follows: one takes a piece of alder, burns it completely and allows the charcoal to cool. Then the latter is pulverized. The natives do this by rubbing it between their hands. Then one puts this powder into a vessel, adds water to it, and allows it to stand until it is well saturated. When they wish to paint some figures on the body, they draw first with a piece of charcoal the design which they desire to have painted. Then they take a needle, made somewhat like a fleam, dip it into the prepared dye and with it prick or puncture the skin along the lines of the design previously made with the charcoal. They dip the needle into the dye between every puncture; thus the color is left between the skin and flesh.[1] When the wound has healed, the color remains and can never be obliterated. The men told me that in the beginning when the skin is pricked and punctured, it is rather painful, but the smart gradually diminishes and at the expiration of a day the smart and pain has almost ceased.

Every morning as we started to row one of the soldiers read the litany to the Virgin Mary which is found in the "Livre de Vie" for Saturday, if I remember correctly. This was never overlooked. It was always read in Latin and the other soldiers always answered either with the "misere nobis" or the "ora pro nobis" according to whether the prayer was directed to one of the saints or to the Virgin Mary. It was amusing to hear them read the Latin so zealously in spite of the fact that they did not understand the language. I also noticed that each and all of my companions hardly ever failed to kneel in prayer on arising and again at night on retiring.

October the 18th

A storm continued throughout the day and prevented us from making any headway. We had to stay on that account in a place

[1] This is obviously a method of tattooing.

where there was no hunting because of the absence of animals.

Here and there in the woods was found the moosewood (or leatherwood) which the French generally called *bois de plomb*. It was used as cord for binding. One of the men said that if the root of this shrub was boiled in water and drunk, it acted as a strong purgative.

In the afternoon the storm abated somewhat and people began to venture forth on the lake. We saw first a few natives sail by us, and shortly thereafter the two English boats with the Englishmen on board who had escorted home the French prisoners captured in the last war with the English. This made me indignant, as I had left three days before they reached Montreal. But the treacherous weather made my journey to begin with a slow one, and now we had put our boat in such an inlet that the weather prevented our leaving it. We were compelled to remain until toward evening when the gale moderated and we eventually got away from there.

Wild ginger (*Asarum Canadense*) was found here and there in the woods where we stopped. The root had a strong aromatic odor and was said to be good to use in food. The French called it *gingembre*.

The *bearberry*[1] was also found here. The French gathered it and mixed it with the tobacco which they smoked.

Finally we left there in the evening and rowed far into the night until we came within six miles of Fort St. Frédéric. On some parts of the shore the rocks were perpendicular and in others they sloped slightly. In still other places where the rocks consisted of the black stone the water had worn away the rock at the base so that the upper part extended over the water. I paid particular attention this evening to the rocks and noticed that the height of the water in this lake at this time could not be more than three to four feet higher than at present, since the water had worn smooth a horizontal band at the latter height. From this it was possible to see how high the water rose when at its greatest height. Above this line the rocks were overgrown with lichens, and everywhere were trees, especially arbor vitae and pines.

In many places the rocks on the shore were composed of black limestone which generally lay like slate in strata. In some places

[1] *Arctostaphylos uva ursi.*

these did not lie horizontally but sloped, so that the southern part was lower and the northern higher, or they appeared as if they had been placed there during a south wind. In one place the black stone had separated into thin flag-stones, like a slate shingle, but as they had been exposed to the air for some time they could not be used for roof covering. Monsieur de la Galissonnière showed me in Quebec pieces which had been taken from this place.

In the evening we stopped for the night on the shore of the lake; the night was rather cold.

October the 19th

We continued our journey from the place where we had passed the night. The sky was now almost entirely overcast and it looked as if snow might fall. It was quite cold. At 9:30 A. M. we arrived at Fort St. Frédéric. Before we reached the shore the soldiers accompanying me gave their customary salute with their muskets after which they called *vive le roi!* As soon as I reached the shore and stepped out of the boat, they gave a salute of five or six guns from the fortress in my honor, and all the officers together with the barefooted monk, Père Hippolite, came toward me on the shore and conducted me up to the commandant who was now Monsieur Herbin,[1] as Monsieur Lusignan, who was in command this summer when I first came here, had now been released from his post. I was received here with all possible good will and graciousness. As the Englishmen who had come hither yesterday afternoon and had left here this morning had taken away with them a large part of the bread supply, there was not enough left for my four men and myself for our whole journey. I was forced to postpone my departure to the following day, until fresh bread had been baked, which was indeed better for us. The officers here fared very well, for not only were all bodies of water full of ducks and game which now flew over in large numbers, preparing themselves for their fall migration, but it was also the season when the natives staged their hunt for roe deer in the large neighboring wastes and woodlands, whence they very frequently came to the

[1] Cf *Bulletin des Recherches Historiques,* XXII, 381, for Lieutenant Herbin's part in an exploit of November, 1747.

fortress with fresh meat to be exchanged for gunpowder, bullets, shot, bread and anything else they needed.

Natives. The officers told how the natives had learned to know almost every part of the forests. He (the native) goes out to hunt far from his home all day long, and at dusk he returns home in almost a straight line. Occasionally when he has killed more game than he is able to carry home, he leaves a part of it in the depths of the woods. On arriving home he tells his wife that he has left it in such and such a place and that she should go in such a direction to reach it. The wife does this and rarely fails to find it. Sometimes the Indians go through the woods when it is pitch dark, yet they find their way home safely.

October the 20th

Collecting the Seeds of the Sugar Maple. In the morning when everything was ready we continued our journey. I went a little ahead down the country road in order to gather seeds from the sugar maple. Some of them had fallen from the trees. My men rowed along the bay which extends up to the Woodcreek. The commander, Monsieur Herbin, and the other officers followed me part of the way along the road and when I was a short distance outside of the fort I was honored with a salute of from five to six guns which were fired at the fort for my sake. When I had said farewell to the officers, I walked quickly to the last Canadian farm on this side. It was about three miles south of the fort. I stopped there until 2 P. M. to gather the seeds of the sugar maple which had fallen and were lying on the ground. There were indeed several trees which still retained their leaves, likewise a considerable numbers of their seeds, but they had such a weak hold that when I had several of these trees felled, they received on falling such a blow that all the seeds fell off; not a single one remained on the tree. It is worth noting that these sugar maples are among the highest trees found in Canada. I had all the trees cut down on the side where they were to fall, so that the seeds and leaves of the maples might not be damaged by falling against other trees. I had the maples cut only to the point where they gradually started to lean and sink downward, but still I could not prevent the trees from falling heavily upon the ground and scattering all the seeds, so that I was forced to hunt for them on the ground where they had been strewn. But here I ran into a difficulty, because half the seed pods of the sugar maple were empty. I had to squeeze every seed between my fingers to find out which were the solid ones and which the soft ones; the former were good and the latter useless. I could not burden myself with worthless seeds; I did not have room for such. Then among those which had seeds, some were unripe, others decayed or worm eaten. All of the latter had to be discarded. In addition to the seeds from the sugar maple which had just fallen, were also the newly fallen seeds of the American hornbeam (*Carpinus Caroliniana*), of the linden (*Tilia*) and of the ash (*Fraxinus*), but not of any other trees.

I tried to get the ginseng, root and all, and had a soldier for that purpose who knew where it grew, but the ground was so covered with leaves that neither I nor he could procure a single plant.

We rowed from there at two o'clock and in the evening stopped for the night on a promontory about twelve miles from Fort St. Frédéric.

I shall call a certain long grass for the present "melic" grass (*Melica*). It grew here in the rich soil in the less dense woodlands. It had many soft leaves. I gathered seeds under this name and preserved several specimens.

I took some seeds from the brome grass (*Bromus*). A specimen was preserved previously on my journey from Saratoga to Fort St. Frédéric.

I called a certain grass here in Canada "Indian grass", of which I also gathered seeds to-day. A specimen was prepared a few days ago.

The woodlands which covered the stony and worthless promontories and islands ordinarily consisted of spruce, pine and arbor vitae, and sometimes also birch.

October the 21st

In the morning we continued our journey and after rowing one and a half French miles we turned to the right and took the course which led to Lac St. Sacrement (Lake George). On both sides we were inclosed by rocky hills or mountains, steep on nearly all sides and fairly high. They were valueless for purposes of cultivation, but no matter how rocky and useless they were, like our worst wooded hills, they were everywhere overgrown with the arbor vitae. Such hills must have been its native habitat. The channel from Lac Sacrement became indeed narrow, hardly a gunshot across, and also so shallow that the boat could scarcely proceed. After rowing three miles we came to the portage where we had to carry the canoe and our goods overland for a distance of a mile and a half. Here is a waterfall over a cliff of eighteen to twenty-four feet sloping height, and furthermore, above this same fall, it is so rocky and narrow that no boat can proceed.

Bustards and a few *ducks* lay in large flocks, swimming about at the entrance of the channel where it flows into Lac Sacrement. Hundreds of them rose into the air as we approached. At this time of the year the natives travel along the rivers and inlets killing large numbers of these birds.

Chestnuts with their burrs were lying here and there below the falls which Lac St. Sacrement made at the first portage. It was a sign that they either grew there near the shore or in the neighborhood of the lake. The soldiers at Fort St. Frédéric informed me that a chestnut tree is occasionally found between the afore-mentioned fort and Lac St. Sacrement.

We travelled overland a mile and a half carrying our belongings and the canoe. A native and his wife, both Iroquois Indians, followed us with their canoe, which the native without any effort carried on his head the whole distance. The region was slightly elevated, but yet fairly level and everywhere overgrown with woodland, which consisted largely of spruce and pine with no rocks of any considerable size. At 11 o'clock we came to the beginning of Lac St. Sacrement itself where we put our belongings aboard, after we had pushed our boat out into the water.

Lac St. Sacrement is a long, narrow lake which extends mostly from northeast to southwest, but with a few small bends. To begin with it was so shallow that we could scarcely proceed with our boat, which was somewhat heavily loaded. After rowing a distance of three or four gunshots we came to a place where the water flowed over a cliff and was only about twelve feet wide. We pulled the boat over this however without having to unload it, but we had to pull it six to twelve feet only before the water again became deep and the lake wider. On both sides of the lake are high, quite steep mountains covered with forests which send up into the air one high peak after another. The forests consist partly of pine and partly of leaf-bearing trees. There are a great many small islands scattered about in the lake. Bulrushes grow in many places in the middle of the lake, also near the shore. The islands are tops of small mountains or rocky formations. Some of the mountains are unusually high, especially on the northwestern side where they are more separated and not in a range as on the southeastern side. It was sometimes so shallow where the rushes grew that we could scarcely float along. The common reed also flourished here. The water was clear and had a pleasant taste. In some of the shallow places there were pebbles and sand. The width of the lake is half an English mile, now a little more, now a little less. In some places there were neither islands nor rushes to be found. The land on both sides between the mountains seemed to be of such a character that it would not be worth while cultivating, since it would not yield much of an income. He who settled here would doubtless have to live very frugally so far as grain was concerned, but he could have plenty of game, since here is where the natives start their hunt for the roe deer.

Juniper was found here and there on the stony islands and rocks of this lake. They were as large as our ordinary ones and grew in the same way.

The shore is filled with rocks, both large and small. The sides just above are rather steep. In some places are found pines, firs, and arbor vitae. The birch also is found now and then on the sides of this steep, rocky shore. The water is clear, so that it is possible to see the bottom even though the depth is great. The mountains on the edge of the lake are in some places very steep, extending out into the water where it is quite deep.

An Edible Lichen. Finally I came to know the *tripe-de-roche* which I had read about in Père Charlevoix's description of Canada. I had talked with several persons and asked them to show me this [lichen], but although they either were acquainted with it or at least said they knew it, they could not show it to me, since it was not to be found in the locality where they were. They who pointed out the lichen to me did not know it, because others and also the natives said afterwards that it was somewhat similar to it. All the men who now accompanied me had travelled much in Canada and often eaten it. They had frequently talked of it on our journey and to-day when they discovered it, they called out in chorus: "Look, there is *tripe-de-roche*"! It grew upon the sides of the steep mountains mentioned above. It is a foliaceous lichen . . . and is somewhat like the crustaceous variety. . . . It is one of the largest of the foliaceous lichens, often twice the breadth of a palm in length and width and even more. It is turned up on all sides and is shaped almost like a skullcap or bowl. . . . The under side is pitch black, almost furry, the whole resembling black cordovan, or the black shoes which are like velvet on top. On top within the folds and creases it is grayish with a little olive coloring, even and smooth. When it is broken, the inside appears white and it is thin as a wafer. The upper side of some is a somewhat dirty black. This is the lichen which the natives sometimes eat and once in a while even the French travellers, though it happens only in the case of extreme necessity when they have no other food. This *tripe-de-roche* is boiled in water and is improved by the addition of some fat (suet). This moss swells so that it becomes almost half an inch thick. Most of the water [in which it is boiled] is poured off and it is eaten with a little of . . . It is a food which neither nourishes nor satisfies hunger, nor has it a pleasant taste. It serves only to sustain life. One of our men said that on one occasion he had subsisted for five days on this lichen and prunes (sour). If white fish is added thereto, it is fairly good. Salt should be added when it is boiled.

Toward evening quite a strong wind blew against us. It gained force from the high mountains situated on both sides of the lake and consequently became stronger and the waves grew higher. On much of our course we had many small rocky islands now over-grown with fir and sweet gale, which was especially plentiful on the lowlands. These islets or cliffs sloped gently on all sides. Pines were the trees most frequently found on them and they were not exceptionally tall. We sailed between and in back of the islands so that the wind and waves might not trouble us so much. The water was everywhere quite clear. The shore was sandy in places. We followed the northwest shore.

The *red juniper* was found now on the mountains and now in the clefts of the mountains. In some of the inlets where the moun-tains were farther apart, the land was level and it would not have been so hard to cultivate, for it was mostly overgrown by deciduous trees.

There were *chestnut trees* here and there near the water's edge.

At a certain place the lake was about two English miles in breadth, but it grew narrower again. On the northwest the land for a while remained level and not so high, but on the southeast it was just the opposite, high and steep, one high mountain after the other. The shore was rocky, yet not precipitous. Pines, firs (the *perusse*) and hardwood trees flourished on the plain. The width of the lake was one English mile.

The *reindeer moss* flourished on the rocks in the woods.

We encamped for the night back of a point northeast of the high mountains located on the southeast shore of the lake. These are almost the highest of the mountains. The length of the lake was said to be twenty-four miles and we estimated that we still had nine miles to its end. The neighboring forest consisted mostly of birches (?) which still retained their leaves. Next in number was the water-beech (*Carpinus*) and the mountain maple (*Acer spicatum*), but I could find no seeds of the latter. To be sure I found seeds here and there under the trees, but I am not certain that they were of this particular tree. Yet I gathered them.

Wild Geese. I inquired of my companions if they knew of any places in Canada where the wild geese remained throughout the summer. They answered that the geese laid their eggs and hatched the young in those places in Canada where the Illinois Indians lived. They also said that swans were found there; yet swans were said to be found on rivers in several localities in Canada.

The Saulteurs were said to be the worst of all natives in Canada. They live mostly near Lake Superior and Mishillmakira (?), but are also found scattered about the whole region. When they come across the French in the forests they treat them with kindness, bring game to them, etc., but when they [see an opportunity] . . . they kill them and carry off all that they have with them.

Linden or *basswood* (*Tilia*). The French called the kind of linden which grew abundantly in the woods *bois blanc* (white wood). They had with them bags in which they carried their food, which were made by the natives from the bark of this tree. The Indians take the bark, boil it in lye and pound it to make it soft. It becomes like a coarse hemp. They weave it in such a manner that the lengthwise threads along the side are broad and scarcely twisted at all, while the crosswise threads which they have twisted on their thighs are about the size of small hemp cords and are not woven in with the lengthwise threads, but wound around as on a hamper.

October the 22nd

The wind was so strong that we could not proceed in our boat, which now was very heavily loaded with food which the men had taken along to have on their return journey. We had to stop for a considerable time. The wind came in gusts.

The *witch-hazel* grew here and there in the woods under the deciduous trees. Although the leaves had fallen from the majority of them, a few still remained, but so loosely attached that they fell off when one approached the bush. The size and shape of this shrub is much like that of the hazel, yet I believe that the hazel grows taller. It was now in full bloom, although the leaves on some of the bushes which had ceased blooming had already fallen. . . . The French who accompanied me did not know this shrub.

Seeds which had recently fallen from the trees were to be found upon the ground, namely, those of:

The sugar maple, which had lost its leaves, with the exception of the small new seedlings which had green leaves.

The mountain maple, which still had its leaves, though it had changed color.

The water-beech still had leaves, but had begun to fade.

The beech still had its leaves.

The leaves of the ash had fallen.

The walnut had no leaves.

The oak still had leaves.

The linden . . . had lost its leaves.

Red sand was found here and there on the shore of Lac St. Sacrement. It was made up of very small, round and shiny grains, just as if they had been either rubies or garnets. I never found it alone, but always mixed with the coarser grains of the light-colored quartz; yet I found it more often on top of the coarse sand than mixed with it. I discovered also the fine sand resembling iron on the shore.

A fog as thick as smoke arose to-day from the high mountains so that at a distance it appeared like the smoke from a burning charcoal kiln.

Bois de calumet (wood for pipes) is what the French called the *dogwood,* which had green branches and stem and not red like the [European] cornel. It sent forth each year long, narrow, even shoots without branches which had a fairly large pith. Both Indians and French took these shoots or branches, removed the outer bark so that they became smooth, then bored out the soft kernel and used the branches for their pipestems. The bowls of the pipes were made of a kind of limestone which was blackened so that they appeared pitch black. The process is described in my notes for my journey to and from Bay St. Paul.[1] The pith of this tree is so soft that a long narrow stick cut from the same tree will force out the kernel when it is pushed through a branch of it. This is not true of the red dogwood, the core of which is harder. On paring the bark from this branch people loosened it so that it remained attached at one end. Then they put the other end of the cutting in the ground a little distance from the fire and allowed the bark to dry a bit, whereupon they placed it in their tobacco pouch and smoked it mixed with their tobacco. They also used the bark of the red dogwood in the same manner. I inquired why they smoked this bark and what benefit they derived therefrom. They replied that the tobacco was too strong to smoke alone and therefore was mixed [with bark]. This is one of the customs which the French learned from the natives and it is remarkable that the Frenchmen's whole smoking etiquette here in Canada, namely, the preparation of the tobacco, the tobacco pouch, the pipe, the pipe-stem, etc. was derived from the natives, with the exception of the fire-steel and flint, which came from Europe and which the natives did not have before the French or Europeans came here. Red dogwood the French called *bois rouge* and knew of no other use for it than the one mentioned above.

The French call the water-beech (*Carpinus*) *bois dur.* They know of no other use for the wood than that it is good for fuel, and since it is hard and durable, it is used for cart axles.

I shall call that part of Canada a wilderness which lies between

[1] See page 498.

the French farms at Fort St. Frédéric and Fort Nicholson on the Hudson River, where Mr. Lydius and other Englishmen have their farms. Not a human being lives in these waste regions and no Indian villages are found here. It is a land still left to wild animals, birds, ect. At this time in the autumn Indians come hither from various localities; even natives who sometimes wage war against one another. They live here for several months by hunting alone, especially for roe deer which are plentiful in this vicinity. When this hunt is finished, they begin the hunt for beaver, which takes place toward spring. About the time when the beaver hunt begins, the meat of the roe deer loses its flavor, as it is the mating season. The skins are not as good as during other seasons, so that it seems as if nature itself commands them to cease the hunt. From the skins of these animals the natives as well as the French in Canada make their shoes which they use on their journeys and which somewhat resemble the Finnish (elastic) boots.

The so-called tea bushes (*Chiogenes hispidula* or moxie plum) were found where we were staying, growing along the ground in large numbers. I have also found them everywhere about the lake.

Fir (*Abies*). The hemlock spruce (*Epinette blanche*) which was used in brewing is found generally on the shore among the rocks.

Fir (*Abies*). The pine called *perusse* was also very common. I took a sufficient quantity of the seeds which were mature.

The pine with leaves in twos from the same sheath, with cones egg-shaped, were commonly found on the shore in a sandy and poor soil. The seeds were ripe, but had disappeared.

The pine with leaves in threes from the same sheath, with prickly cones, was also quite common and growing in the same kind of soil. The seeds had dropped from the cones.

Shepherdia grew here and there on the shore, but I could not find any fruit on it, except the small buds which are said to be the fruit. Might it be a kind of Myrica? I don't believe it.

Maple (*Acer spicatum*). The mountain maple was abundant here, but I do not know if it was the seed of this that I found under the tree. The French called it ordinarily *bois noir* (black wood) because the bark had black lines in it, especially when it was somewhat older. They also said that it was called by some *bois d'orignal* (wood of the elks) since these animals eat many of the twigs of this tree during the winter. When one passes through the forests at this season, one can tell from these trees the whereabouts of the elks, because the twigs are chewed off these trees. It is possible to see in what direction they have gone since the twigs have been broken off, hang and are bent in the direction in which they have gone.

Tales of Horror. During the evenings my companions were busy telling one another how they had gone forth in the last war to attack the English; how they had had Indians along and how they had beaten to death the enemy and scalped him. They also told how the natives often scalped the enemy while he was still alive; how they did the same thing with prisoners who were too weak to follow them, and of other gruesome deeds which it was horrible for me to listen to in these wildernesses, where the forests were now full of Indians who to-day might be at peace with one another and to-morrow at war, killing and beating to death whomsoever they could steal upon. A little while ago there was a crackling sound in the woods just as if something had walked or approached slowly in order to steal upon us. Almost everyone arose to see what was the matter, but we heard nothing more. It was said that we had just been talking about scalping and that we could suffer the same fate before we were aware of it. The long autumn nights are rather terrifying in these vast wildernsses. May God be with us!

October the 23rd

We continued our journey from this place at dawn, inasmuch as both the weather and wind were less severe.

The lake had at this point about the same appearance as before described; namely, on both sides high, fairly steep and wooded mountains. I do not know on which side they were the higher. There were small rocks here and there in the lake. The shore in some places was covered with a light sand, in other places with stones or bed rock. The water was clear and pleasant to the taste. Here it would have been impossible to paddle a canoe at night because of the many rocks along the shore. The lake extended from northeast to southwest with an occasional short curve. It seemed that the curves bent more toward the west than the east and left side, along which we were now proceeding.

Indians. One of the natives had put up his tent, if I may so call it, on the shore. He was one of those who had set out to hunt. The canoe lay upside down on the shore, as was the custom, and a short distance above in the woods, the Indian had made his hut which was constructed in the following way. He had placed pieces of birch bark and other bark on top of slender rods as a roof over himself where he lay and had hung an old [blanket?] . . . .[1] to protect himself on the sides from the wind and storms. His companion had done likewise on his side and their fire was between them. The wife and children were also sitting before the fire. The native had killed a great number of roe deer and hung up the flesh on all sides to dry. The skins were also stretched out to dry. At the fire's edge sticks were set into the ground perpendicularly and at the tip of these were pieces of meat for cooking. Indians discarded the horns, yet these were sometimes used for making knife handles and the like. The native men had pulled out the hair from the front part of their heads as far up as the part above their ears, so that the whole of this part of the head was bare, which gave them the appearance of having rather high foreheads. When they sat down they crossed their legs in front of them, one leg in front of the other. After we had bought a little of the meat of the roe deer we continued our journey.

Lake George. The lake had the same appearance as before mentioned with fairly high mountains on both sides, the one mountain piled upon the other so to speak. We followed along the northwest shore and in one place came upon a terribly high and steep mountain, which at the top, on the side toward the lake, was almost perpendicular. Below and next to the water there was a rather steep hill which was composed of large and small stones that had fallen down from the mountain. It was awe-inspiring when we rowed at the foot of the mountain and looked up, for it seemed as if the mountain hung right over our heads as we proceeded. The lake at the shore was very deep. The shore now for a while consisted of either stones or bed-rock and beyond that point the shore was precipitous. The mountains were everywhere overgrown with forests. The wind began to blow in strong gusts against us and the waves were fairly high, so that it was not especially pleasant to sit

[1] Word omitted in original, probably from defect or illegibility of manuscript.

here in the heavily loaded boat and write. I should not have wished to have the boat founder here, since it was so fearfully deep that no one of us could have saved his life. Fog was rising from the mountains in many places just like the smoke from a charcoal kiln. There were many islands here and there in the lake.

The trees about this lake had not so generally lost their leaves as those had in the neighborhood of Montreal and Lake Champlain. Many trees still retained their green leaves and the farther south I proceeded the greater was the difference that I perceived. These high mountains surrounding all sides prevented the cold from being felt as early as around Lake Champlain, where the mountains are not as high.

The *red juniper* grew here and there in the crevices of the rocks and cliffs. Such are natural places for their growth. Some of them doubtless had berries.

The shores for about six miles consisted of bed rock only or large boulders, and was so steep that in case of a storm a birch-bark canoe could not make a landing there. It is possible that one might be able to save one's life with difficulty, but hardly the boat, and one's belongings. In many places the rocks were so steep that it would have been impossible to have escaped alive, if the waves had forced one to seek land in this locality. Judging from the moss on the rocks the water in this lake when at its full height is from two to three feet higher than at present. Above this distance the mountains and rocks were covered with lichens and mosses.

The pitch pine was generally found growing about the lake in sandy and poor soil and in the crevices of the mountains. Also the scrub pine flourished in similar places.

The arbor vitae was plentiful in some places and grew much under the same conditions as the red juniper, even among the rocks.

Birch likewise grew in the region about the lake and often in the crevices in the bed-rock. The leaves were yellow, but had not fallen from their branches. The opposite was true at Lake Champlain, where most of the trees had lost their leaves.

The mountains everywhere on both sides were fearfully high, one close to the other and often quite steep, although covered with forests of trees and pines. Here and there in the lake was a small rocky island with a few trees upon it.

The wind was about the same as it had been the days before;

it blew in gusts, at times very strong, again more gentle and almost calm. Then it would change.

Juniper (the Swedish) also flourished here and there in the crevices of the rocks. The oak (the white, red and black) was found to grow on both sides of this lake.

There was an island in the lake of some considerable size in proportion to the other small islands. It consisted of a long low rock overgrown with shrubs.

The trees which grew among the stones that had tumbled down from the mountains were, among others, the following:

Birch in considerable numbers and even flourishing luxuriantly where the soil was the poorest: firs, both perusse and epinette; pines of all kinds; arbor vitae, in natural habitat; and a fair quantity of red juniper.

Note. Nearly all of these were trees which throve in the clefts of the mountains and grew quite rapidly. If anyone should wish to make use of such crevices, he should plant these trees and others which he finds will thrive there. Yet I did not find very many firs in the clefts of the bed-rock, but all varieties of pines, red and ordinary junipers, sometimes also the birch.

There were at this place in the lake several small islands covered with woodland and situated close to one another. All were of bed-rock covered with trees, mostly firs.

The lake was nearly everywhere about an English mile across, sometimes it grew a little wider and again it became narrower. Just about southwest of these islands it seemed to be quite broad. We rowed down toward the southeastern side and followed it. The lake now became about a couple of English miles broad with small islands here and there and surrounded by high mountains on all sides.

At noon the wind became so strong, augmented no doubt by the greater breadth of the lake here, that we could no longer continue our rowing. We were forced to seek the shore until it calmed down somewhat. We landed in back of a peninsula formed by a mountain. This was a barren place as far as herbs were concerned, as nothing much was to be found on a mountain. The rarer kinds found were these:

Bearberry plants covered the mountains in many places and flourished here in the same kind of places as ours [in Sweden].

The crevices in the mountains were full of them. Sweet fern (Myrica) grew everywhere in this locality. The so-called tea bush (moxie plum) was also common. The white pine (*Pinus alba*) throve in this region and grew to an unusual height. The Andromeda was found here also. The juniper was everywhere in the crevices. Indian grass, so called in New Sweden, was commonly found here. I do not recall that I have seen it farther north.

At 2:30 o'clock in the afternoon the weather became more calm and we set forth from here.

Flies now began to follow us, since we had received fresh venison which we were carrying along with us. They were the ordinary house flies.

The wind had nearly died down when we set out from the shore. It has not been very windy since. A squall accompanied with rain came, and suddenly there was such commotion in the lake that it looked almost like a boiling kettle. The waves went crisscross and were so large that we were in great danger. Yet it was almost calm. As soon as the squall and rain subsided, the strong agitation of the waters ceased. When the movement of the waters was at its height, we were in such a place that it would have been an impossibility for us to reach the shore, since it consisted of precipitous mountains reaching to the water's edge. The commotion of the waters was greatest about the promontories.

Arbor vitae flourished most where the shore consisted of fairly large rocks covered with a little soil or moss. The red American larch, which is used in making a beverage, grew here as did the perusse, its companion.

Lichen. Reindeer moss was everywhere abundant on the wooded mountains.

The shore on the southeastern side of the lake where we were rowing to-day consisted either of bed-rock, more or less steep, or it was covered with fairly large stones which could hardly be called cobble stones. The lake was very deep at the edge of the shore. We saw scarcely any other trees than varieties of pines and firs with an occasional birch among them.

The lake on which we had travelled this afternoon was nearly everywhere two English miles broad, if not a little more. My companions guessed that it was three miles or more in some places. Near the shore there was occasionally an island, but seldom any in the middle of the lake. There were mountains on both sides, yet they did not seem as high as those we had seen before, although some were as high if not higher: a mountain just now appeared on the southeastern side where we are rowing, and almost in front of it, in the center of the lake, an island is located. The rocks and mountains about the lake are of granite, and nowhere in this region have I found the black limestone. The sunshine about 3:30 P.M. was quite warm, yet the thermometer did not rise higher than fourteen degrees above 0°C. Toward the end of the lake were large islands covered with woodland. The forests were mostly pine or fir, a sign that the seeds had been carried there by the wind, but not so the seeds from hardwood trees. The seeds of the latter which had been carried by the water had either become decayed or had not reached land because of the rocky shore. There was a birch here and there. Perhaps the firs had come here first, grown up and when the leaf-bearing trees had come later, they did not thrive, since the firs had had the upper hand and had stifled them. This theory seems strengthened by the fact that oak and other trees are to be found here, but they are rare and few in number, small, miserable specimens, surrounded and crowded by the firs.

We had now approached the end of the lake. Now again I encountered the same difficulty as last summer when I travelled to Canada through these wastes, namely, that the person who was to be our guide could find neither the way nor the portage a second time. We then had to begin by following along the shore, thus hunting for the road again. If we do not find it, things will go wrong. Thank God, we have enough food, but if bad weather should set in, there is no pleasure in being in these vast wildernesses.

I have seen very few sugar maples in the vicinity of this lake, and in most places none at all.

The lake divides into two branches at its end, one toward the right and the other toward the left, or one toward the W.S.W. and the other S.S.W. We are now following the left, but after we had gone to the end of the bay and found only a small brook which ran out of a swamp or morass, but no trail nor sign of a portage, we had the pleasure of turning back to see if we were to be more successful along the other branch, which flows W.S.W. or toward the right when one comes from Canada. The land between these two branches or bays is a long peninsula of about a quarter of a mile or so. It is a lowland, mostly overgrown with fir. Arbor vitae is especially plentiful on the shore of this inlet and next to it in number is the perusse. After four o'clock it was extremely calm and the water in the lake very smooth. Once in a while there would come a slight gust of wind.

The wolves were howling fearfully in the bay which we had just left. It was said that they had just torn to bits a roe deer over which they rejoiced, or they had killed one and were calling the others to the feast.

A *black squirrel* was shot. It was the female who was not all black. It was said that the male was entirely black. This one resembled our squirrels, and its size was about the same, but the color varied. The greatest part of the hair on its body was black with the outermost ends light gray. Its head between the ears was almost entirely black. The cheeks . . . and stomach were almost a brown, the whiskers long and black. From the nose and head along its back there was a darker stripe. The tail and legs were the same color as the rest of the body, namely, the end of the hair light gray with the rest black. The tail was broad and fluffy just like that of our squirrels.

We continued our journey farther in search of the right place where one goes ashore and passes over to the English provinces. We discovered smoke coming from a place on shore, toward which we rowed in order to come in touch with people from whom we could get all information [we wanted]. There were three boats of Abenaquis Indians who had set up camp at this place. The greater part of them were out hunting, so that there were a couple of men only and a few children left with the boats. The men were almost drunk, since the Englishmen who had travelled through here a couple of days before, had given them rum in payment for the meat of the roe deer which they had given them. As soon as we came ashore, they put on the pot to boil meat for us. According to their wishes we must of necessity stay over night and eat with them. But as the natives when they are intoxicated are often very troublesome and even dangerous, we decided that it was wisest to proceed from this place, especially since we learned that the portage was at the end of the bay on which we were now rowing. We continued a little farther on and set up our quarters for the night, as usual, on the shore.

October the 24th

We continued our journey from this place, rowing about a quarter of a Swedish mile before we came to the end of Lake St. Sacrement. Here the shore became sandy and sloping. My companions left their boat and the greater part of their food on the shore, since they had all they could do to take care of my belongings. They had to carry the latter fifteen miles if not more over land from the aforesaid lake to an arm of the Hudson River. The mountains at the end of the lake did not seem as high as those we had seen before on both sides of the lake. In the beginning, and almost along the whole way, we had mostly pine woods around us, though here and there was a clump of oaks. The pine woods consisted in part of red and white pine and cypress. Both white and red oak grew abundantly among the pines, but they were small.

Sweet fern flourished everywhere under and among the pines along our path. I have found the same variety in similar barren localities from Fort Nicholson way up to Albany. Most of the trees had now shed their leaves, but this shrub still had its leaves.

Tea bushes, so called in New Sweden, grew everywhere abundantly in the same barren places as just now mentioned.

The ground over which we were travelling was not exactly level, but slightly hilly, yet the hills were not especially high. Nearly everywhere there were numerous large pines which had fallen over and were left to decay. These had often fallen right across our path and hindered much those who were carrying the luggage. Now and again we saw a mountain in the distance. The region sloped uphill as a rule and was full of tea bushes. The greater part of the forest consisted of the pines and shrubs mentioned above.

We found *chestnut trees* in some spots in the forests. The soil in which they grew was in some places rather poor, hence it is easy to see that we could get this tree to grow in Sweden provided we could get some good nuts for seed.

Witch-hazel grew in many places in the forests, especially on level ground. Its flowers were now at their height and were very attractive from a distance since the leaves had fallen. The flowers were yellow.

We found that chestnut trees were among the common trees of the forests. The men who were with me said that the roe deer were very fond of the nuts after they had fallen from the trees. They eat acorns also.

Smuggling. When we had gone half the distance [to the English territory] we met a couple of Englishmen who came from Boston and had been sent to Quebec on the prisoners' behalf. They had several Iroquois with them as guides, some of whom carried their luggage, others their canoes. Merchants in Albany who carried on questionable business with those of Montreal, took advantage of this opportunity to send a lot of forbidden wares to Canada in exchange for which they were to receive the skins of beaver and other animals. In one spot we came upon an oak forest. We also saw now on the right and again on the left a high mountain some distance from us. We then passed over a mossy region. After that we travelled mostly through a sparse pine forest, the pine being mostly like ours [in Sweden].

I found several small larches growing in a swamp or morass. In Canada it grew more often in level and dry regions; yet also there in a swamp, e.g. between Fort St. Jean and Prairie de la Magdaleine.

The ground laurel (trailing arbutus) grew in some places in the same poor soil as the so called tea bushes, but I found no seeds on them.

On the journey from Lac St. Sacrement to Hudson River I did not see a single sugar maple. Wherever there were any, they were few in number, as the soil was too dry and poor.

Kalm Reaches the Hudson. About three o'clock in the afternoon we reached the Hudson River where on the downward journey one boards a boat, while on the upward journey one starts to carry it overland. We had no boat and still had eight English miles to the place where Fort Nicholson had stood and where was now the first English habitation. As it was late and the road poor and wooded, we had to postpone our journey to the following day. We took our quarters for the night under a cliff which projected out from the mountain and formed a room as it were beneath it. The natives had rested here before, otherwise it looked awesome when one looked up at the large cliff below which we were now lying, and contemplated that several rocks from the same steep mountain had but lately descended. We made a large fire under the rock. We stopped here so as to escape the rain if it were to come during the night. In summer this region is said to be infested with rattlesnakes, but the cold had long since driven them into their holes so that we did not need to have any fear of them.

The mountains of which the cliff mentioned above was a part, as well as the bottom and both the high steep sides of the river, were of the black limestone which lay here in strata. These were of varying thicknesses, some only six inches or less, others up to four feet or more, still others in between the two. Arbor vitae grew in the crevices.

October the 25th

In the morning I went with three of the men from our night quarters to Fort Nicholson, where the first English colony was located. Our object was to procure a boat to take our things to Saratoga, since it would have been difficult for the men to carry the canoe through the woods from Lac St. Sacrement to this place. It was reckoned as eight short English miles from the place where we first reached the river to Fort Nicholson. We walked along the side of the river which was overgrown with woodland. The forest consisted mostly of tall white pines, with oaks here and there. The ground was generally level, sometimes slightly hilly. We rarely saw any rocks, let alone mountains. Now and then there were swamps or morasses, yet they were wooded. The most common tree growing here was the arbor vitæ, sometimes the fir called perusse.

To-day I saw *sassafras* for the first time since I turned south from Canada. I have not seen any native to Canada, only that which had been planted on the Isle de Magdaleine at Montreal. The plant which I found to-day was but eighteen inches tall and still retained its leaves. It grew in a barren place.

The *witch-hazel* (*Hamamelis*) grew in surprising numbers under and between the widely separated trees. It was in full bloom. . . .

At 10 o'clock we arrived at *Fort Nicholson*. In the last war the house of the English here was burned by the French and the latters' native allies. Now they had built a new house. Colonel

Lydius who had lived here during the last war had not yet had his rebuilt, but intends to have it done next winter. The reason why the English, or rather the Dutch who live in Albany, build and live here is because of the trade with the natives. The English as well as the French also settle here. The natives bring their fur products to this place and in return receive almost all the wares from the English at a better price than from the French, with the [additional] advantage that they may buy and drink as much rum as they wish, a thing which is not permitted by the French traders. Here we see the reason why the native, even he who belongs to the French, prefers to sell his products to the Englishman rather than the Frenchman. There is nothing in the world which an American Indian prizes so highly as French brandy and rum, though he always prefers the French brandy. In Canada it was said that if an Indian saw brandy before him and the proposition were made to him to receive a good drink under the conditions that he was to be killed thereafter, he might well reflect a little thereupon before giving consent, provided he, somehow, before could still his longing for the brandy. The English know how to take advantage of this in a polite way, that is, to get from the Indian his products for next to nothing, under the pretext that they have only a little brandy or rum left, which they need themselves. A native who lives far to the west of Montreal might travel more than 200 to 300 Swedish miles, past the French colonies, to Oswego with his fur products, there to sell them at a low price, just for the satisfaction of once becoming drunk from rum.

October the 26th

Letter writing consumed a good part of the day. I had four men with me from Montreal; namely, one sergeant and three civilians who had acted as guides the whole way and were therefore paid by the French government, as the original bill of my Canadian expenses which the French Crown had incurred on my behalf testifies. They had orders to go with me as far as Saratoga or four Dutch miles farther (N.B. A Dutch mile is equal to four English; a French lieue or mile, 3 English), but since head winds had delayed them and the season was so far gone, that the river might freeze over and they would encounter difficulties in returning home, I allowed them to turn back from here. I then wrote letters to some of my hosts and friends to thank them for their favors and friendship which they had shown me during my stay there. The letters I wrote were to the following:

Governor-general Marquis la Jonquière;
General de la Galissonnière, in France;
Commissary Bigat;[1]
The Governor of Montreal, Baron Longueuil;
Gauthier, the Royal Physician;
The Commander at Fort St. Frédéric, Mr. Herbin;
The Barrack-sergeant in Montreal, Mr. Martel;
Monsieur Renet de Couagne;[2]
Mademoiselle Charlotte de Couagne;
The Superior of the Jesuit [priests] in Quebec;
The Rev. Father Saint Pé, Society of Jesus, Montreal;
The Commander at Fort St. Jean, Monsieur de Ganne;[3] and
Barrack-sergeant, *ibid.*, Monsieur La Croix.

I continued my journey from Fort Nicholson in the afternoon. I had difficulty in starting out, as the natives who were there demanded so much [pay] that I could not afford to take them along. Finally the man who had charge of the trading post accompanied me with his boat. I thought I could make the descending journey easily, as he had another boat below the uppermost and largest waterfall. But when we arrived there and had begun to carry my luggage overland (about half a mile), we found that some one had carried off his other boat, and I was forced to leave all my goods in the unprotected forest during the night, since it was impossible for the two of us to get the first boat down the fall. Then we had to travel on foot along the country road a Dutch mile to Saratoga: a wretched road it was.

---

[1] This is possibly François Bigot (1703-c. 1777), who had been appointed commissary at Louisbourg in 1739 and intendant of New France, 1748. Because of gross fraud in Government funds he was thrown into the Bastile in 1759 and after a sensational trial was finally exiled from French territory.

[2] It has proved impossible to learn any details about the life of Renet de Couagne. The Couagnes in Canada were very numerous, all being the descendants of Charles de Couagne (1651-1706), who had emigrated to Canada in the seventeenth century.

[3] Probably either Michel de Gannes, sieur de Falaise (1702-52) or his brother, Charles-Thomas de Gannes (1714-65). Both were military officers in Canada in Kalm's day.

October the 27th

Kalm in Saratoga. I occupied myself during the whole day with getting my things home from the waterfall, where I had left them the evening before. It became necessary to remove them, to prevent the Indians who might happen that way from carrying off my property, if they should take a fancy to it. Besides, the place where my goods were left under the open skies demanded that they be removed from there, if they were not to be damaged. The road was very poor for travelling with horses, but I finally succeeded in getting my belongings away. Part of the time afterwards I was busy wiping some of them dry, as the unusually heavy rain which had fallen the night before had penetrated some of them, although we had attempted to protect them well before we left them the previous evening.

Dutch Food. The whole region about the Hudson River above Albany is inhabited by the Dutch: this is true of Saratoga as well as other places. During my stay with them I had an opportunity of observing their way of living, so far as food is concerned, and wherein they differ from other Europeans. Their breakfast here in the country was as follows: they drank tea in the customary way by putting brown sugar into the cup of tea. With the tea they ate bread and butter and radishes; they would take a bite of the bread and butter and would cut off a piece of the radish as they ate. They spread the butter upon the bread and it was each one's duty to do this for himself. They sometimes had small round cheeses (not especially fine tasting) on the table, which they cut into thin slices and spread upon the buttered bread. At noon they had a regular meal and I observed nothing unusual about it. In the evening they made a porridge of corn, poured it as customary into a dish, made a large hole in the center into which they poured fresh milk, but more often buttermilk. They ate it taking half a spoonful of porridge and half of milk. As they ordinarily took more milk than porridge, the milk in the dish was soon consumed. Then more milk was poured in. This was their supper nearly every evening. After that they would eat some meat left over from the noonday meal, or bread and butter with cheese. If any of the porridge remained from the evening, it was boiled with buttermilk in the morning so that it became almost like a gruel. In order to make the buttermilk more tasty, they added either syrup or sugar, after it had been poured into the dish. Then they stirred it so that all of it should be equally sweet. Pudding or pie, the Englishman's perpetual dish, one seldom saw among the Dutch, neither here nor in Albany. But they were indeed fond of meat.

October the 28th

Leaving Saratoga. Early in the morning I resumed my journey from this place. It was indeed possible to go down the river in a boat, but as there were many places where there were cliffs and rocks in it, and as the water was very low, I chose the ordinary means of travel, by wagon along the country road. We crossed the river for the first time some distance above the Saratoga redoubt. The water was so deep here that it went over the front wheels of the wagon and the horses were just on the verge of swimming. The next time was just before we came to the place where Fort Saratoga had formerly been located. Here I was shown the spot where in the last war the French through an artful trick had taken a couple of hundred Englishmen prisoners right in view of the garrison of said fort. The story runs thus: Monsieur St. Luc, a Canadian officer with whom I have the pleasure of being well acquainted,[1] was ordered to make a sally. He stationed his men during the night in the woods not far from the fortress. In the morning, after daylight, he sent forth a few natives who were to shoot or take prisoners any who might leave the fort. The English shot at these, who pretended that they had been wounded and so could not run. Soon three hundred men rushed out of the fort to take them prisoners, ran along the field located on the northern side of the fort, and before they were aware of it, they were cut off from the fort and surrounded by the French. They saw no other way out than to give

[1] This is undoubtedly M. St. Luc de La Corne (Lacorne) (1712-1784), later legislative councillor of Quebec, who was an officer in the Canadian army about this time. He "took part in the defense of the province against Americans in 1775-76; and he commanded the Canadians and Indians in the campaign of 1778, under Burgoyne." The attack on Saratoga had taken place in May and June, 1747; but the attacking force seems to have been the larger, for St. Luc had two hundred men at his command, and only forty English prisoners were captured. Apparently the victory had been considerably exaggerated by the Frenchmen who told the story to Kalm. Cf. article on St. Luc de La Corne in the *Dictionaire Générale du Canada.*

themselves up as prisoners.[1] Shortly thereafter we drove by the fort, close to its gate, the fort being on our right and the river on our left. The fort, however, was now in ruins. Later places were pointed out to me where the French in the last war had killed the English or stolen upon them when they were out chopping wood or doing something else.

Triticum [a genus of cereal grass, wheat] was found at Saratoga and also in other places.

The poison ivy seen to-day had lost its leaves.

I saw a considerable amount of speedwell (*Veronica*) to-day in the woods and in the fields, but I do not remember seeing this plant farther north.

In the beginning of my journey there was an abundance of pine woods, mostly low pines. In other places there were many scrub oaks which thrive in barren spots.

The majority of the trees have now lost their leaves. I make an exception of the oaks whose leaves remain on the trees until spring, but they were either brown, reddish or brick colored.

We saw apple orchards, some of considerable size, on nearly every farm. It was a common custom in the English colonies, when anyone paid a visit to a house, to bring in a large dish of apples and invite the guests to partake of them. In the evening when we sat warming ourselves before the fire, a basket of apples was carried in and all in the house ate of them according to their desire. A large quantity of cider was made on every farm.

We took lodgings for the night about ten English miles from Albany. Ordinarily it is possible to drive from Saratoga to Albany in one day, but because so many of the bridges had fallen into decay on account of the war, we were of necessity greatly delayed on the road. New houses had been built this summer in many places to replace those burned by the enemy in the last war.

October the 29th

Approaching Albany. In the morning we continued our journey. The weather was not especially pleasant for travelling. It snowed fairly hard all day. The snow was the more disagreeable because it

[1] Substantially the same story was told by Kalm under the date of June 24. See p. 358.

was very wet. The greater part of the forest at the beginning of our journey consisted of low pines which grew in the barren low-lands. We now drove through all three branches of the river. There was here neither ferry nor bridge, so one drove through. The current in all three was indeed strong, but it did not come higher than the belly of the horse, except in a single place where it reached the side of his body. The river bottom was stony and uneven so that both the horses and the wagon swayed. The branches were a short distance apart. When the water is high as a result of heavy rains, it is difficult and even dangerous to get across. We saw now fields and now farms on the rest of the journey to Albany, where I arrived at 11 A.M.

Dutch Customs. I think that on the occasion of my former visit to this city I cited a few customs of the people in this locality. Inasmuch as I have not my notes with me, I shall record a few items here. With the tea [at breakfast] was eaten bread and butter or buttered bread toasted over the coals so that the butter penetrated the whole slice of bread. In the afternoon about three o'clock tea was drunk again in the same fashion, except that bread and butter was not served with it. At this time of the year since it was beginning to grow cold, it was customary for the women, all of them, even maidens, servants and little girls, to put live coals into small iron pans which were in turn placed in a small stool resembling somewhat a footstool, but with a bottom . . . upon which the pan was set. The top of the pan was full of holes through which the heat came. They placed this stool with the warming pan under their skirts so that the heat therefrom might go up to the *regiones superiores* and to all parts of the body which the skirts covered. As soon as the coals grew black they were thrown away and replaced by live coals and treated as above. It was almost painful to see all this changing and trouble in order that no part should freeze or fare badly. The women had however spoiled themselves, for they could not do without this heat.

October the 30th

The snow now entirely covered the ground, but the sun to-day melted a considerable part of it on the southern sides of the hills. On the northern sides it remained untouched by the sun.

Wood. Hickory was considered the best wood of those trees which grew here, but it was said to have one disadvantage, namely, that when burning it caused great injury to the eyes. So far as excellence was concerned the sugar maple was said to be almost as good as the one just mentioned.

Dogwood. The wife of Colonel Lydius told me that when she had arrived in Canada she had suffered from pain in her legs as a result of the cold. It became so severe that for a period of three months she could not use one leg and had to go about with a crutch. She tried various remedies without avail. Finally, a native woman came to the house who cured her in the following manner. She went out into the forest, took twigs and cuttings of the dogwood, removed the bark, boiled them in water and rubbed the legs with this water. The pain disappeared within two or three days and she regained her former health.

Iris. Colonel Lydius related how the Indians make use of the iris root as a remedy for sores on the legs. This cure is prepared as follows. They take the root, wash it clean, boil it a little, then crush it between a couple of stones. They spread this crushed root as a poultice over the sores and at the same time rub the leg with the water in which the root is boiled. Mr. Lydius said that he had seen great cures brought about by the use of this remedy. It is the blue iris, which is extremely common here in Canada, that is used for this purpose.

Sassafras. He also told me that the natives consider the sassafras very valuable in the treatment of diseased eyes. They take the young slips, cut them into halves, scrape out the pith or the medulla, put it into water, and after it has been there for some time, wash the eyes with the same water. When the natives from Canada formerly came to his house at the . . . ying[1] place they were very anxious to hunt for sassafras. They cut the stems in two, took out the pith and preserved it and took it home with them to use as described above.

OCTOBER THE 31ST

Pumpkins. A certain kind of oblong and large gourds were called "pumpkins" in Dutch. They were much used by the Dutch, the

[1] The first part of the word is omitted in the original.

English, the Swedes and others here in America. The French in Canada had also some use for them, but not as much as those mentioned before. They ripen in the middle of September here in Albany and are able to stand a fair amount of cold. The natives both in Canada and in the English provinces plant a considerable amount of them, yet I think that they came here first from Europe. They were delicious eating and they were prepared in various ways. Sometimes they were cut in two or more parts, placed before the fire and roasted. Frequently they were boiled in water and then eaten. It was customary to eat them this way with meat. Here at Albany the Dutch made a kind of porridge out of them, prepared in the following way. They boiled them first in water, next mashed them in about the same way as we do turnips, then boiled them [again] in a little of the water they had first been boiled in, with fresh milk added, and stirred them while they were boiling. What a delicious dish it became! Another way of preparing these which I observed, was to make a thick pancake of them. It was made by taking the mashed pumpkin and mixing it with corn-meal after which it was either boiled or fried. Both the gruel and the pancake were pleasing to my taste, yet I preferred the former. The Indians do not raise as many pumpkins as they do squashes. Some mix flour with the pumpkins when making the porridge mentioned above, others add nothing. They often make pudding and even pie or a kind of tart out of them. Pumpkins can be preserved throughout the winter until spring, when kept in a cellar where the cold cannot reach them. They are also cut into halves, the seeds removed, the two halves replaced and the whole put into an oven to roast. When they are roasted, butter is spread over the inside while it is still hot so that the butter is drawn into the pumpkins after which they are especially good eating.[1]

Discord had taken a firm hold among the inhabitants of Albany. Although they were very closely related through marriage and kinship, they had divided into two parties. Some members of these bore such strong aversion to one another that they could scarcely tolerate the presence of another member, nor could they even hear his name mentioned. If a visit was made at the home of one of them, you were then hated by his opponent, even though you had

[1] Cf. diary for September 19th, 1749.

visited him previously. It was interpreted that you were not satisfied with his friendship alone, but also wished to enjoy that of the other or that you liked him better.

November the 1st

Petroleum. Both Colonel Lydius and Mr. Rosbom and many others told me to-day about a kind of earth-oil which is to be had from the Indians allied with the English, in a place about three days' journey from the country of the Senecas. There is said to be there a small lake, about a musket's shot in length and breadth, upon the surface of which continually floats an oil. The natives go there in the spring when they find the lake almost covered with leaves which have fallen into it during the previous autumn and which are now saturated with oil. They set fire to them and allow them to burn up. The fire then continues to burn over the lake until all the leaves and all the oil on the water is consumed and only the water is left. They then take twelve-foot poles which they force down into the lake's bottom and an oil comes welling up from the bottom and the holes they have made therein. The bottom is not hard, but very soft. They skim with a spoon or scoop the oil which floats upon the water and preserve it in some vessel. Colonel Lydius accompanied me to the home of a man or gentleman who had some of this oil in a bottle. It was real petroleum and not at all different from what is sold at the apothecary's, except heavier; but I learned that this condition resulted from allowing it to stand several months in an open bottle, that is, without a stopper in the bottle. Its color is dark brown with a tendency toward red. The smell is absolutely that of petroleum or Jew's pitch (asphalt). The gentleman who had it in a bottle related that nothing was better for wounds than this. If a little were applied to the sore, it would heal within a short time, which he had experienced on his many journeys. He had a horse last year that injured itself badly by kicking one leg with the horseshoe on another foot. The wound was smeared with the oil and within a few days it was healed. Another horse had some time thereafter suffered a bad injury and was likewise cured within a short time. It is said to be this oil which Père Charlevoix tells about in his *Histoire de la Nouvelle France*.[1]

[1] Tome I, page 422.

The hawkbit (*Leontodon*) was found on the hills outside of the city in full bloom, although the ground had but recently been covered with snow.

Salt Springs. In Philadelphia Mr. Bartram related that he had seen several salt springs in the country occupied by the native allies of the English and that the Indians prepared salt from them by the boiling method. He also showed me a piece of it. In Albany several people told me the same story, and that when travelling from here to Oswego the place is on the left, and about a day's journey from the road. The brine is boiled by the natives in copper utensils. Salt springs are found there in more than one place. In one location is a salt spring from which they obtain salt, and on each side of it, hardly a musket's shot away, is a spring of fresh water in which it is impossible to detect a trace of salt. There is in Albany no one who is able to let me have a little of this salt. It is about these salt springs which P. Charlevoix tells in his *Histoire de la Nouvelle France.*[1]

Cabbage Salad. My landlady, Mrs. Visher, prepared to-day an unusual salad which I never remember having seen or eaten. She took the inner leaves of a head of cabbage, namely, the leaves which usually remain when the outermost leaves have been removed, and cut them in long, thin strips, about 1/12 to 1/6 of an inch wide, seldom more. When she had cut up as much as she thought necessary, she put them upon a platter, poured oil and vinegar upon them, added salt and some pepper while mixing the shredded cabbage, so that the oil etc. might be evenly distributed, as is the custom when making salads. Then it was ready. In place of oil, melted butter is frequently used. This is kept in a warm pot or crock and poured over the salad after it has been served. This dish has a very pleasing flavor and tastes better than one can imagine. She told me that many strangers who had eaten at her house had liked this so much that they not only had informed themselves of how to prepare it, but said that they were going to have it prepared for them when they reached their homes.

The gourds or melons raised by the Indians were generally called *squashes* by the English, *cascuta* by the Dutch and *citrouilles* by the French.

---

[1] Tome I, p. 421.

Melons d'eau so called by the French in Canada, were called "watermelons" by the English and "*watlemone*" (*watermeloen*) by the Dutch. The majority of those raised here were said to have red meat. As they had gone by, it was impossible now to see one, as they do not keep them but eat them only when they are fresh. It was said here that those obtained in New York are better and tastier than those in any other place in North America.

Mulberry trees are still found scattered through the forests about this town. I requested several people in this locality last summer to gather some of their berries when they became ripe so that I might have some seed. The majority had forgotten about it and I would have lost out entirely, if young Mr. Rosbom, a barber-surgeon, had not preserved a few for me. He said that he had gotten them with great difficulty and that they were the only ones he had seen during the summer, as they had been scarce. Formerly there had been more mulberry trees, but they had been wasted by cutting them down for the purpose of making dowels for boats and yachts. This wood withstands rot for an unusually long time. On a farm twenty English miles from this town on the way to Saratoga lived a farmer who told me that here and there in the woods in that region one might find a single mulberry tree. In Saratoga, which was twenty English miles further away, or forty miles from Albany, nearly due north, all the farmers said that they had never come across nor seen a mulberry in the woods. Colonel Lydius, however, assured me that he had seen such a tree in Saratoga and that Saratoga was the place farthest north where he had found these trees growing. He has lived for several years at the old Fort Nicholson, only twelve English miles north of Saratoga, and on a hundred journeys through the woods there he had never seen one of these trees. The fruit of the mulberry which grows here, he said, is much smaller than that of the trees planted in Holland, but as far as taste is concerned it is as pleasing and agreeable as the Dutch variety.

Oswego which is the trading center of the English on Lake Ontario is situated at 44° 47′ according to the observations made by a French engineer.

The Dutch began, they said, to settle New York in 1623.[1]

A kind of seed was called by the French *piment* (if I remember

[1] New Amsterdam was settled by the Dutch in 1614, and New Jersey colonized by them in 1617-1620.

rightly, by botanists called *Capsicum* or *Cor indum*) [*Cardiospermum halicacabum*]. This is used very commonly here to improve the flavor of food. It is used very much as pepper is. Nearly everyone here sowed the seeds, which were said to lie in the ground six weeks before they cracked open or germinated.

November the 2nd

Walnuts. All the varieties of walnuts found in Canada flourish here with one additional kind. The oblong are called oil-nuts (butter nuts). *Noix amères* are called "bitter nuts"; *noix dures* are called "hickory nuts" (*kiskatom*, an American Indian name). In addition to these three there is a fourth kind which is smaller than the last mentioned, the shell being less hard yet the kernel sweet. When you paid a visit to any home a bowl of cracked nuts was also set before you, which you ate after drinking tea and even at times while partaking of the tea. A plate filled with large sweet apples of no particular name was also often added.

Ginseng (*Aralia Canad.*) was called . . . *wurtzel* by the Dutch. The old surgeon, Mr. Rosbom, said that the root not only is invaluable for wounds but taken internally it also affects the urine and is fine for stones in the bladder.

The *moccasin flower* (*Cypripedium*), which is found quite generally in the woods here, is said to be rather good for women in the throes of childbirth. I refer to the decoction made of the root.

Houses. Last summer on my journey up to Canada, on the day I arrived at Saratoga, I described how the houses belonging to the Dutch living in the country were built. They first put up the framework upon which the rafters and both roofs rested and then filled in the framework with unfired bricks. The inner side was brushed over with lime and whitewashed so that from the inside it looked like a stone house except where the perpendicular timbers which supported the rafters were visible. On the outside the houses were generally covered with clapboards so that the unfired brick might not be damaged by moisture, weather and wind. As a rule they did not have more perpendicular supports in the walls than they had cross beams, from three to five on each side or long wall. On the walls of the gables they had two or three upright beams, doubtless for the sake of firmness and strength when they put the bricks between them. The ceiling was horizontal and beamed. The roof was either of boards or shingles; there were several rooms under one roof. There was nearly always a cellar under the house and usually this was large enough to extend under the whole house. The walls of it were of masonry, yet the ceiling was not vaulted but made of plain boards or of planks with the floor of the house upon them. The fireplace in the houses in the country was built in an unusual way and it was nearly always placed in the wall on the gable end opposite the door. The fireplace for about six feet or more from the ground consisted of nothing more than the wall of the house which was six to seven feet wide and made of brick only. There were no projections on the sides of the fireplace, so it was possible to sit on all three sides of the fire and enjoy the warmth equally. Instead of forming this figure ⌒ or that one ⌴, namely of three sides, as our fireplaces do, the fireplaces here formed this one —, or a straight line without projections, corners or embellishments. Above, where the chimney began the bricks rested upon rafters and cross beams on three sides which had been arranged so as to support them. As the chimney was some distance above the floor they had put boards about these rafters, or as was more common, they had hung short curtains extending downward [and outward] to prevent the smoke from coming in. But in spite of this and because the fireplace had no sides, it frequently happened when the door was opened that the smoke was driven into the room. The hearth itself was always even with the floor. The fireplace was ordinarily six to eight feet in width. Occasionally a shelf had been made above it upon which teacups, etc. were placed. In Albany the fireplaces had small sides projecting out about six inches made of Dutch tiles with a white background and blue figures. As they knew nothing about dampers the fire burned during the winter all day, both in the houses in the towns as well as in the country. No tile stoves were used here. The houses in Albany were built in the same style as those in the country except that the gables which were always toward the street were constructed of fired brick so that, as previously noted, a stranger walking along the streets and not paying close attention might easily conclude that all houses here were built of fired brick. The floors of the houses were kept quite clean by the women who sometimes scrubbed them several times a week, and Saturday was the day especially set aside for that task. In many houses in the town they had partitioned off the part of the room where the beds stood by placing large doors before them, [like cupboards], and thus completely concealing the beds from view. Every house in the town as well as in the country had a bedroom or an attic where they kept miscellaneous household goods. This was generally reached by a ladder or staircase within the house. The ovens for baking which were made of brick were placed on a hill [apart from the house].

Decrease in the Amount of Water. Among the signs that the water in early times had been higher than at present, the following may be noted: Colonel Lydius told how just a little north of this town a whole tree had been found buried in the ground at a depth of twenty-five feet and not far distant from the river. A few English miles below this town there has been found on digging into the ground some very large teeth, almost like those of elephants, but shaped like human teeth. There has also been discovered here a fairly long tubular bone, so that it has been concluded that this skeleton was not that of an animal but of a human being. I inquired whether one had found any oyster shells or other shells in the ground. He answered that he did not know that any had been found in this locality, but farther down the river large quantities of them had been discovered in some places. It is not known whether there is a decrease of water in this river, but it is a fact that small banks are shifted from place to place, which generally occurs during the spring floods.

Rattlesnakes. Colonel Lydius has seen these in the clefts of the rocks on Lake Champlain. When the snake attacks a human being the blood in the beginning spurts high into the air from the wound. If it should strike certain parts of the body, as an artery, etc., the wound was believed to be fatal. It does not usually attack anyone if it is not trampled upon. On the other hand, it is possible to approach quite close to it without the snake minding it, though it looks sharply at a human being and, as it were, marvels at him. It cannot extend itself further than its own length; it floats like a bladder upon the water; it usually rattles before it strikes, yet not always. It is rare that one hears of anyone being attacked by it, despite the fact that the people travel so much about the woods; so it must be true what is said about it being stepped upon, etc. It is known that it has crawled over the stomach of a sleeping person without doing harm to him. The rattlesnakes first come forth from their dens into the sunshine of the early spring, sun themselves the whole day long and then in the evening they crawl back again. This they do from two to three weeks early in the spring until it becomes warmer. Colonel Lydius once found a great number of them in one place and he shot sixteen of them. His companion beat several to death, and the rest of them lay under a large stone and rattled vigorously. When he pointed toward them with his cane, they threw themselves toward him and then hastily drew back, just like a flash of lightning, and they repeated this at frequent intervals. The rattlesnake does not withdraw when it hears a human being approach, but lies still.

November the 3rd

I shall soon become tired of having to remain here and wait so long before I can get away. The morning after I arrived in this town two yachts departed, but it was impossible for me to get ready to leave on them, as I must procure various kinds of seeds of the walnut, chestnut, squash and other useful plants, a thing which I could not do in a hurry. People assured me that another boat was to leave as of Saturday last, but it is still tied up here, even though there is plenty of wind to-day.

The inhabitants of this town [*Albany*] are as a whole all Dutch or of Dutch extraction, descended from those who first came to settle this part of the country. Both sexes dress now very nearly like the English. In their homes and between themselves they always speak Dutch, so that rarely is an English word heard. They are so to speak permeated with a hatred toward the English, whom they ridicule and slander at every opportunity. This hatred is said to date back to the time when the English took this country away from the Dutch. Nearly all the books found in the homes are Dutch and it is seldom that an English book is seen. They are also more thrifty in their homes than the English. They are more frugal when preparing food, and seldom is more of it seen on the table than is consumed, and sometimes hardly that. They are careful not to load up the table with food as the English are accustomed to do. They are not so given to drink as the latter, and the punch bowl does not make a daily round in their households. When the men go out of doors, they frequently have only a white cap under the hat and no wig. Here are seen many men who make use of their own hair, cut short without a braid or knot, as both of the latter are considered a mark and characteristic of a Frenchman. The vast majority, in fact almost everyone here, carries on a business, though a great many have in addition their houses and farms in the country, close to or at some distance from the town. They have there good country estates, several sawmills and in many places even flour mills. The servants in this town are nearly all negroes. The children are instructed in both the English and Dutch languages. The English accuse the inhabitants here of being big cheats and worse than the Jews.

November the 4th

Corn. From Mrs. Vischer with whom I had lodgings, I learned one way of preparing corn so that it would be especially suitable for cooking and eating. They take the corn before it is ripe, boil it in a little water, allow it to dry in the sun and preserve it for future use. Corn prepared this way is then boiled with meat, etc. when it, as well as the soup in which it is boiled, is good to eat. The younger the corn is when picked, the better it is, provided, however, that it is not too young. When the corn is prepared this way it is not necessary to remove the hull as this is not yet hard. When it is boiled, it is done in the following manner: The whole cob is placed in the saucepan and when it is ready, the kernels are removed from it.

The departure from Albany took place at 2 P. M. on Capt. Wilj. Winov's boat. We had a favorable wind, but the captain had to stop for more freight at a couple of places, so we did not make as good headway as usual. Yet before sunset the freight was all on board.

The *Province of New York*, it was said, was not nearly as well populated and cultivated as the other English provinces here in America. The reason was said to be that for the most part the inhabitants were Dutch who in the past had acquired large stretches of land. Most of them are now very wealthy and their feeling of jealousy toward the English prevents their selling a piece of land unless they are able to get for it much more than it is worth. As the English as well as persons of other nations can buy land at a far more reasonable price in the other provinces here, they gladly allow the inhabitants of New York to keep their land. Moreover, in Pennsylvania there are greater advantages to be gained than here in New York. Because of this it so happens that people come from Germany every year to settle in this country. They come sometimes from London by boat to New York, but they hardly come ashore before they leave here for Pennsylvania. It is a fact that Pennsylvania alone has now almost more inhabitants than Virginia, Maryland and New York put together, due to the generous privileges which the sagacious Penn wrote into his wise laws and constitution for Pennsylvania. The inhabitants of New York console themselves with the thought that when once all the land in Pennsylvania has been filled up with inhabitants, the remainder will of necessity come hither, and then they can set whatever price they wish upon their land. But it is the general opinion that they will have to wait a long time.

The *red stone* which is found about New Brunswick, and of which I spoke last spring when I was there, is said to be useless for building purposes. Even though it seems firm and strong when in the ground, it happens that when it is taken up and exposed to the air for a time, it disintegrates and crumbles. One of the townsmen here had a house made of this stone, but it began so to crumble on the side which was exposed to the air that he was forced to have boards put on the outside to prevent its falling to pieces. There is, however, one use for this stone: when crushed and spread upon the fields it is a good fertilizer, and weeds do not thrive where this has been put upon the farm land and the kitchen gardens. Mr. Wiljams, a townsman, who had long lived in New Brunswick told me all this.

November the 5th

To-day we made very little headway on our journey, as there was hardly any wind, and in addition the captain was delayed in one place.

The *best cider* in America is said to be made in New Jersey and about New York, hence this cider is preferred to any other. I have scarcely ever tasted any better cider than that from New Jersey.

Squashes. The best squashes are said to be found in the region around Albany. Those from New York are said to be not so good and those from New Jersey, and Pennsylvania even less so. For this reason those living in New York procure their seed from Albany. The squashes are good then for three years, at the end of which period they must send for new seed. The squashes can be kept in good condition up to the month of March, if stored in cellars where the cold does not reach them. But the squashes keep longer than pumpkins. The latter are called *pumponen* by the Dutch. I was told about this at a large gathering of old men and women last evening.

Watermelons were also kept in Albany a part of the winter. They could even be eaten by the sick without harm, so far as the people of Albany knew.

Roofing slate, very beautiful and not affected by air, rain, etc., is said to be found in Highland which is a part of the Province of New York, situated about halfway between New York and Albany. The place where it is found is about five English miles from the Hudson River. They are thinking of roofing the houses in New York with it.

There is said to be a *sulphur spring* in the Mohawk country about seventy miles from Albany. The water comes down a mountain and when it reaches the gravel below it leaves so much sulphur that the deposit of it can be swept up from the ground. A man drank of the water with the result that his body swelled up therefrom. Mr. Wiljams from Brunswick has seen the place.

Iron ore is said to be found everywhere and in great quantities in this country, which everyone with whom I have conversed on this subject has confirmed as with one voice.

Ash. The baskets here in which food and the like was sold, were generally made of ash; the withes were thin and about a digit broad. The natives make quantities of these baskets and sell them to the Europeans.

Catskill Mountains. At two P. M. we reached the Catskill mountains which are also called the Blue Mountains. They form the long range in this part of America. They were north of us and their tops were all covered with snow, although on the ground not a bit of snow was to be seen. It had no doubt been washed away by the heavy and prolonged rain. These mountains tower above all the others, are visible as far as Albany, and as clear as if they were quite close, when in fact they are situated forty miles from the place just mentioned. There is found on these mountains many herbs and trees which are not to be found on the plains round about them and not any nearer than Quebec and the northern part of Canada. These mountains are said to be infested with rattlesnakes.

Limestone is found in abundance on both sides of the Hudson River. Formerly they burned the lime here and shipped it down to New York to be sold. Now, however, this is not done as much, for a considerable amount of lime is made in New York from oyster shells.

White pine is commonly used for roofs of houses in Albany, and such a roof is said to last forty years. This white pine is not found in New Jersey, so they say.

November the 6th

Our journey proceeded rather slowly as we had contrary winds and we did not advance faster than the tide carried us.

Charles XII. The passengers on the boat had the *History of Charles XII* [*King of Sweden*] with them to help while away the time.[1] It was the life written by Voltaire which had been translated into English. In London I saw it in nearly every bookseller's shop and bookstall. I had noticed during my travels that it was widely circulated and was one of the most read and best-known of books. They also had descriptions of his life by other authors. All marveled much at this great king and all had acquired a singular regard for him. But there was one thing of which nearly every one disapproved and that was his behavior at Bender when the Tartar Khan attacked him and he began to defend himself.

Commerce. We saw one boat after the other tied up at the shores of the river as we proceeded, all being loaded with wood, flour or

---

[1] Voltaire's *Histoire de Charles XII* appeared in 1731, and became immensely popular. It was immediately translated into English, and at least three editions were published in London during the year 1732. By 1740 seven English editions had been printed.

something else. The country was inhabited in most places, especially nearest the river. The captain told me that the soil on both sides of the river was said to be inferior, especially on the southern side, and on that account the inhabitants there were poor.

The *skulls of the American natives* are said to be much thicker and harder than those of the Europeans. This is a fact, according to what I learned from several Frenchmen with whom I talked in Canada and [from what I heard] also from Englishmen in this country.

The products which came from New Brunswick were said to be grain, flour in quite large quantities, bread, linseed in considerable amounts, and various utensils. All of these were sent on small sailing boats to New York, which is the only trading center, and to which wares are shipped. New York is situated forty English miles from New Brunswick.

The *Seneca Country* and the surrounding region is inhabited by Indians only. Formerly because of wars they lived in large villages surrounded by stockades, but now the natives have scattered, one living here, another there. In some places they do not allow unrestricted sale of rum, but only certain ones may buy it and sell it again in small quantities to prevent the harm occasioned by drunkenness. This was told by someone who had spent three years among them for purposes of trade.

There is said to be a street in New Brunswick which is inhabited by the Dutch only who have come from Albany and is therefore called Albany Street. These Dutchmen call upon one another, but seldom visit any other residents of the city; they keep themselves apart.

Prognosticon Tempestatum (Signs of the weather). In Albany it was considered a certain sign, especially in the fall when there had been two or three days of southeast or southwest wind with rainy weather, that a strong northwest wind would follow with a clear and cold atmosphere.

The boats from Albany were now making their last trip to New York for the season, since it was the general opinion that they would not be able to make another trip before the river was frozen over. Most of the boats are tied up for the winter at Albany, but a few remain at New York, since they continue their trips as long as possible, and then dare not go up the river, since it is frozen over earlier at Albany than at New York. The reasons for this are not only that Albany is situated farther north, but that the tide is not as strong there and the water is not salt but fresh.

November the 7th

The journey was to-day also a very slow one. We had to depend upon the tide, and when we came opposite the home of young Mr. Colden we had to come to anchor, as the tide was low, and besides the wind coming down from between the high mountains just below us was so strong against us that it was impossible for us to proceed. We were forced to remain here until half past twelve, when the tide turned and we again could continue our journey. The wind also changed and became favorable, but it was too gentle to be of any assistance to us. We floated along past the high mountains where we anchored late in the evening, as the skies did not look especially promising.

The chestnut oak grew near the home of Mr. Colden.

Wheat (*Triticum*) with waving spikes was also visible there.

That part of the country is called "Highlands" which is situated just below the high mountains, and the latter are right below Mr. Colden's house. The Highlands are located on both sides of the river.

Wild geese and ducks were flying about here and there in the neighborhood of the river, and large numbers were also swimming about in the water. Yet we were not permitted to shoot any of them.

November the 8th—In New York

The current prices on products in New York for this autumn were those given in the New York *Gazette* of November 13, 1749, as follows:

Wheat, per bush.	6 s.	Molasses,	1 s. 9 d. per Gal.
Flour, per C.	18 s.	Westindia Rum	3 s. 9 d.
Milk bread,	39 s.	New England d:o	2 s. 6 d.
White d:o	29 s.	Beef, per Bus.	36 s.
Middling,	24 s.	Pork	21.18 s.

Brown	18 s.	Flax-Seed	10 s.
Single refined Sugar	16 d.	Bohea Tea,	6 s. 6 d. per Box
Muscovado Sugar, per C.	50 s.	Indigo	7 s.
Salt, per Bush.	2 s. 6 d.	Chocolate	2 s. by the doz.

The *copper mine* which is located nine to twelve miles from New York on the side toward Philadelphia is as yet the only one known in this country. At least none of that ore is mined elsewhere. Nearly all the owners of it live in New York. The mine was worked previous to the last war. The ore from it is said to be of the best obtainable. Following an act of the Parliament in England the inhabitants here are no longer allowed to smelt and refine the silver and copper ores which they find, but are obliged to send them to England in their original state and have them smelted there. This ore has been so rich that they have sent it to England and sold it at a profit in spite of the high freight charges incurred. During the last war the work in the mine was abandoned, as it was not safe to send the ore over the seas to England for the privateers to seize. It was at this time that the mine became full of the water which is still in it. Now the owners intend to send to London for engines to pump the water out of it.

November the 9th

The Dutch Church, the Service, etc.—*Hunc diem perdidi.* (This day I have spent uselessly) I can in a certain way say about this day because I spent the morning in one of the Dutch churches and the afternoon in the other. I listened to two sermons, one of which lasted two hours, the other two and a half, and in neither instance could I understand much, because I was so far away from the pulpit; and even if I had understood all of it, I could not have remembered so long a sermon. But nevertheless, for the mere pleasure of it, I shall here record the ceremonies which are used here, a description of the church, etc. The morning service began exactly at ten A. M. and was to-day especially noteworthy, inasmuch as it was fifty years to-day since the minister preached his first sermon in that church. The fact was also stressed that the sum total of those who had heard his first (installation) sermon and were now present in the church, could not be very great. The church service was conducted as follows: first, the bell was sounded two or three times (they had here [in America] only one bell in each church tower) and this was rung by means of a rope which reached way down to the floor of the church, where it was customary to stand when ringing, or more correctly speaking, when tolling the bell. Then the cantor began to sing one of David's psalms rendered into verse. Only a few stanzas thereof were sung, and it seemed to me that nearly all of their hymns had similar tunes. While this hymn was being sung the minister mounted the pulpit, hung his hat on a peg, took off his robe and also hung that there, since the old fellow was eighty years of age and could not endure wearing his robe for two hours. He sat down on a chair and when the singing ceased, he stood up and preached for a while; then he read a lot of prayers, after which he began preaching again. At that time two men came forward, took the [longhandled] contribution bags and placed themselves before the pulpit, holding the handles upright, with the bags at the top. After the minister had mentioned a few facts about church work, they went about taking the offering. It was the custom here that everyone contributed, but no one was asked to contribute more than once during the same service. When the minister had finished the sermon, the clerk [or cantor] passed a note to him. He had put the note in a split at the end of a cane and thus passed it up to the minister from the place where he stood below the pulpit, saving himself the trouble of going up into it. It was the license to have the banns of marriage announced of those who were to be married. Finally he finished his sermon with the Lord's Prayer, announced the hymn to be sung, which was another [rhymed version] from the Psalms of King David, and when he had finished, the cantor began to sing it. They had also in the church several boards on which were indicated what hymns were to be sung before the sermon: e. g. to-day the following words were on the boards: "Psalm 18 pause"; but no post-sermon hymn had been indicated. The minister remained seated in the pulpit during the singing, and when this was finished he pronounced the benediction, after which everyone departed. No mass was said and no altar devotions with gospel, epistle, etc. were held. Even if they had had this chanting service at the altar, it could not have been performed since there were no altars in the Dutch churches here, but just a pulpit.

The church where I to-day attended morning worship was the so called New Dutch Church. I shall now describe the building. It is a large structure built of stone like the ordinary church, with a tower and bell. The church is located nearly in the direction of S. S. W. and N. N. E. with the tower at the latter end. Here I wish to note that the churches in this town are not constructed to face any particular direction, one standing east and west [for instance] in the customary way, as was almost true of the English church. The others were set north and south, the direction generally followed by the Lutheran and Reformed or Presbyterian churches, etc. Quite a large churchyard surrounds the temple, and about it are planted trees which give it the appearance of an enclosure. The church has several doors and large, high windows on all sides. Within there is not a sign of a painting or figure, only white walls and ceiling and an unpainted pulpit. The church is not vaulted, but has a ceiling built of boards in the form of a vault. There is no balcony in it. There are pews everywhere in the church, also several aisles; but both are rather narrow. The pews are made like ours except in the manner hereafter described. The backs of our pews extend perpendicularly from the top to the floor, while here the back extends from the top to the seat only. Then there is another perpendicular partition which is not in line with the back of the pew but placed a little ways under the seat of the pew so that a person can place his feet under the pew in front of him. The number of the pew is painted on its door. There are no chandeliers and no candlesticks in the church. There is no sacristy and no other room to take its place.

The ministers are dressed in black with a gown and collar like those of our ministers except that the gown is not quite as long. During the week when there is no service in the church, the shutters are closed. In the tower is a clock, which strikes the hour, the only one of its kind in the city, as far as I know.

The church was filled to-day with people who came out of curiosity to hear a man begin his fifty-first year as minister. For this reason more had flocked hither than usual. The men were dressed like the English; the majority wore wigs, but a few had their own hair, which was not very long, was not powdered, and had no more curls than nature had bestowed. A few elderly persons had worsted caps on and three or four wore hats. Perhaps they were Quakers whom curiosity had driven hither. The women as a rule had black velvet caps which they could fasten on by tying at the ears. Others wore the ordinary English gowns and short coats of broadcloth of various colors. Nearly everyone had her little container, with the glowing coals of which I have spoken before, under her skirt in order to keep warm. The negroes or their other servants accompanied them to church mornings carrying the warming pans. When the minister had finished his sermon and the last hymn had been sung, the same negroes, etc. came and removed the warming pans and carried them home.

In the afternoon I went to the so called Old Dutch Church. The service began there at two P. M. and was performed in exactly the same manner as the morning worship which I attended. There was singing before the sermon and the latter was preached in the same way. It ended with the Lord's Prayer, followed by a final hymn and the benediction. I observed one ceremony here which I did not see during the morning service, namely, the christening of a child. The service was as follows: as soon as the sermon was finished and the minister had read a few prayers, a woman carrying a child on her arm came forward to the pulpit. Then the minister began to read from there all the prayers [for the occasion] contained in the Dutch prayer book, and when he had finished she took the child to the rector, who sat in his pew, and let him christen it. This was performed, as we do in baptism, by putting water upon the child's head. After this the other minister in the pulpit talked briefly on the significance of the christening and then ended the service with the Lord's Prayer.

This church was also of stone with two pillars in the center which supported the wooden roof built in the form of a vault. Some of the windows had colored glass, but they were not old, since "City of New York" was inscribed upon them. There was a small organ, the gift of former Governor Burnet. The pulpit and vault were painted, but without any figures on them. There was not a single figure or painting in the church, only several coats-of-arms had been hung there. There were large balconies, and it was the custom here that the men sat there while the women occupied the ground floor, except the pews against the walls which were reserved for the men.

There were many chandeliers here. The examination in the Catechism was held late at night by candle-light.

Peter Kock. I learned very unexpectedly this afternoon of the death of Mr. Peter Kock, the merchant. I was telling those with whom I was staying that I visited him here last year just about this time, when they informed me that he was dead. He had first served with the admiralty and then for several years sailed to foreign ports. He married in Cusassov (?) a wealthy Dutch woman and in this marriage had several children, two of whom, daughters, are still living. He became a widower, then came to New York to build a ship in which to return to Sweden. He married here a woman of the Dutch family Van Horne and then took up his residence in Philadelphia. There he became one of the wealthiest of merchants. He was a pillar of the Swedish congregation there; he warded off the attempts of the United Brethren to get a foothold. It was to him that I took my bill of exchange. He received me with extraordinary kindness and performed unusual favors for me, so that it is not at all surprising that his death should affect me so deeply. He was quite well acquainted with the theological writings.

November the 10th

Letters. To-day I received letters from several of my friends; namely, two from Mr. [Abraham] Spalding in which he advises me not to leave here in the fall, inasmuch as I might experience difficulty on my arrival in London so late in the season to get an opportunity to continue [my return journey] to Sweden. Therefore he considered it wisest to postpone my return until next spring. In addition I received a letter from Mr. Collinson in London, also one from Jungström who described his whole journey from Albany, etc. From the gazettes which were printed here in town I discovered also that Mr. Franklin, the postmaster at Philadelphia, my very special friend, had had printed in the Philadelphia gazette an extract from the letter which I had written to him from Quebec, Canada. This extract had been reprinted in the gazette which was issued in this town. All of the French were especially pleased at this, as they had [in this article] been given considerable honor for their learning.

French Refinement vs. the Dutch. I have already told in my journal of the good breeding of the French in Canada. Now I must emphasize one item before I forget it: namely, that the inhabitant of Canada, even the ordinary man, surpasses in politeness by far those people who live in these English provinces, especially the Dutch. I just recently came from Canada and left behind me in the vicinity of Saratoga the French who had brought me to the English colonies. When I reached Saratoga and came in contact with the first English inhabitants who were of Dutch descent, I noticed a vast difference in the courtesy shown me in comparison with that shown me by the French; it was just as if I had come from the court to a crude peasant. Yet I must grant that although they [the Dutch] showed a lack of breeding in their speech, their intentions were of the best. I noticed that when they believed, or were persuaded to believe, that a person did not understand Dutch, they amused themselves by censuring the manners which differed [from their own]. The women, especially in the towns, had this habit, for they did not like the French mode of living at all. But never have I seen folks more ashamed than they when a person let them know that he understood every word they said. Some of them did not dare show themselves again.

The King of England's birthday was celebrated in town to-day, but the people did not make a great fuss over it. A cannon was fired at noon and the warships were decorated with many flags. In the evening there were candles in some windows and a ball at the governor's. Some drank until they became intoxicated, and that was all.

November the 11th

Dutch was generally the language which was spoken in Albany, as before mentioned. In this region and also in the places between Albany and New York the predominating language was Dutch. In New York were also many homes in which Dutch was commonly spoken, especially by the elderly people. The majority, however, who were of Dutch descent, were succumbing to the English language. The younger generation scarcely ever spoke anything but English, and there were many who became offended if they were taken for Dutch, because they preferred to pass for English.

Therefore it also happened that the majority of the young people attended the English church, although their parents remained loyal to the Dutch. For this same reason many deserted the Reformed and Presbyterian churches in favor of the English.

Lodgings, food, wood etc. were in this town much more expensive than in Philadelphia. The rooms were said to have grown more costly since so many people from Albany had lived here during the last war and thus brought about a shortage of living quarters. The prices paid then were still in force. Food was high-priced because so much flour, corn and other food stuffs had been exported in great quantities to the West Indies, and even to New England. The farming region round about here is not so well populated that it can supply the town in such quantity as is the case in Philadelphia, where the whole countryside is thickly inhabited by the Germans and others who have settled there. On the other hand, it is the general opinion that broadcloth and other merchandise can be had here at a more reasonable price than in Philadelphia.

November the 13th

My correspondence kept me occupied all day. Now that I had resolved to stay one more year in America (under H. E.'s[1] protection), since it would [otherwise] be impossible for me to accomplish all that the Academy of Sciences demanded of me and that I ought to accomplish, I wrote to several friends in Sweden about the matter. My letters were to be sent by a boat which was to leave New York for London within a fortnight. I wrote to the following:

1. His Excellency Count Tessin. I told him that I had written from Canada; also that I had resolved to stay another year, and about the receiving of my allowance.
2. The Royal Academy of Sciences. I wrote about my Canadian journey and what I had procured there. I related how I had been refused permission to go through Fort Frontenac, etc. I also told why it was necessary for me to remain another year; I told of . . . [2] from Mr. Spalding. I informed them

[1] H. E. = His Excellency (Count Tessin).

[2] The rest is omitted in the original.

that I would soon begin to send them seeds, and that I had received permission from the King of France.

3. The Vice-president, Baron Bielke, and the archiater, Dr. Linnæus, a letter together. I wrote them about the same things as I did the Academy, except at greater length. This was to be read by the Academy.
4. Mr. John Clason.[1] I told him that I had asked Mr. Spalding for a new draft.
5. The Secretary of the Academy of Sciences. I asked him to seal my letters and mail them.
6. Cousin Wilh. Ross. I requested him to countermand my house rent.
7. Mr. Abr. Spalding. I wrote him of my reasons for staying another year and asked for a draft of sixty pounds sterling. I told him of the letters sent, etc.

November the 14th

Dutch Customs. In New York I had lodgings with Mrs. van Wagenen, a woman of Dutch extraction whose dwelling was opposite the new Dutch church. She as well as everyone in her house was quite polite and kind. It is true that the Dutch both in speech and outward manners were not as polite and well-bred as the English, and still less so than the French, but their intentions were good and they showed their kindly spirit in all they did. When a Frenchman talks about a man in his presence and even in his absence he always uses "monsieur": e.g. "*donnez á monsieur* etc.". An Englishman says: "give the gentleman etc." while the Dutch always said: "*giw dese man*". The women were treated in the same way without ceremony, and yet the Dutchman always had the same good intentions as he who used more formality. If several persons of Dutch extraction should come into a house at this time of the year, as many as could be accommodated would sit down about the fire. Then if any others should happen in, they pretended not to see them. Even though they saw them and conversed with them, they did not consider it wise to move from the fire and give the others a little room, but they sat there like lifeless statues. The

[1] Johan Clason (1704-1790), wealthy merchant of Stockholm. He had become a member of the Swedish Academy of Sciences in 1745.

French and English always made room by moving a little.—When one spoke of refinement as the word is now used, and in applying it to the French and Dutch, it was just as if the one had lived a long time at the court while the other, a peasant, had scarcely ever visited the city. The difference between the English and the Dutch was like that of a refined merchant in the city and a rather crude farmer in the country. But it is well to remember that there are exceptions to every rule.

I have lived now for almost a week in a house with a good-sized family. There was the same perpetual evening meal of porridge made of corn meal. (The Dutch in Albany as well as those in New York called this porridge *Sappaan.*) It was put into a good-sized dish and a large hole made in its center into which the milk was poured, and then one proceeded to help oneself. When the milk was gone, more was added until all the porridge had been consumed. Care was usually taken that there should be no waste, so that when all had eaten, not a bit of porridge should remain. After the porridge one ate bread and butter to hold it down. I had observed from my previous contacts with people of Dutch extraction that their evening meal usually consisted of this "Sappaan". For dinner they rarely had more than one dish, meat with turnips or cabbage; occasionally there were two [dishes]. They never served more than was consumed before they left the table. Nearly all women who had passed their fortieth year smoked tobacco; even those who were considered as belonging to the foremost families. I frequently saw about a dozen old ladies sitting about the fire smoking. Once in a while I discovered newly-married wives of twenty and some years sitting there with pipes in their mouths. But nothing amused me more than to observe how occupied they were with the placing of the warming pans beneath their skirts. In a house where there were four women present it was well nigh impossible to glance in the direction of the fire without seeing at least one of them busily engaged in replacing the coals in her warming pan. Even their negro women had acquired this habit, and if time allowed, they also kept warming pans under their skirts.

My departure from New York for New Brunswick took place at 2:30 P. M. New Brunswick was said to be about forty English miles distant. The weather was fine, but the wind from the south was not very strong. There are two ways by water from New York to New Brunswick: namely, around the outside of Staten Island or on the inner side of the same. If the weather is good and there is some wind, the journey is much shorter by way of the outer passage. If, however, there are signs of a storm or bad weather, the inner passage is to be preferred. We now took the outer one. After we had sailed for a time, the wind died down so that we proceeded very slowly. Yet we crept along with the tide and the gentle wind which aided us until midnight, when the low tide forced us to drop anchor. This journey is made usually in rather small boats from New Brunswick. These have a small cabin, and as there was now a considerable number of passengers and the night was very cold, we had to crowd into the cabin. No one can expect to get a wink of sleep where there is scarcely room enough to sit. Larger yachts cannot be used, as the river which they navigate up to New Brunswick is in some places very shallow.

Signs of Future Weather Conditions. An elderly farmer who was with us on the boat prophesied that the winter was to be a very cold one this year. The reason for this he gathered from the fact that there were more squirrels this year than for many a year before and that they had been very busy gathering nuts and other things for preservation in their holes as food during the winter. As a result it happened that even though there had been plenty of nuts, it was difficult to obtain any quantity of them, because the squirrels had carried them off to their hiding places. He assured us that this was an old sign of a cold winter. He even intimated that when the winter was to begin, it would come suddenly and in a hurry, since the fall had been so beautiful.

To Prevent Cracking of Punch Bowls by Warm Water. Often when warm punch was poured into a porcelain bowl, even though the latter felt warm, it would crack. A Jew, who also was a passenger on the vessel from New York, said that if a new porcelain bowl is boiled for a time in water in which there are some husks ("brains" i.e. bran, hulls or chaff) the bowl will not be so apt to crack when a warm liquid is poured into it.

Jews. The Jew just mentioned was a rather good-natured and polite man and it would scarcely have been possible to take him for a Jew from his appearance. During the evening of this day which ushered in his Sabbath he was rather quiet, though he conversed with me about all kinds of things, and he himself often began the discourse. He told me that the Jews never cook any food for themselves on Saturday, but that it is done on the day before. Yet, he said, they keep a fire in their houses on Saturdays during the winter. Furthermore, he said, that the majority of the Jews do not eat pork, but that this custom does not trouble the conscience of the young people when on their journeys, for then they eat whatever they can get, and that even together with the Christians.

November the 15th

The journey was continued in the morning after seven A.M. when the tide began to go up the river. The wind was right against us, but it was so gentle that it could neither hinder nor help us. We had to steer the boat as we rowed. I have described this river and the region about it in my notes for the month of May of this year [1749].[1] Now I just wish to add that the salt water runs almost up to New Brunswick, yet it never goes so far up that it actually reaches the town. In the river there are several oyster beds in which oysters are found and it is these which the boats run into when the water is low. The people living near the river have no evidence that it has grown less deep. It was shallow along its banks, but deep beyond them. We arrived at New Brunswick at ten A.M., where I met with an unusual piece of good luck. Just as I jumped off the boat and was about to run up into the town to procure lodgings until next week when the mail vans were to leave here, a man from Trenton who had just brought some travellers here and was about to return, came down to the boat and inquired whether there was anyone who desired to go to Trenton. I immediately took advantage of such a favorable opportunity, had my belongings put aboard his wagon and we set out from New Brunswick exactly at twelve o'clock noon. We had a distance of twenty-seven English miles to Trenton with good, dry roads. I have before described the whole of this road[2] so that I do not find anything of special interest to add at this time. We stopped at one place only, where we ate our dinner and fed our horses. Then we continued our journey. It became almost dark before we were halfway, because the roads are not as good in the vicinity of Brunswick

[1] See also pages 121, 122 and 323

[2] See pages 117 ff. and 322.

as they are nearer Trenton. The night was somewhat cloudy, yet whatever moonlight there was, assisted us. The gazettes had contained during the summer various accounts of how many travellers in the region about Philadelphia had been robbed on the roads of all their money. These facts made this journey by night particularly unpleasant for me, but thank God! we arrived safely in Trenton at a quarter of twelve.

Pennyroyal began to be found in large quantities at the sides of the roads about Brunswick and likewise along the road from there. It had entirely dried up.

The hawkbit (*Leontodon vulg.*) was everywhere along the road in full bloom, even though the ground here had already once been covered with snow. Compare also the [entry for the] 21st of October. This, i.e. the hawkbit, had short and somewhat broad leaves.

There was one plant of the chamomile (*Anthemis*) still in bloom. There were no signs of any other herbs in bloom at this season.

The products of the soil were said this year to be especially fine in New Jersey and Pennsylvania. There was a greater amount of wheat and corn than for several years past. Likewise there was an abundance of buckwheat, more than had been harvested for many years. It is often injured here by the frost in the autumn, before it has had a chance to ripen, but this year there had been no frost during this whole period, so that it had had an opportunity to mature. There had also been a rather plentiful crop of apples, but the spring sowing, such as of barley and oats, had failed, as had also the hay crop. The crop was small here, but still smaller in the vicinity of Boston. Boats loaded with hay had been sent there from Philadelphia. It was said that hay had been shipped there this year even from England.

The Bravery of the French Nation. I entered into conversation with an Englishman who had been a privateer for several years during the last war. While on these expeditions he had twice been taken prisoner by the Spanish and four times by the French. He had assisted in the capture of many ships, both French and Spanish. I asked him his opinion concerning the bravery of these two nations on the seas, their treatment of prisoners, etc. He answered me that if, for example, the French had a vessel with sixteen guns and a hundred men and the English likewise one with just as many men when they started to fight, the result would ordinarily be that they would have to leave one another, each having received equal injury. They cannot conquer one another, since the French are said to be merciless fighters on the seas. If a person is made a prisoner by the French, he is handled with extreme politeness and no prisoner can be shown greater kindness and consideration than that which the French nation allows its prisoners to experience. If an English vessel with sixteen guns and a hundred men should happen upon a Spanish man-of-war with the same number of guns and men, the English is usually able to conquer the Spanish within one half to an hour's time. If the English should remain at a distance and shoot at the Spanish, then the latter would be able to withstand the attack for a longer time and would return shot for shot. But if a sudden attack is made and an attempt made to board the ship, the Spanish cannot long withstand it. He said that an Englishman with such a ship and manned as described above much prefers to capture from two to four such Spanish ships than one of the French. If a man is taken prisoner by the Spanish, hardly a Turk or a heathen treats him worse than the Spanish do. The prisoner is treated as if he were a negro or a slave. They rob the prisoners, take off their clothes and leave them almost stark naked. Very frequently after they have captured an English vessel and made all on board prisoners, they attack them all with their swords and hew them down. If they obtain a prisoner whose life they wish to spare, they throw him into chains. They send the majority to their silver mines or to work as galley slaves. I remember when I was in England that they made the same distinction between the French and Spanish in their way of handling the prisoners, and they could not be particular enough in portraying the cruelty of the Spanish nation in this. Even neutral nations like the Swedish and Danish were quite rudely handled by the Spanish privateers during the last war between Spain and England, when they sometimes happened to come aboard some Swedish or Spanish (evidently an error in writing for *Danish*) ship. They robbed them of almost everything they owned, beat the sailors under their bare feet to make them say where they had concealed their money, etc. which tales I heard in London from those who had been forced to experience these things.

NOVEMBER THE 16TH

Farmers sow a considerable amount of buckwheat in this locality. It is used especially in preparing cakes similar to pancakes (griddle cakes). As these cakes come hot from the pan they are covered with butter which is allowed to soak into them. These cakes while they are still warm and prepared as above, are eaten in the morning with tea or coffee. The buckwheat straw is said to be useless; therefore it is left lying on the ground after it has been threshed. Neither the cattle nor other animals will eat it; only in cases of extreme necessity when the ground is covered with snow and they cannot obtain anything else will they occasionally chew a little of it.

Commerce. The products of this region are as a rule to be had at a lower price in Philadelphia than New York. The result is that many who live in the neighborhood of Trenton and also further down towards Philadelphia send the products of their farms, as butter, flour, etc., overland to Brunswick and then on to New York, there to be sold at a better price than they can obtain in Philadelphia. Throughout the summer the wagons travel between Trenton and New Brunswick.[1] They are loaded with flour and various kinds of merchandise which are being sent to New York. Many are carrying passengers only. On their return trips they have not much to transport except that they are frequently full of travellers of all classes and nationalities who during the whole summer do nothing but travel back and forth between Philadelphia and New York. The products mentioned above are brought in boats from Philadelphia here to Trenton and then overland from here to New Brunswick wherefrom they are carried on boats to New York. It is possible to travel this way from the Delaware River up to Trenton, but not further than here, as there is a fall or a cataract in this same river over which a small boat can scarcely pass, let alone a somewhat larger boat. The current above this fall or cataract was said to be quite strong. Yet it is possible to come down with boats from a distance of one hundred English miles from here, and therefore great quantities of the products of the region are transported here by boats and even to Philadelphia. But whereas it takes only one day to come down, it requires from ten to fifteen to go back against

[1] Cf. diary for October 28 and 29, 1748.

the current. The tide goes up the Delaware River to the falls here near the town of Trenton, but it can go no further on account of this cataract. At this time there were in here, in Trenton, four boats, each of which made a weekly trip to Philadelphia and back again. It often happens that on the return trip from New Brunswick, the wagons are loaded with all kinds of goods which are being sent to Philadelphia since sometimes, as I have mentioned before, certain articles are more expensive in Philadelphia; at other times in New York, and it is this fact that the merchants in the two towns take into account. Yet the man with whom I had lodgings, and who was a carter and continually carried these wares over the road, said that generally more goods were sent from Philadelphia to New York than *vice versa*. The country on both sides of the Delaware above this town is populated for about a distance of one hundred and fifty English miles. The land on this side of the river is more sandy and on the Pennsylvania side it consists of clay soil, especially a short distance from the river. About a hundred English miles further up in this New Jersey Government is a tract of land which is called Minisiek [Minisink or Meenesink].[1] It is lowland and rather thickly populated. The whole of this province, as well as Minisiek and other places on the western side of the road, is said to be nearly everywhere thickly inhabited, mostly by the Dutch or descendants of old Dutch families. From all of these places comes considerable corn, flour, hemp, linseed, lumber (since the country is full of forests and sawmills), butter, together with other products of the region, which are then sent either down the Delaware River to Philadelphia or down other rivers to New York. In various places in the country the merchants have stores to which they send, and where they sell, all kinds of merchandise. Besides these, there are to be found in every large village one or more stores.

The savin (*Juniperus sabina*) is said to grow as a common thing on the banks of the Delaware. It is found only as large shrubs and never grows to the size of a tree.

---

[1] Concerning this settlement see John W. Barber and Henry Howe, *Historical Collections of the State of New Jersey*, edition of 1855, p. 506. According to the map in this volume it seems to have been located in the southwestern part of what is now Delaware County, New York, near the Pennsylvania border, about where Cadosia and Hancock are now situated. It will be remembered that the Delaware River runs far up into New York State.

The red juniper, so called by the Swedes in New Sweden, or the Englishman's red cedar, grows abundantly further down New Jersey in sandy and dry places, according to what they told me.

Hvita cedern so called by the Swedes and *white cedar* by the Englishmen was to be found, as I was told, in quite a number of places in the deep swamps further down the province. In English such morasses were called *cedar swamps.* I described one of them last spring on my journey to Rapaapo in the month of May. The savin is also said to grow wild here, and I have likewise found it in the same state in the region about Albany.

Trenton. When I travelled through this little town last summer toward the end of May, I gave a short description of the same. Now I just wish to add one or two things. The soil upon which the town is located is mixed with sand, hence the location is considered very healthful. The old gentleman with whom I had lodgings, told me that when he came to this town some twenty years ago, there was not more than one house here, and since that time the town has grown so that there are now almost a hundred houses. These are built so that the street passes along on one side of the buildings, while on the other are kitchen gardens of varied dimensions, and in the furthermost corners of the gardens the privies are located. A little back of the houses in the gardens, also, are the wells, ordinarily with a draw bucket. The houses are two stories high with a cellar beneath them and a kitchen underground, next to the cellar. The houses are divided within into several rooms by thin board partitions.

November the 18th

Thermometrical Observation. I had my personal belongings, including the thermometer on board, so that I could not take a temperature reading before the men entered the boat, when I removed the thermometer and at half after nine exposed it for a long time in the open air to see how low the mercury might go. But no matter how long I kept it in the air, I could not get it lower than one degree above zero C., which appeared strange to me, since the wind was blowing so cold that it was impossible to stay outdoors very long before one became thoroughly chilled through; and when I held the instrument between my fingers, they became so stiff that I could scarcely move them. From this it may be seen that the body feeling of cold does not always correspond with the reading on the thermometer.[1]

The yachts [or boats] which were now en route from Trenton to Philadelphia were loaded with timber, wheat flour, several barrels of linseed, and many ditto of pork. The timber was of white and red or black oak.

Oakwood in Trenton cost seven shillings per cord; hickory, nine or ten.

At ten in the evening we left Trenton by boat.

November the 19th

Slow Sailing. The continuation of our journey proceeded very slowly to-day, for the wind was contrary, becoming a biting storm. The boat, too, was worn out and so loaded both above and below the deck that it sank down almost to the deck itself. When the tide was against us it was impossible to move, so that we had to cast anchor; and when the tide went out the storm was so violent and the waves so high that we dared not sail in the old, heavily-laden vessel much before evening, when the wind died down. We then tacked slowly forward and about seven o'clock in the evening reached Burlington, where we lay the following night until one in the morning, when the tide changed again and began to go out. The skippers confided that in the river between Trenton and Philadelphia are hidden rocks, but that one can easily dodge them with a boat.

Tobacco. The English chewed tobacco a great deal, especially if they had been sailors. Not an hour passed when they did not take as much cut tobacco as they could hold in the fingers of the right hand and stuff it in the mouth. Young fellows of from fifteen to eighteen years of age were often as bad as the older men.

The barrels that carried the flour (which constituted a large part of the cargo) were made of white oak, but the hoops were of hickory, and the ends of pine.

[1] A paragraph on watermelons which follows is but a repetition of earlier accounts and is therefore omitted here.

November the 20th—At Philadelphia

Oysters were sold in large quantities at this time in Philadelphia.[1]

November the 21st

Swedish Clergymen. Since I learned in my quarters this morning that two ministers had arrived from Sweden, I went out to find and talk with them. They were lodged at the home of Mr. Hesselius and were the Rev. Mr. [Israel] Acrelius, appointed dean of the Swedish congregations and pastor of the Christina parish, and Rev. Unander, who expected to be made pastor of the parishes of Penn's Neck and Raccoon. Both came from Roslagen [Uppland], had left Sweden in June, and had been on the way between Gravesend and Philadelphia more than six weeks.

In the forenoon, also, I was busy unpacking my goods from the boat, and in the afternoon in visiting some acquaintances.

The Library in this city was established a few years ago. Mr. [Benjamin] Franklin was one of the first to lay foundation for it, and was the real cause of its origin. A number of gentlemen then came together and each one gave forty shillings for the purchase of books, so that one hundred pounds were acquired for the purpose. Those who now are directors of the Library give annually so much that books may be bought for 50 pounds. Several others in the town have followed their example, and there are now severa branches here, where books may be borrowed for a small fee. A rich gentleman from Rhode Island was here, and when he had the opportunity of examining this institution he liked it so well that when he had returned home he persuaded some gentlemen in that state to build a house for a library, to which he made a gift of 500 pounds sterling for books. Proprietor [Thomas] Penn presented the city with a piece of land for the building, with some telescopes, two globes, and a machine for [display of] electricity.[2]

---

[1] Here follows an article on oysters which is omitted in our translation, since the substance of it has already been given by Kalm in other entries of his diary. See especially entry for October 31, 1748.

[2] We must not forget that this is in 1749.

NOVEMBER THE 22ND

This morning we left Philadelphia for Raccoon, N. J. Jungström had gathered several seeds there, and at my departure last spring I had asked several of the Swedish people to collect seeds for me of various kinds. It was now time for me to find out whether they had heeded my request. At four in the afternoon I reached the Raccoon parsonage, where I found the wife[1] [now a widow] of the Rev. Mr. Sandin tolerably well. Jungström and the rest were also well.

Diseases. Last summer there was a great deal of sickness in New Sweden, and the reason is not so easy to understand. All asserted that the heat had been more severe than in many summers before, and according to Mr. Evans and others had been the worst on the Sunday I left Fort Anne.[2] They spoke of several who had fallen dead from the heat while walking. The epidemic began soon thereafter, and people over almost the whole country were ill, even those who otherwise had the strongest of health and were born in this land. They had to stay in bed a long period. Mr. [John] Bartram and most members of his household were sick; in Wilmington there was much illness too; also in Philadelphia. But nowhere was the epidemic so severe as in Raccoon and Penn's Neck. These two places and Salem had the reputation of being the most unhealthy localities in the whole country. There was scarcely a person who did not reiterate that these were the worst. Last summer there was hardly a house without several sick members, and it lasted a large part of the season. On some farms almost every person was ill, even the little children; and those who had been able before to keep their health there had this summer been obliged to take to the bed. Chills and fever [malaria probably] had been the predominating disease, but with several symptoms differentiating one person's illness from that of another. However, only a few died from it, and toward the end of October the disease had for the most part left. Mrs. Sandin had had a particularly difficult attack, from the beginning of June up to that time.

[1] This lady became, in the following year, Mrs. Kalm.

[2] June 29, 1749. Kalm in the entries for June 28 complains of the heat at Fort Anne. It must have been much hotter in New Jersey.

She had often been so weak that she thought her last hour had come.

November the 23rd—At Raccoon

There are many *black walnut trees*[1] in this locality, but not many further north. I never saw them north of New York. In planting them old van Neeman considered it best to bury the nuts where the trees were to stand permanently, because they would then better withstand the cold. Once he had planted some black walnuts, and the following summer transplanted some of the young trees, and allowed others to stand where they had come up. The next winter was unusually cold, and the majority of those transplanted froze off right near the roots, but those he had left untouched were not damaged.

Mulberry Trees. Old Kijhn (Keen) told me that sometimes during unusually severe winters one-year old mulberry trees freeze close to the ground. But this was not surprising, he claimed, for when a mulberry tree comes up it may grow as much as eight feet the first summer. In that case the pith is loose, juicy and large, and so is the tree itself. It can therefore not endure the frost so well. Nevertheless, this does not happen except in extreme cases, and only with young trees, and in the following spring new shoots will sprout from the old roots. But when a mulberry tree is two or three years old, no cold can damage it, no matter how severe the winter may be. These mulberry trees are of the black variety and grow wild here and there in the woods. None around here had ever seen any white mulberry trees.

November the 24th

Squashes are planted in large quantities everywhere in New Sweden, but they do not last after the end of October and consequently cannot be kept over winter. They become ripe in September, when

[1] By the Swedes of Raccoon called *svartnöttbom;* Ger., *Schwarznussbaum;* Dutch, *zwartnootboom*, translating the stems literally. The influence here is Dutch probably rather than German.

they are delicious; but later their shell hardens so that it becomes almost like wood, and the pulp and seeds decay. They have to be left out in the fields therefore, or thrown away, if not eaten in September or October.

November the 25th

In the morning I returned from Raccoon to Philadelphia, where I arrived at sunset. The roads were dry and the weather fine for the season. All hardwood trees here had already lost their leaves, and the forest looked everywhere wintry.

The Harvest. I have already, in my diary for the 15th of November, referred to the season's harvest in New Jersey and Pennsylvania, and all people I have met lately have confirmed what I said there. Farmers everywhere complained about the poor hay-crop; it had not been so poor for years they said, and the reason was the terrible drought which had prevailed all summer in this locality. Many a husbandman did not know how he was to feed his cattle over winter. The crop of apples this autumn was bountiful, but many had the tendency to rot after they had been brought indoors. A large number of nut trees had borne no fruit this year, or at most very little. This was particularly the case with the various kinds of walnut and hickory trees, the chestnuts, and the many varieties of oak. There was so little fruit of the latter that the farmers of the place, who otherwise let their pigs feed and fatten on acorns, either by collecting them or allowing the pigs to run wild among the oaks, this year had to feed their swine on corn and other grain. This, again, despite the abundance of corn, has made the latter quite expensive. Nor has this season the crop of beechnuts been any larger than that of acorns; Jungström could not obtain any when he arrived here. On the other hand, there was an abundance of them in Canada this year: the trees were full of them. There was also a large quantity there of walnuts and hickory nuts and an appreciable crop of acorns. The hay crop, too, was heavy, excellent. The wheat harvest there was exceptional: they had not had one like it for many years. But in Albany and New York State in general the complaint was commonly heard that the wheat crop this season had been one of the poorest.

November the 26th—At Philadelphia

Remarks about Plants. To-day I met Mr. John Bartram for the first time since my return from Canada. He reiterated what he had often told me before, namely that all plants and trees have a special latitude where they thrive best, and that the further they grow from this region, whether to the north or south, the smaller and more delicate they become, until finally they disappear entirely. When I apprised him of the fact that I had found sassafras trees at Fort Anne, far to the north, but that these trees had been very small and without seeds, he answered that all such trees had grown there from seeds carried by birds, and that their slender stems showed it was not their natural habitat. In the Blue Mountains he had found the *Abies balsamifera* (balsam spruce), which generally grows in Canada. Near Oswego, he had come upon a mulberry tree, and one single specimen of *Arbor tulipifera* [*Liriodendron tulipifera*], which the local settlers called Old Woman's Smoke, and whose leaves were held to be a remedy for gout. He discovered an aloe plant in the northern part of Virginia, the farthest north that it grows.

Fol. Avoine. When I talked with him about Fol. avoine (*Zizania aquatica*, Indian rice), he told me he thought this to be that tall, thick grass which grows here in brooks and other bodies of water and has long, grain-bearing seeds. The Indians had formerly gathered these seeds for food. Now they are eaten by a bird which is described and pictured in [Mark] Catesby's Ornithology [1] and is called the ricebird (bobolink). This bird remains here until the seeds of the plant have fallen out, and then he moves on to Carolina, where the rice is ripe about that time. Mr. Bartram believed this wild Indian rice to be a good food, but encountered a difficulty in its gathering, since it ripens very unevenly and not all simultaneously. It begins to ripen in the beginning of August and continues to do so the rest of the month.

November the 27th

An Apple Beverage. Mr. Hesselius's daughter related how some

[1] Either *The Natural History of Carolina, Florida, and the Bahama Islands* (1731-1743, with appendix, 1748) or a paper on birds presented in 1747 before the Royal Society of London and printed in its *Philosophical Transactions* that same year. In this monograph Catesby "gave examples from among the South Carolina birds."

colonists in this vicinity made a pleasant beverage of apples, as follows: some apples—which need not be the best—and apple peelings are taken and dried. Half a peck of this dried fruit is then boiled in ten gallons of water and when removed from the fire the solid part taken out. Then yeast is added to the water, which is allowed to ferment, whereupon it is poured into vessels like any other drink. One who has not tasted it before would not believe that such a palatable beverage could be prepared from apples. It was said to be better than that made from persimmons, because it retains its quality longer and does not get sour.—I forgot one thing: when the apple ale is made, some bran should be added to the water.

Varieties of Stone. To-day Benjamin Franklin showed me several varieties of stone, which he had in part collected himself and in part received from others. All were formed in the English provinces of America and consisted of:

1. A rock crystal, the largest I had ever seen. It was four inches long and of a diameter of three fingers' breadth. I regretted it was not transparent but of a dingy, watery color and opaque texture. All six sides were smooth as if ground, and had been found in Pennsylvania.

2. *Asbestus stellatus,* with fibers radiating out from the center, as described in Wallerius's *Mineralogy,* page 145. Its color was a very dark gray, mostly blackish, and felt oily to the touch. It came from New Engand, where it is found in big stones that are utilized for fireplaces, because it does not change or crumble in the least from the action of fire.

3. *Stalactites.* These were discovered in a cave near Virginia and were of two kinds: the *stalactites conicus* which had depended from the roof of the cavern, and the [stalagmite] that had been deposited like a round, uneven, scraggy fungus on the floor of it, where the [calcareous] water had dripped from above. In color they resembled an unclean white.

November the 28th

Graphite. To-day I received in my quarters a piece of graphite that had been discovered some 20 English miles from here [Philadelphia]. It was dark blue, very soft so that it could be cut with a knife, and had an undulating surface. One could hardly touch it before the hands became black; it left a mark everywhere.

November the 29th

In the evening I accompanied Mr. John Bartram to his estate in the country.

The Falling Trees. I told Mr. Bartram about my observation of falling trees the summer before in the wilderness between Albany and Canada, describing how on calm nights I heard the trees crack to the ground. He offered the [highly fanciful] [1] explanation that it was due to a difference in atmospheric pressure, and that in his experience such a phenomenon was a sign of rain.

November the 30th

Mr. Bartram had found some *Salicornia* (glasswort) and *Potentilla anserina* L. (silver-weed) growing near salt springs. He had discovered some arbutus in the sand in western New Jersey, and found the taxus in several places near the Hudson, Delaware and Susquehanna Rivers, though always dwarfed specimens of it.

The persimmon is a very hard wood. Hammers and clubs are made of the sour gum tree. *Filipendula* (meadowsweet or dropwort) are also called tea bushes. There is a birch [sweet birch] which is chewed like sugar.

In the evening I returned with Mr. Bartram's son to the city. We met a man on the road who was complaining bitterly that two culprits had just attacked him and robbed him of 50 shillings and his overcoat, and had then beaten him and run away.

December the 1st

Swedes. A man from Chester called on me last Friday morning. He was of old Dutch extraction whose ancestors had been in America since the Hollanders captured the land from the Swedes, but he was married to a Swedish woman. He gave me much information about the old Swedish settlers. They had been entirely satis-

[1] Words in brackets are not in the original text, of course.

fied to come under Dutch rule, he believed, because they had not heard anything for a long time from their mother country, nor received any aid from there. And more particularly because the Governor [Printz] of the Swedes had been rather severe, and treated them mostly as slaves.[1] A short time before the Dutch had captured New Sweden, a Dutch vessel appeared, loaded with all kinds of goods which they knew were needed by the Indians. At their arrival the Dutch asked the natives for permission to tie their ship with a rope to a tree on the shore, since otherwise the current might carry it away. When the Indians had permitted this, they were presented with various gifts and were promised more in return for a permanent permit to fasten their boat there and for a piece of land at the spot, where the Dutch might put up their tents and cook their food, it being so troublesome always to stay on the ship. The natives allowed this too: *hinc causa belli inter Suecos et Batavos.*[2] When Penn came he took much land away from the Swedes and the Dutch who lived near Philadelphia and gave it to the Quakers under the pretext that the Swedes had more land than they needed and that otherwise it would lie waste and uncultivated. Much land was taken away by fraud, and whenever [the higher authorities] got hold of the old deeds, describing how much land had been apportioned to each one, the Swedes and Dutch never had theirs returned.

December the 2nd

Violent Storms in North America (quotation). "The *American Weekly Mercury*, No. 231, Philadelphia May 31, 1724. On Saturday last, being the sixteenth of this instant, we had a violent storme of wind, wid very hard shower of rain and claps of thunder, which lasted about eight minutes, it has done very considerable damages to our orchards, and has killed several Cattle by the fall of the Trees, and in some parts of the country they had very large hail-stones, which they say has destroyed whole fields of corn.

N:o 243. Philadelphia aug. 13, 1724. On the third instant, about the hour of 12 (at New Garden in Chester) there began a most

[1] There is, possibly, some truth in this Dutch contention.

[2] "Hence a cause for war between the Swedes and the Dutch."

terrible and surprizing Whirl-wind; which took the roof of a barn and carried it into the air, and scattered it about 2 miles off, also a mill that had a large quantity of wheat in it, and has thrown it down and removed the millstones, and took a lath of the barn, and carried it into the air; it also carried a plough into the air." . . . Both were hurled down later; one into an oak lid and the other deep into the ground. A flock of geese and three or four hawks had to go along. From the *Philadelphia Country* I obtained the following: At Plymouth the whole roof was pulled off a big barn and carried out in the lot; a woman's skirt sailed seven or eight miles through the air, and grain stacks were strewn about the fields. "It took up almost all the apple trees in the orchard by the root and carried them some distance." People were in danger of being carried off right in their houses. From Bucks' county [came the report] of houses, fences and trees being torn down. In No. 245 it goes on to speak of this whirl-wind as follows: "Philad. aug. 27. We have a farther account of the Whirl-wind, or Hurrican, from the Great-Valley, where it took up very large trees by the roots; but particularly one about 3 foot over, and carried it up in the air a great height, so that when it fell, stuck very deep upright; it made clean work where it went, took up all as tho it had been pulled, and where it went across the roades, it laid trees so thick that it is very difficult to travel; it made a road of about 40 pole in breath, and in some places it parted, and then met again about two miles off.

N:o 244. Philad. Aug. 20. 1724 on th 17th, 18th and 19th of this instant, we had a violent storm of wind and Rain, which has caused such a fresh in our Creeks, and river, that it has broke several of our best mill dams, in this and our Neighbouring Provinces, and the fresh is so strong, that our Vessels in the road have not winded; Such a fresh has not been known this 20 years.

December the 4th

To-day I went to Raccoon.

Cheese. In almost every place that I have been this summer, both in the English provinces and in the French ones in Canada, the residents have made cheese; but there has been an appreciable difference in the quality of it, which has varied with the locality. In Canada cheese was made only in certain places, such as on the

Isle d'Orleans below Quebec. The cheeses there were very small, round and thick, but not very good, about the poorest I have eaten in America. They tasted dry, lacked nutrition, and only common people for the most part were seen eating them; most of the better class left them alone. These bought cheese imported from France, which came in large, thick and round forms and tasted well, but smelled like sour herring. In New York State several kinds of cheese were made. Most of them appeared in an average, suitable-sized mold, were round, reasonably thick; and some tasted pretty well; but most varieties were poor and manufactured from sour milk. In Esopos [Kingston] good cheese was made in the shape of small globes that were flattened a little on top and bottom; one had added spices to it to improve its taste. Here in Pennsylvania we got cheese of both kinds, good and bad; but in general better cheese was made here than in any other place in America that I visited. The cheese made by the Swedes of Raccoon was especially good and looked very appetizing. It was molded in round, thick forms of from nine to twelve inches in diameter, and was the best made in this part of the world. Some of it could rival the English variety.

December the 5th

Centenarians in America. The *American Weekly Mercury*, No. 196, in a communication from Boston, September 2, 1723, mentions an old Indian by the name of John Quittamog, "living in Nipmug Countrey near Woodstock in New England. He is reckoned to be alone one hundred and twelve years old; he still confirms, that he was at Boston when the English first arrived; and when there was but one cellar in the place, and that near the common, and then brought down a bushel and a half of corn upon his back." He is still in good health and mind, has a retentive memory, "and is capable of travelling on foot ten miles a day."

The Pennsylvania *Gazette*, No. 138, in a note from New York of July 5, 1731: "We hear that about 26 miles above Albany there died lately one Johannes Legrange, aged 106."

"Dedham in New England June 5, 1732. This day died here the famous Sam Hide, Indian, in the 106th year of his age." He was well known for his "running jests and uncommon wit." He had slain 19 other Indians and made the same number of notches or dents on his gun.

No. 343. Boston, June 16, 1735. Thomas Curries died at Colchester, Connecticut in the 110th year of his age. He left 5 children, 39 grandchildren, and 28 great grandchildren; "some of the last are married." His hair was not gray, nor was he bald. He walked over six English miles a day during his last years.

The Germans [of the locality] were said to be very industrious and in good circumstances. Others lived mostly from hand to mouth.

December the 6th

Scurvy in the Mouth.—An Englishman by the name of Wood has offered the following remedy for scurvy in the mouth: one takes that sole of an old shoe which is nearest the foot in walking, burns it to a coal in a fire, and grinds it into a powder. Burn some alum, powder it, and add it to the shoe-sole powder. Take some of the mixture in a rag or cloth and rub the teeth and gums with it several mornings in succession.

Squirrels. "*The American Weekly Mercury*, No. 302. Boston, September 20, 1725. We have advice from Connecticut and several Towns in this province, that the Squirrels enter their fields, in droves, and destroy their corn; and that in some places they even run into the farmers' houses for food."

I left Racoon to-day at half after twelve, arriving in Philadelphia at exactly six o'clock.

December the 7th

Quakers. To-day I attended service in a Quaker meeting-house. I was once present at such a service in London, and once in this city, where there are two meeting-houses or churches. It is known that the Quakers are a religious sect that arose in England during the last century, of whom the majority is found in Pennsylvania, since this province was granted to Penn, who was a Quaker and who brought his religious followers to this territory. But I should like first to describe the service that took place in their church today, and then speak about their faith and customs.

The Quakers in this town attend meeting three times on Sunday—from ten to twelve in the morning, at two, and finally, at six in the evening. Besides, they attend service twice during the week, namely on Tuesdays from ten to twelve and on Thursdays at the same time. Then also a religious service is held in the church the last Friday of each month, not to mention their general gatherings, which I shall discuss presently.—To-day we appeared at ten, as the bells of the English church were ringing. We sat down on benches made like those in our academies on which the students sit. The front benches, however, were provided with a long, horizontal pole in the back, against which one could lean for support. Men and women sit apart. (In London they sat together). The early comers sit on the front seats, and so on down. Nearest the front by the walls are two benches, one on either side of the aisle, made of boards like our ordinary pews, and placed a little higher up than the other seats in the church. On one of them, on the men's side, sat to-day two old men; on the other, in the women's section, were four women. In these pews sit those of both sexes who either are already accustomed to preach or who expect on that particular day to be inspired by the Holy Ghost to expound the Word. All men and women are dressed in the usual English manner. When a man comes into the meeting-house he does not remove or raise his hat but goes and sits down with his hat on.—Here we sat and waited very quietly from ten o'clock to a quarter after eleven, during which the people gathered and then waited for inspiration of the Spirit to speak. Finally, one of the two old men in the front pew rose, removed his hat, turned hither and yon, and began to speak, but so softly that even in the middle of the church, which was not very large, it was impossible to hear anything except the confused murmur of the words. Later he began to talk a little louder, but so slowly that four or five minutes elapsed between the sentences; the words came both louder and faster. In their preaching the Quakers have a peculiar mode of expression, which is half singing, with a strange cadence and accent, and ending each cadence, as it were, with a full or partial sob. Each cadence consists of from two to four syllables, but sometimes more, according to the demand of the words and meaning; i. e. my friends//put in your mind//we can//do nothing//good of our self//without God's//help and assistance//etc. In the beginning the sobbing is not heard so plainly, but the deeper and farther the reader or preacher gets into his sermon the more violent is the sobbing between the cadences. The speaker to-day had no gestures, but turned in various directions; sometimes he placed a hand on his chin; and during most of the sermon kept buttoning and unbuttoning his vest. The gist of his sermon was that we can do nothing of ourselves without the help of our Savior. When he had stood for a while using his sing-song method he changed his manner of delivery and spoke in a more natural way, or as ministers do when they read a prayer. Shortly afterwards, however, he reverted to his former practice, and at the end, just as he seemed to have attained a certain momentum he stopped abruptly, sat down and put on his hat. After that we sat quietly for a while looking at each other until one of the old women in the front pew arose, when the whole congregation stood up and the men removed their hats. The woman turned from the people to the wall and began to read extemporaneously a few prayers with a loud but fearfully sobbing voice. When she was through she sat down, and the whole congregation with her, when the clock struck twelve, whereupon after a short pause each one got up and went home. The man's sermon lasted half an hour. During the sermon a man would get up now and then, but in order to show that he did not do so to speak he would turn his back to the front of the church—a sign that he did not arise from any spiritual inspiration. There were some present who kept their hats off; but these sheep were not of this flock, only strangers who had [like Kalm himself] come from curiosity and not because of any special prompting by the Spirit.

The meeting-house was whitewashed inside and had a gallery almost all the way around. The tin candle-holders on the pillars supporting the gallery constituted the only ornaments of the church. There was no pulpit, altar, baptismal font, or bridal pew, no prie-dieu or collection bag, no clergyman, cantor or church beadle, and no announcements were read after the sermon, nor were any prayers said for the sick.—This was the way the service was conducted to-day.

But otherwise there are often infinite variations. Many times after a long silence a man rises first, and when he gets through a woman rises and preaches; after her comes another man or woman;

occasionally only the women speak; then again a woman might start, [followed by a man], and so on alternately; sometimes only men rise to talk; now and then either a man or woman gets up, begins to puff and sigh, and endeavors to speak, but is unable to squeeze out a word and so sits down again. Then it happens, also that the whole congregation gathers in the meeting-house and sits there silently for two hours, waiting for someone to preach; but since none has prepared himself or feels moved by the Spirit, the whole audience rises again at the end of the period and goes home without the members having accomplished anything in the church except sitting and looking at each other. The women who hope to preach and therefore sit in a special pew generally keep their heads bowed, or hold a handkerchief with both hands over their eyes. The others, however, sit upright and look up, and do not cover their eyes. The men and women have separate doors, through which they enter and leave the church.

I shall now say something about their clothes and manners, in so far as they vary from those of others. The women have no clothing that differs from that of the other English [ladies], except that I do not remember having seen them wear cuffs, and although they censure all adornment I have seen them wear just as gaudy shoes as other English women. But the men's clothes differ somewhat from those of other gentlemen. For instance, they have no buttons on their hats, and these are neither turned up entirely nor turned down, but just a trifle folded up on the side and covered with black silk, so that they look like the headgear of our Swedish clergymen. They wear no cuffs; they never take off their hats, neither when they meet anyone nor when they enter a stranger's house or receive friends at their own home; they make no bows and hate all courtesies; and the only form of address is "My Friend" or "thee and thou". A son addresses his mother with "thee". In the plural for "they" one says "I and the Friends", naming them. Although they pretend not to have their clothes made after the latest fashion, or to wear cuffs and be dressed as gaily as others, they strangely enough have their garments made of the finest and costliest material that can be procured. So far as food is concerned, other Englishmen regard the Quakers as semi-Epicureans; for no people want such choice and well-prepared food as the Quakers. The staunchest Quaker families in the city are said to live the best. Yet in the matter of drinking they practice restraint. The majority of colonists did not look upon them as any *societas pia*, as they at first represented themselves, but as a political body. They cling together very close now, and the more well-to-do employ only Quaker artisans, if they can be found. If a skilled workman, laborer or someone else of their faith backslides and joins another church, they have no more to do with him. The Quakers have a general fund from which they lend money to their poor, at little or no interest, according to the circumstances of the borrower. Often when one of their tradesmen gets into financial difficulties, the Friends will collect a sum and present him with it. They have overseers who go about and see how the brothers live, which may be seen from their Code of Discipline.

When two members wish to marry they attend the monthly meeting together, rise and announce jointly and loudly that they expect to take each other as man and wife. This is repeated at three such meetings in succession, so that it requires three months before the banns are duly proclaimed. For further details, see the marriage code.

December the 8th

The Newly Invented Pennsylvania "Fireplaces." Although this city [of Philadelphia] is situated at 40° north latitude and should be relatively warm in winter, when we compare it with European towns of the same latitude, experience has taught that sometimes it is as cold here as in old Sweden. It is necessary only to examine the meteorological observations in this travelogue for this year, the supplementary remarks here and there in the diary about weather conditions and temperature, and what Mr. Franklin records in his *Poor Richard's Almanac Improved*. It has therefore been imperative to make some provisions for heating the houses for a period of several months in winter, and for that purpose various types of stoves have been used. I shall not attempt to describe them all at this time, because they are not only described but their good features and serious faults elaborated in the book which Mr. Franklin published in Philadelphia, 1744, under the name of *An Account of the New Invented Pennsylvanian Fire-Places* etc. But since all these had their faults, and the Englishmen liked to see the fire burn instead of confining it in a stove, Benjamin Franklin invented a new type of stove, which not only provides plenty of heat, saves fuel and brings fresh air into the room, but is so constructed that the flame may be seen. An extensive description of it is found in the above-mentioned book by Mr. Franklin, so that I have but little to add, especially since Mr. [Lewis] Evans has in addition made copious marginal notes about it in his presentation copy of it to me [and which may be consulted]. Mr. Franklin invented the stove and Mr. Evans made the drawings and figures. The bottom should not rest flat on the floor or ground, but be elevated a little, and an opening or passage left to the air-box, so that the air which is heated beneath the bottom plate can get out into the room. There are several types of this stove: some have dampers, which in the description are called registers, and others have not; some have a front plate near the bottom which can be moved up and down, and when it is moved down the draft becomes stronger and the wood begins to burn quicker; others have a hole on the frontal plate with a small trap-door which when opened allows air to enter and fan the fire, for there is a narrow passage leading to this trap-door, either from the room below or from a space under the floor connected with the outside. Under the air-box is an opening through which fresh, cold air enters it, and here it is heated and sent into the room through side openings. If there is a cellar beneath the room with the stove, a hole is made to it, so that the air can come from that direction; but otherwise it must pass under the floor to the air-box from some side of the house, preferably an entrance hall, where a small opening may be made in the wall near the floor, through which fresh air may enter. Several who could not afford to purchase these stoves imitated them by making them either of brick or white Dutch tile [or brick], making only the top of iron. But these were not so warm.

December the 9th

More about the Franklin Stoves. I wrote yesterday about the newly invented fireplaces [stoves]. I shall now add a few more details. They are made or cast of iron, i. e. the iron which is obtained in this province. The size varies and so does the price.

Despite their usefulness they have been criticised. It was held that if the chimney could not be swept [as in big chimneys with open fireplaces] there was danger of fire; some thought the stoves gave too much heat, and since the Englishmen were not accustomed to this they liked open fires better; and the Germans preferred *their* small, oblong, square iron stoves, which are constructed in the same way as the stoves in Bohuslän and Norway, because they give more heat and cost less. Nevertheless, there were many who used these new stoves, both in Philadelphia, New York and elsewhere. In the country where they had plenty of fuel they used the large fireplaces, since many believed these new ones expensive, and did not reckon the cost of wood. Mr. Franklin loaned me one of the stoves for the winter. It kept the house quite warm, but then one had to use short wood in it. It proved often unnecessary to have a fire in the kitchen, and one could prepare chocolate and other food in the little stove. Also, it proved possible, by suspending a cord from above in front of the fire-box, to roast meat or fowl attached to it and turned. And curiously enough it was roasted better and quicker than in a regular kitchen fireplace, since the room was warm and heat came both from the fire and the iron of the stove. The chimney is seldom cleaned more than once a year, but Mr. Franklin was in the habit of setting fire to a sheet of paper every fortnight and let it pass through the flue leading to the stove and so burn off the soot there also. If the stove is narrow it is not so easy to sweep the chimney, after everything is closed up by masonry; but Mr. Franklin had a brick removed beside the stove, let a man pass down through the chimney, clean it, and when he reached the bottom near the stove had him force the soot through the hole made by the removal of the brick. When this was done the brick was replaced. Where the hearth is broad the stove is placed on one side of it, and a door made on the other through which the chimney sweep can enter and do his work. To get fresh air into the stove of a house with no cellar, and where no outside air is wanted, Mr. Franklin this year had had the stove in his own room set on a rim of masonry six inches from the floor, with an opening through the bricks on one side to let the cold air near the floor enter, pass through the air-box, where it was heated, and then pass through the holes on the iron sides of the stove into the room, etc. This brought about a constant circulation of air.

December the 10th

Remarks about the Heat. After dinner I talked with Mr. Evans about the weather here in Philadelphia last summer, especially about the intensity of the heat. He said the severest heat had come on June 26th and 27th (N. S.), and more particularly the latter, when his thermometer showed 86° F., which he found to be the same as the temperature of the human body.[1] The heat on that day would have been unbearable, had it not been for a gentle breeze. According to a Fahrenheit thermometer owned by Mr. Franklin, and believed to be more accurate than Mr. Evans's the mercury on that day rose to 100°. No one remembered a hotter summer. [Then follows a brief comparison of the Fahrenheit and Centigrade (Celsius) thermometers, which is unnecessary here]. By referring to my meteorological observations for the 26th and 27th of June [1749] and the days immediately preceding and following one can notice the difference between the temperatures of Philadelphia and of Fort Anne, where I stayed on those dates. Strangely enough Mr. Evans reported that about the 14th of June, according to the newspapers, both England and France had had an unusually cold spell, and that there had been ice in England of the thickness of a silver dollar. In France the cold had damaged the grapevines and other fruit trees and vegetables.

December the 11th

The Cold. It is now becoming quite cold. Yesterday especially it felt very penetrating, in particular when walking against the wind. All small streams and inlets are now covered with ice so that one can pass over on most of them without danger. The Schuylkill River was well frozen too. On the Delaware River there was so much ice that it was completely frozen this morning near the city. Yet the ice was very thin and disappeared as soon as the tide appeared. Otherwise there drifted so much ice in the river that at Gloucester's Ferry it was impossible to cross with a horse. One

[1] Mr. Evans's thermometer was not very accurate, as we can see. The temperature of the human body is about 98.4° F. It is possible, too, that Kalm's original manuscript showed, or meant to show, 96° instead of 86°, which still leaves an appreciable error.

could now cross only in small boats between the pieces of ice. People who had brought wood to town on boats expected to leave them and return by the highway. Ships that intended to sail hastened down the river before they would be frozen fast, and those only half loaded began to consider remaining where they were over winter, wherefore some took their boats into dock or to some safe place where it would be protected from the ice, both now and in the spring. The price of wood went up rapidly, because before that one had been able to buy a cord of hickory for 22 shillings, but now it had gone up to from 25 to 27 shillings per cord, and even then one had to hurry and take it lest it be snapped up by someone else. Oak wood rose from 16 to 19 and 20 shillings per cord, and one was glad to get it at that price. All wood bought now was green, cut this fall, for although the country people cut much wood in the spring and summer, which they let dry and cart into town in the early autumn, it is bought up at once by the wealthy and thoughtful, who then get their supply at a lower price. Later only green wood can be found on the market.—The following days the weather changed and became milder, so that by the 13th of December there was hardly a piece of ice to be seen near the Gloucester ferry; it had either melted or been carried away by the water.

December the 12th

Newspapers in North America. There were said to be at present seventeen different newspapers published in the various English colonies in North America; for every week dailies or weeklies appear in Boston, New York, Philadelphia, Virginia and Carolina, and on the islands belonging to the English. Occasionally there are several printed in one city: in New York are both English and Dutch newspapers; in Philadelphia two English and one German. The first publication of this type in Philadelphia was *The American Weekly Mercury* which made its appearance on December 22, 1719. In the beginning it was only half a sheet, folio, in size, and printed by Andrew Bradford [1686-1742]. It has continued up to the present time.[1] In 1728 the second type of newspaper appeared, called

[1] Isaiah Thomas, *History of Printing*, II, 327, believes it was discontinued "soon after" 1746.

The Pennsylvanian Gazette. On October 1, 1728, the following announcement of it was circulated: "Whereas several gentlemen in this and the neighboring provinces, have given encouragement to the Printer hereof, to publish a paper of intelligence; and whereas the late[1] Mercury has been so wretchedly perform'd, that it has been not only a reproach to the province, but such a Scandal to the very name of printing, that it may, for its unparallel'd blunders and incorrectness, be truly stiled Nonsence in Folio, instead of a serviceable News-Paper." The first number by Samuel Keimer [1688-c. 1739],[2] was published on December 24, 1728. Mr. Keimer was a so-called Sabbatarian, who never designated the months as January, February, March, etc. but as the first, second, third month etc., and March was the first month in his system.[3] In place of Monday, Tuesday and so on, he put first day, second day etc. He was very queer and satirized everything so that it is not surprising that he had many enemies. On September 25, 1729, Mr. Franklin and Mr. [Hugh] Meredith began to publish this gazette, since Keimer left and gave up his position to the men just named. Newspapers then received a still better reputation, and as soon as the new editors started to print, they used selected paper and type. Januarius, Februarius, and Monday, Tuesday, etc. were recalled from their exile. The first two years the *Gazette* contained a weekly necrology, but later this section was given up.

December the 13th

Prices Current in Philadelphia 1719, 1720 etc. From *The American Weekly Mercury* I have copied the following prices current in Philadelphia during the years 1719 and 1720, when newspapers here first appeared. The first list is printed on December 29, 1719 [in the second number].

---

[1] The *Mercury* had not in 1728 been discontinued of course, and was not, incidentally, by any means as black, relatively, as it was painted.

[2] See articles on Samuel Keimer and Andrew Bradford in the *Dictionary of American Biography.*

[3] This was common in the old Colonial style of reckoning.

Flour	9s. 6d. to 10s. per hundr.
Middling Bread	14 s. per Hundr.
Brown Bread	12 s. p. H.
Tobacco	14 s. p. H.
Muscovado sugar	40 to 45 sh. p. H.
Pork	45 sh. p. Barrel
Beef	30 sh. p. Barrel
Rum	3 sh. 9 d. p. Gallon
Molasses	1 sh. 6 d. p. Gallon
Wheat	3 sh. 3 d. to 3 sh. 5 d. p. Bushel
Indian corn	1 sh. 3 d. to 1 sh. 8 d. p. Bushel

THE 26th OF JANUARY 1720

Flour	9 to 10 sh. p. H.
White Bread	18 sh. p. H.
Middling Bread	14 sh. 6 per C.
Brown Bread	11 sh. 6 per C.
English salt	3 sh. per Bushel
Tobacco	14 sh. per H.
Muscovado sugar	30 to 45 C.
Pork	45 sh. per Barrel
Rum	3 sh. 8 d. per G.
Molasses	17 to 18 d. per G.
Wheat	3 sh. to 3 sh. 3 d. p. Bush.
Indian corn	1 sh. 6d. to 1sh. 8d.
Bohea Tea	24 sh. p. Pound.
Madera Wine	16 to 20 l. Pipe. [2 hogsheads]
Pitch	16 to 17 sh. p. Barrel
Tar	10 sh. p. Barrel
Turpentine	8 sh. p. H.
Rice	16 sh. p. H.
Pipe staves	3 l. per Thousand
Hoggshead stav.	45 sh. per Thousand
Barrel stav.	22 sh. 6 d. p. Thousand
Gunn Powder	7 l. 10 sh. p. Barrel
Brown oznabrigs	12 d. per Ell.
High coloured Malt	3 sh. 3 d. to 3 sh. 6 d. p. Bush.

DECEMBER THE 14TH

Peas. I have previously mentioned in these *Travels* that but few peas are planted in New Sweden, because worms eat them up before they are ripe. But this is not all. I was shown a lot of peas that had been grown here during the summer of 1748 and had first been left untouched by worms, but which during the past winter, while lying dry in storage, had been so infested by vermin that hardly a pea had escaped. The kernels were so thoroughly consumed that there was hardly anything left but the shells. There was scarcely a pea that had not thus been hollowed out and spoiled.

December the 15th

Peach Trees. Old Kijhn (Keen) gave me the following observation and advice about planting peach trees in a cold country: they should be set out on high land or hills, where they are exposed to the wind. The reason for this idea came from his own experience on his farm in Raccoon, from which he had a view over a large part of the country thereabout. He noticed that one summer night, when frost had damaged almost all peach trees in the whole land so that little or no fruit was obtained that year, that the flowers on his own peach trees had not been injured at all. He received an excellent peach crop that autumn. People came from near and far to buy peaches from him. Those who lived in the valleys right below his own farm as well as others in the country, had their peach blossoms frozen. It is to be noted that Mr. Keen's estate lies on a very high elevation, which consists to a large extent of sand or loam, while in the vales here and there in the locality are marshes with gently flowing water. There is one kind of peach which is ripe two or three weeks before any other, and this is considered the best, both in smell and taste. The later varieties are not deemed so good.

December the 16th

To-day's prices of goods, from *The Pennsylvania Gazette*:

Flour	15 sh. 5 d.
Wheat	5 sh. 6 d.
Indian corn	2 sh. 7 d.
Shipbrood	17 sh.
Midling br.	26 sh.
White bisket	30 sh.
Beef	30 sh.
Pork	60 sh.

Pipe staves	9 l.
Hogshead staves	6 l.
Barrel staves	4 l.
Madera wine	27 l.
Westindia Rum	4 sh.
New England Rum	2 sh. 6 d.
Muscovado sugar	60 sh.
Molasses	20 d.
Coarse salt	2 sh.
Fine salt	3 sh.
Rice	20 sh.
Tobacco	18 sh.
Pennsylvania Loaf Sugar	1 sh. 5 d.
Indigo	7 sh.
Powder	10 l.
Hemp	3 d. halfpenny
Flaxseed	12 sh.
Barley	5 sh. 6 d.

Mortar. When some of the Swedes were building the fireplace in the Raccoon parsonage they mixed fresh horse manure with the mortar, maintaining that this made it better—it held together longer and did not crack. The mortar used here contained more sand than ordinary clay, so that it did not stick well together. In setting up the new Franklin stoves it was the custom at first to mix horsehair in the mortar to make it stick better, but now this has been given up, since it was found that the stoves smelled from it for a time. Perhaps horse manure in the mortar has the same effect, although it is not noticeable where there is no damper and most of the heat goes up through the chimney.

December the 17th

Prices Current in New York 1720. From *The American Weekly Mercury* of January 12.

Flour	14 to 15 sh. p. H.
White Bread	20 to 21 sh. p. H.
Midling Bread	18 to 19 sh. p. H.
Wheat	4 sh. to 4 sh. 6 d. p. B.
Indian corn	2 sh. p. B.
Pease	5 sh. p. B.

Beef	36 to 38 sh. per Barrel
Pork	56 sh. to 3 l. per Barrel
Logwood	12 l. per ton.
Rum	3 sh. 6 d. per G.
Molasses	1 sh. 6 d. per G.
Muscovado Sugar	40 to 45 sh. per H.
Madera wine	24 to 25 l. per Pipe
Pitch	16 to 17 sh. p. Barrel
Tar	13 sh. p. Barrel
Spanish plate	8 sh. 6 d. to 9 sh. per ounce
Pistoles	28 sh. p. piece
Addenda d. 28 Apr.	
Indigo	7 sh. p. P.

December the 18th

Prices Current in Boston 1720. From *The American Weekly Mercury* of February 23.

Pitch	11 sh. p. H.
Tar	22 sh. p. Barrel
Turpentine	12 sh. p. H.
Train oil	36 pound p. Tun, and falling
Fish merchantable	23 sh. 6 p. per Quintal
D:o Jamaica	18 sh. per Q.
D:o Barbadoes	15 sh. per Q.
Barbadoes Rum	5 sh. per G.
Molasses	2 sh. 4 d. p. G.
Cocoa	7 pound p. H.
Beaver skins	3 sh. 10 d. p. pound
Buck and Doe Skins in oil	8 sh. 6 d. per P.
D:o indian dress	5 sh. per P.
D:o in the hair	1 sh. 8 d. per P.
Pine Boardes	55 sh. per Thousand
Flour	28 sh. p. H.
Bread coarse	25 sh. p. H.
Wheat	7 sh. 6 d. p. Bushel
Indian corn	4 sh. p. Bushel

Addenda of the 28th of April

Hops	4 d. halfpenny p. l.
Mackeril	35 sh. per B.
Isle of May salt	24 sh. p. Hog.
Whalebone	4 to 5 sh. p. l.

December the 19th

List of Births and Deaths in Philadelphia, 1722. I found in *The American Weekly Mercury* for the latter part of the year 1721 and the whole of 1722 a list of births and deaths of both sexes for each month during that period, not only of the members of the English Church, of the Quakers and of the Presbyterians, but also a list of the Negroes dead, the number of deaths from accidents, i. e. "casualties", and the number buried in the Strangers' Burying Ground. (See list on next page).

December the 20th

Weather Forecast. The weather to-day was glorious, and later in the evening we had starlight, with the moon appearing in its pure white radiance. Still there were signs of some change, because just before and at noon it was very warm, and not long after it the hens went to roost, as though it had been late in the afternoon. The following night it began to rain right after twelve o'cloock. See the meteorological observations for next day.

December the 21st

The Shortest Days. We were now passing through that time of the year when the northern hemisphere has its shortest days. But we learned that there was a large difference in this respect between the Old and the New Sweden. The sun rose here at 7:15 and set at 4:45. It was possible to read a book without artificial light at seven in the morning and five in the afternoon. But it was very dark at six, both morning and night, and before 7 A.M., and after 5 P.M. it was almost impossible to read by natural light. The frost had not yet had any serious results, for the ground was bare everywhere; but little snow had fallen so far, and what little had come had melted the same day it came, or the day after. The River was still open, and the smaller streams that had been covered with ice a little more than a week before had by the sunshine and recent rains been cleared of the ice again. This neighborhood now had about the appearance of a Swedish countryside in the beginning of October. The cattle went outdoors continually seeking their food, and but little was fed

	English Church				Quakers				Presbyterians				Buried in the Strangers' Burying Ground	Casual- ties	Negroes Dead
	Born		Died		Born		Died		Born		Died				
	Boys	Girls	Men	Women	Boys	Girls	Men	Women	Boys	Girls	Men	Women			
Year 1721 fr. the 21st of July to 25th of Dec.	16	18	26	22	—	—	—	—	—	—	—	—	—	1	—
Year 1722															
Ianuarius	2	1	2	—	—	—	4	3	2	—	1	—	2	—	—
Februarius	3	1	1	2	12	10	2	2	—	1	—	—	1	2	5
Martius	4	4	2	2	10	15	2	4	2	4	1	2	2	—	6
Aprilis	2	3	3	1	9	9	5	1	3	2	—	—	4	2	—
Maius	2	2	4	4	12	16	1	4	1	—	—	—	1	5	3
Iunius	—	1	—	—	13	12	6	2	—	2	—	—	1	1	1
Iulius	2	3	4	2	11	15	2	5	—	—	1	—	1	—	3
Augustius															
September	4	4	9	6	20	18	7	4	4	2	2	3	1	1	3
October	4	2	5	0	13	16	1	4	2	1	1	1	4	3	3
November	2	3	5	2	4	9	3	1	2	1	1	2	—	—	2
December	2	1	2	2	9	11	1	3	1	3	3	2	—	—	3
Year 1723															
Ianuarius	4	3	2	1	18	12	5	1	—	1	1	—	2	—	2

them at home. It may be seen from the meteorological notes that, with few exceptions, the weather had for a long time been fair and pleasant.

December the 22nd

How to Prevent Candles from Dripping. I asked several people how to prevent candles from dripping. I was told [at first] that no remedy was known for it, but that a frequent cause of it was the adulteration of the tallow by lard. Mr. Franklin admitted that he had seen such candles, but that he had found no other remedy for the dripping than to wind a strip of paper round the candle. This would prevent the tallow from running. The paper will burn of course as fast as the tallow but not faster, since the tallow itself hinders it. We tested the suggestion and found it to be true. The candle will burn as brightly as otherwise, but it is necessary from time to time to remove the charred paper. Five sheets of paper suffice for twenty tolerably large candles. This remedy applies only when the candles are stationary; when they are being carried the hot tallow may easily, with an unsteady hand, run down over the paper and fingers and burn them. But a paper-wound candle is not consumed any more rapidly than a bare one of the same size.

Weather Forecast. To-day and for a couple of days following the sunsets colored the clouds in the west red. Mr. Turner, a merchant, said that in this city [Philadelphia] it meant fair weather for the next day, perhaps for several days, with relatively high temperatures. No winds followed as in the old countries.

December the 23rd

Candles from Spermaceti. This evening I had the opportunity to see some candles made of spermaceti. They looked like ordinary tallow or white wax candles, very white, translucent, with a cotton wick. They are made of the fat formed in the brain and head of the whale, a substance which in itself is said to be white and hard; but it needs some scientific knowledge to make these candles. There is a man in town who makes them, but he will not teach others the secrets of the process. The current price in the city for these candles is two shillings, nine pence per pound, while the tallow ones may be had for only eight pence per pound. The spermaceti candles burn as brightly as those of tallow, if not brighter. To-night we made an experiment to see which one of the two kinds would be consumed the faster, using candles of the same size, and learned that they burned about equally fast. The advantages which the new candles have, beside their rarity, are: (1) that they hardly ever need to be snuffed, but continue burning brightly, and the wick bends of itself and drops off as soon as it is burned; (2) they never drip, and though it may be warm in the room, such as when they are used in the summer time, they do not become soft, nor do they melt or soften in the hand when being carried, as tallow candles may do. Mr. Franklin and several others in the city made use of these candles for reading. It was said that there was plenty of them in London, and that they were not so expensive there as here, because more spermaceti was brought to England than to America. Consequently, candles are manufactured here (1) of spermaceti, (2) of various kinds of tallow, (3) of wax, and (4) of *myrica* or candleberries (bayberries), although it is claimed that now a lesser number of the latter are made than formerly. There is said to be a tree in the East Indies that yields wax or tallow from which candles are made.

December the 24th

Raven quills are said to be the best of any to use as picks for plucking the strings of a spinette.

The following *current prices* of goods in the city have been taken from *The Pennsylvania Journal or Weekly Advertiser* for December 12, 1749. It will give a general idea of how to read and understand price lists that have previously appeared in this journal.

By the Hundred		By the Barrel	
Flour	15 sh.	Beef	30 sh.
White Bisket	28 sh.	Pork	60 sh.
Middling d:o	26 sh.	Pitch	14 sh.
Ship d:o	17 sh.	Tar	11 sh.
Muscovado sugar	60 sh.	Powder	10 l.
Rice	20 sh.		
Tobacco	18 sh.		
Turpentine	18 sh.		

By the Pound		By the Bushel	
London Loaf sugar	2 sh. 6 d.	Wheat	5 sh. 7 d to —
Pennsylvania sugar	1 sh. 4 d.	Indian corn	2 sh. 7 d.
Cotton	20 d.	Fine salt	3 sh.
Indigo	7 sh.	Coarse salt	2 sh. 6 d.
		Flax seed	12 sh.

By the Gallon		By the Thousand	
West India Rum	4 sh.	Pipe staves	10 l.
New England Rum	2 sh. 6 d.	Hogshead ditto	6 l.
Molasses	20 20 d.	Barrel ditto	4 l.

		By the Pipe.	
Butter one lb.	1 sh. à 10d.	Madeira Wine	27 l. to 30
Turnips ½ peck	4 d.		

December the 25th

Extracts from *The American Weekly Mercury*. From the library in the city I brought home the first old newspapers of Philadelphia to make extracts about North American events that might have special interest.

Governor William Keith, Esq. of Pennsylvania issued on August 10, 1720, a proclamation concerning the administration of justice.[1]

Governor William Burnet, Esq. [of New York and New Jersey and later of Massachusetts], who entered office as chief magistrate on October 3, 1720, was on that date officially congratulated by the clergy and government offices of New York.

Under the date of February 7, 1721, the paper enters a complaint and protest over the number of criminals sent over from England to settle these colonies.

On February 20, 1721, Governor Burnet visited Philadelphia and was much eulogized everywhere.

New York, February 27. "Five or six weeks ago [came] the deepest snow fall at Albany that has been known for many years, and . . . all the fine Weather we have had, hath been winter-weather there."

[1] Unless given in quotation marks, the extracts are the editor's English translations of Kalm's summaries in Swedish. Only a few are literally copied from the *Mercury*.

Philadelphia, September 28, 1721. "Several Bears were seen yesterday near this place, and one killed at Germantown, and another near Derby." One was killed when he was eating acorns in an oak.

Philadelphia, December 19, 1721. The river is so full of ice that no boats can enter the harbor. A similar item is repeated under the date of the 26th.

1722

Philadelphia, January 2. The river is frozen, so that no ships can move in or out. The same condition is announced under date of January 9th and 16th. In the middle of February boats began to leave.

All numbers for the years 1721 and 1722 are full of reports about pirates. The high seas were full of them, declares the weekly; ships were plundered, and mercantile trade almost stopped. The sea-robbers captured any ships they wanted.

On December 11th we are told that Monsieur Vaudreuil, the governor-general of Canada, had sent messages to the savages north of New England, inciting them to war against the English, and that the Englishmen had secured a copy of the message and sent it to the King of England.

The paper says nothing about the [Delaware River] being covered with ice in December of this year, nor is the matter mentioned during January of the following year, 1723.

1723

On the 30th of July a terrible storm raged over New York, coming first from the northeast and gradually shifting to the southeast. The rainfall was very heavy. "The water came up into the City higher than ever was known before." Many roofs were blown off and landingplaces damaged. "It . . . broke up all the Wharfs from one end of the city to the other, drove all the Vessels (except three) on shore", and spoiled wares stored in cellars. A severe storm visited Philadelphia on the same day, coming from the same direction. Many chimneys were hurled down. The storm "occasioned such a high tide here as has not been known there many years; the storm continued about 2 hours and a half; it has blown down a great many trees and very much damaged the fruit." In the number for August 22nd, it is reported that the storm did not reach Bermuda.

December the 26th

A Continuation of the Extracts from the American Weekly Mercury.

February 1, 1726. No ships can enter the city because of ice. There is still much drifting ice on the fifteenth of the month. —On June 23, Patrick Gordon Esq. governor of Pennsylvania, arrived in Philadelphia. Two weeks previously the newspaper had printed the former Governor Keith's farewell address to the people of Pennsylvania, in which he sharply takes them to task for neglecting the interests of the King.[1]

Philadelphia, February 14, 1727. "We have had very cold weather for these four days past, which has filled our river full of ice." Ships that ventured out were obliged to return.

January 23, 1728. "We have had very hard weather here for near this two weeks past, so that it has froze our river up to such a degree that people go over daily, and they have set up two booths on the ice, about the middle of the said river." On the 30th the river was still frozen, but not by the 6th of February.

New York, January 29, 1728. No ships have passed in or out of the harbor for two weeks, because of the ice in the River. People have walked across on the ice from New York to Long Island.

N. 452. "At Boston upon the Lords day Aug. 11:th, 1728. p. m. a noble Rainbow was seen in the Cloud, after great thundering and darkness and rain; one foot . . . stood upon Dorchester neck, the Eastern end of it; and the other foot stood upon the Town; it was very bright, and the reflection of it, caused another faint rainbow to the westward of it. For the entire compleatness of it throughout the whole arch, and its duration, the like has been rarely seen. It lasted about a quarter of an hour. The middle part of it was dis-

---

[1] This was exactly half a century before the Declaration of Independence. Kalm was one of the first men of importance to note the budding neglect by the English government [of the colonists] and their growing discontent with it, though he did not always sympathize with the viewpoint of the colonists.

continued for a while; but the former integrity and splendor were quickly recovered."

December the 27th

A Curious Phenomenon. The American Weekly Mercury N. 122, Newport, Rhode Island, March 30, 1722. There has lately a surprising appearance been seen at Narraganset, which is the occasion of much discourse here, and is variously represented; but for the substance of it, it is matter of fact beyond dispute, it having been seen by abundance of people, and one night about 20 persons at the same time, who came together for that purpose. The truth, as near as we can gather from the relations of several persons, is as follows. This last winter there was a woman died at Narraganset of the small pox, and since she was buried, there has appeared, upon her grave chiefly, and in various other places, a bright light as the appearance of fire. This appearance commonly begins about 9 or 10 of the clock at night, and sometimes as soon as it was dark. It appears variously as to time, place, shape and magnitude, but commonly on or about the grave, and sometimes about and upon the barn and trees adjacent; sometimes in several parts, but commonly in one entire body. The first appearance is commonly small, but increases to a great bigness and brightness, so that in a dark night they can see the grass and bark of the trees very plainly; and when it is at the height, they can see sparks fly from the appearance like sparks of fire, and the likeness of a person in the midst wrapt in a sheet with its arms folded. This appearance moves with incredible swiftness, sometimes the distance of half a mile from one place to another in the twinkling of an eye. It commonly appears every night, and continues till break of day. A woman in that neighbourhood says she has seen it every night for these six weeks past.

December the 28th

The cold now became noticeably severe; the wind that blew was biting, and the River full of floating ice. A ship went down the River this morning, but encountered so much ice that it was compelled to turn back and either postpone the trip until a milder temperature should drive away the ice, or until spring.

December the 29th

Price list of goods in New York according to the New York *Gazette* of December 22 (Dec. 11 O. S.), 1749.

Wheat per Bushel	6 sh.	Molasses	1 sh. 9 d. per G.
Flour, per H.	17 to 18 sh.	West India Rum	3 sh. 9 d.
Milk bread	40 sh.	New England "	2 sh. 6 d.
White d:o	30 sh.	Beef, per Bar,	34 sh.
Middling	24 sh.	Pork	2 l. 16 sh.
Brown	17 sh.	Flax see	10 sh.
Single refin'd sugar	16 d.	Bohea Tea	6 sh. 6 d. by the box
Muscovado sugar	50 sh. p. C.	Indigo	8 sh.
Salt, per Bushel	2 sh. 3 d.	Chocolate	24 sh. per Doz.

Season Forecast. In the above-mentioned *Gazette* I read the following: "Annapolis in Maryland, October 25. We are told by people from the back parts of this province, that they have had great numbers of Bears and other wild beasts, come down among them this fall; which they look upon as a certain token of an approaching hard winter."

December the 30th

Pastor Dylander. Here I shall record a few facts about the late Johan Dylander, pastor of the Swedish church in Philadelphia, generally called Wicaco [now Southwark, Philadelphia], before the present incumbent of that office, Mr. Gabriel Näsman. When Mr. Dylander first arrived here there was no German minister, though a large number of Germans had settled in this locality. Consequently he was requested to preach to them occasionally, whenever the duties of his own parish would allow it. He travelled about among the Germans, therefore, preaching to them and suggesting, first of all, that they designate certain central localities where they might congregate for divine service. In Philadelphia he gave three sermons every Sunday: first one in the morning, in German; then at ten o'clock a second one in Swedish; and finally, after dinner, a third one in English. At every one of these sermons the church was full, but this was especially the case at the English service, when so many people appeared that some of them had to stand outside the doors and windows of the church. The Germans came to church in a body on Sunday morning with their psalmbooks under their arms. Some of the Swedes disapproved of having the Germans come to their temple, and did not like that Mr. Dylander gave them permission to come. Therefore Mr. Dylander [eventually] persuaded the Germans to hire a big old building in the city, where he attended to their spiritual needs. As a result he had no opportunity to accomplish anything else during the short time he was here from [1737 to 1741] than continually to travel around preaching, baptizing, administering the sacrament of the Holy Communion, etc. Once a month he went up to Lancaster to hold service, although this was far from Philadelphia. He had the German church in Germantown built, and preached in it at least once a month. He often said that in one week he gave sixteen sermons. He was incredibly beloved by all, high and low, rich and poor, by Englishmen, Swedes, Germans, and by members of almost all denominations. The reasons for this were not only his divine teaching and exemplary life but also his social affability, for in intercourse he had a special gift of being pleasant and entertaining with innocent and diverting sallies of wit. Mr. Dylander was so highly respected by the English that not only the most prominent in the city, like the proprietor, governor and others, visited him, but none felt well married unless the ceremony had been performed by him. His last illness came on the same afternoon that he had attended [officiated at?] a funeral. He rode from that to Mr. Kock's but soon felt so ill that he was compelled to go home that night and take to his bed. This was on a Monday. When Mr. Kock visited him a few days later he did not seem to be fatally ill, but he himself was certain that he would never get well, and requested that Rev. Mr. Petrus Tranberg administer the Holy Communion of the following Sunday to those of his church who were entitled to it, for, he said, he would never again perform that service. When he was thus seriously ill a bridal couple appeared to be married by him, and his wife sought to prevent it because of his weakness, but when he saw the wishes of the party concerned he bade them come forward, for he desired in this manner to end his official work on earth. So he had pillows placed behind his back, took the missal in his hands, which he held open, and read the ceremony by heart and gave them his blessing. His sickness was yellow fever, which he had had three months before when an epidemic of it killed so many people in this country, and had almost recovered from it when he again fell a victim to it, and this time he was forced to succumb. His funeral was one of the best attended in the land, and the services were conducted by Mr. Tranberg. There was such a crushing throng in the Swedish Church that many of the foremost ladies of the province lost large pieces of their skirts when they tried to get in; in fact some lost half of their skirts, an apron, or some other garment. Many of the men of rank had to climb in through the windows, and over the ladies, to get to their pews. He had the reputation of having in two years learned to speak better English than Mr. Tranberg, who had been in the country for several years. He spoke so well that it was difficult to notice that he was not a born Englishman or at least born among the English. If he had lived a couple of years longer, he would have returned to Sweden, for he maintained that he did not have the right conscience to remain in this country, where people cared so little for true godliness, especially his Swedish congregation, which cared least of all. A little before his death he lost a little prestige among a few people because of the following incidents. Mr. [George] Whitefield came over [from England] about that time, began to preach, and with his eloquence won many followers. He had accepted the Calvinistic dogma on election of grace, whereupon Mr. Dylander translated the work of Gerardius [Gerhard] [on the same subject] into English and had it published, with his own notes and commentaries.[1] Since this appealed considerably to all right-thinking minds among the Lutheran and English Churches, the Whitefieldians believed themselves offended. Otherwise Dylander was so popular among the English that he sometimes preached in the English Church in the city, and on those occasions the church was always crowded if the congregation knew in time that he was to appear.

Here follows a [brief] notice that was published in *The Penn-*

[1] *Free Grace in Truth: The XXIVth Meditation of Dr. John Gerhard Translated from Latin into English, with Notes for the Better Understanding of the Author's Meaning*. Printed by Benjamin Franklin, 1741.

sylvania Gazette for November 5, 1741 [relative to Mr. Dylander].
—"Monday last died of a lingering indisposition, the Reverend Mr. John Dylander, Pastor of the Swedish Church, at Wicaco, near Philadelphia. His sincerity, affability, and other good qualities make his death lamented by many." [1]

DECEMBER THE 31ST

Pestilence. I have asked several people whether or not there ever had been a pestilence in this country. All agreed that they had never heard of any. It is claimed that certain insects are the cause of pests. This is maintained in a volume published in London in 1738 under the name of *A National Account of the Weather* [2] etc. and where the last article entitled "The Cause of the Plague" contains the following words: "It has been observed that Plagues and the most contagious distempers, have commonly happened in those years, when the Easterly Winds have more than ordinary prevail'd in the spring and summer seasons; then the air comes to be infected, and rarely or never at other times." If it is true that there are but few easterly winds in this region, and the few that exist come over the ocean where such insects necessarily must die—because the sea air is so different from all other,—it would not be surprising if no plague came here. And all other winds in this vicinity are salubrious. . . . The author asserts also that India, China, and the southern parts of Africa and America have never had a real pest, for the same reason. But he is wrong when he claims that none of the plants and insects found in Europe are to be seen in America. These *Travels* can demonstrate the contrary. He is of the opinion that these pestilential insects first came from Tatary, where they are born in the swamps and the stagnant waters of the wilderness.

JANUARY THE 2ND, 1750

The cold was now quite severe. The Delaware River had yesterday been covered by ice near Philadelphia, and to-day the whole river

[1] It should be remembered perhaps in reading this rather long account of Mr. Dylander that Kalm himself was later ordained in the Lutheran Church and eventually became a doctor of divinity.

[2] The title should read *A Rational Account of the Weather,* by John Pointer, 1723; 2nd ed. 1738.

was full of boys, girls and older people moving about upon it, the majority of them skating. But at Gloucester the River was still open, because it is narrower there and the current therefore swifter. On the third of the month several shoppers walked over the River, but on the fifth a large part of it was so open again that horses were being carried across in boats.

JANUARY THE 3RD

Inundations in North America, From *The American Weekly Mercury,* Philadelphia, June 29, 1721.—"It seems the rains of late have been very violent up in the country, especially to the Westward of the Skullkill river, by which there came down such a sudden fresh, that in some places the water rose 20 feet perpendicular from its usual bounds in a few hours time." Many cattle were drowned, mills and bridges carried away, and dams torn down. "A large stone bridge near pennypack mill is wholly destroyed. It is esteemed to be the greatest fresh we have known here these twenty years, & the most suddain & unaccountable, because we have not had such unusual rains."

November 8, 1722. "We have news from South Carolina, that a storm began there the 9th of September last, which continued in all 5 days. The rain was more violent than the wind, doing considerable damage to the corn & rice & carried away some houses and cattle in the country. The water rose upwards of 30 feet more than usual."

JANUARY THE 4TH

Cold Temperature in North America. From *The American Weekly Mercury* of January 15, 1722, a communication from New York. "It is excessive cold, and the river full of ice from the narrows to New York. Yesterday a great many people went upon the ice from New York to the Ferry on Long Island."—Mr. Edw. Holley's thoughts of the reason for the severe cold in North America may be seen in *The Philosophical Transactions,* No. 363.

January the 5th, 1750

(Christmas Day, 1749, Old Style)

To-day Christmas Day was celebrated in the city, but not with such reverence as it is in old Sweden. On the evening before, the bells of the English Church rang for a long time to announce the approaching Yuletide. In the morning guns were fired off in various parts of the town. People went to church, much in the same manner as on ordinary Sundays, both before and after dinner.[1] This took place only in the English, Swedish, and German churches. The Quakers did not regard this day any more remarkable than other days.[2] Stores were open, and anyone might sell or purchase what he wanted. But servants had a three-day vacation period. Nowhere was Christmas Day celebrated with more solemnity than in the Roman Church. Three sermons were preached there, and that which contributed most to the splendor of the ceremony was the beautiful music heard to-day. It was this music which attracted so many people. It must be emphasized that of all the churches in Philadelphia only the Swedish and the Catholic possessed organs. There had formerly been one in the English temple, but it had later become useless, and there had not yet been any measures taken to procure a new one. The organ in the Swedish church had also through improper care become worthless. Consequently an organ was to be heard only in the papal place of worship. The officiating priest was a Jesuit, who also played the violin, and he had collected a few others who played the same instrument. So there was good instrumental music, with singing from the [back] organ-gallery besides. People of all faiths gathered here, not only for the high mass but particularly for the vespers. Pews and altar were decorated with branches of mountain laurel, whose leaves are green in winter time and resemble the lauro-cerasus (cherry laurel). At the morning service the clergyman stood in front of the altar; but in the afternoon he was in the gallery, playing and singing.

[1] In Sweden Christmas was, until recently at least, regarded as a very sacred religious festival. Service is still held in all Swedish churches in Sweden and elsewhere at five or six in the morning on Christmas Day, and no unnecessary work or visiting is done on that day.

[2] Cf. diary for Jan. 5, 1749.

There was no more baking of bread for the Christmas festival than for other days; and no Christmas porridge on Christmas Eve.[1] One did not seem to know what it meant to wish anyone a merry Christmas. However, [after I had written this] I heard several members of the English Church wish one another a happy Christmas holiday. In the English church a sermon was preached in the morning; but after dinner only a prayer meeting was held, and on the day after Christmas again, only a prayer meeting.[2] But, as I have already noted, the Quakers paid not the slightest attention to Christmas; carpentry work, blacksmithing and other trades were plied on this day just as on other days. If Christmas Day falls on a Wednesday or Saturday, which are market days, the Quakers will bring all kinds of food into the market as usual; but no others will, and only Quakers will buy anything of them on such a day. Others make provisions so that purchases will prove unnecessary until the first market day after Christmas. The same custom is observed at New Year's. At first the Presbyterians did not care much for celebrating Christmas, but when they saw most of their members going to the English church on that day, they also started to have services.

JANUARY THE 7TH

Animal Diseases in America. The American Weekly Mercury, in a communication from New York, September 4, 1721, says: "A mortal distemper is got among the horses this way; many hundreds are dead and dying daily. There are 200 and odd dead in the town of Hackinsack, and as many in several other towns. There are 250 dead at Elisabeth town, and there abouts."

New York, September 11, 1721. The distemper among the horses continues, and spreads upon Long Island and Westchester County; & not only horses, but many neat cattle and hogs are dead, and continue to die with the same distemper.

---

[1] In Sweden, where the Christmas season lasted for three weeks, there was enormous gastronomic preparations made of both solid and liquid material for these holidays. Special drinks and special dishes were an imperative part of the menu and general celebration.

[2] In the Swedish church calendar the day after Christmas is also a religious holiday with regular morning service. The same is true of the Easter holidays.

JANUARY THE 8TH

Customs. It was a custom here among the English to express good wishes to a newly-married couple by paying them a personal visit during the first week or first month of their married life. The nearest relatives and friends would come and say "I wish you joy." Men generally made these visits before noon; the women, in the afternoon. The men received each a glass of wine; the women, wine and tea. Upon leaving, each guest was given a piece of wedding-cake [1] done up in clean paper, which he brought home with him. If the cake was not provided, the bride was either considered stingy or ignorant of *savoir-vivre.* The bridal couple was then in duty bound to return all calls by the well-wishers. Failure to return a visit was held as a sign that further visits from the party in question were undesired. In order to receive these visits and felicitations the bride and groom, and especially the bride, were obliged to remain at home for two weeks after the wedding, and to be always dressed after lunch. If she did not stay at home but went out, she was considered afraid of visits and unwilling to receive them. The wedding-cake was made of eggs, flour, butter and sugar mixed and thoroughly beaten, with some sweetmeats added. At baptisms, also, the guests were regaled with tea and wine, and the same kind of cake was served, in the same way, at the exit. The tea and the cake were given out by the woman who attended the young mother; [2] and politeness required that a gratuity of four or five shillings or even a dollar,[3] be given to her. At weddings, however, no tips were given. An average wedding cake cost 30 shillings [or about four dollars].

JANUARY THE 9TH (1750)

The Delaware River was now frozen in most places at Philadelphia. For the last three days there had been a large number of young men

[1] This is very free translation of the Swedish *puderkaka,* made of powdered brown sugar, etc.

[2] The christening took place probably while the young mother was still in bed, since early baptisms were not uncommon. Kalm's Swedish version is not absolutely clear on this point.

[3] The Pennsylvania shilling at this time was worth about 13 1/2 cents, so that it took between seven and eight to make a dollar.

and boys on the ice, some walking but most of them skating. There was still an open place here and there in the middle of the River; nevertheless, to-day at eleven I saw a man successfully driving a horse and sleigh on the ice directly in front of the city. The next day was mild and beautiful, when a section of the ice before the town suddenly broke up and began to move downstream. There were a good many people on this piece of ice: booths had been set up to sell brandy and such things to the skaters, and now they all found something else to do besides enjoying themselves. People rushed away precipitously, and fortunately all reached *terra firma* safely. The ice remained, but for a few days no one dared go out on it. There had been some people on the other side of the river starting to cross when the ice began to loosen, but these were obliged to turn back. On the 13th of this month the river was wholly open again so that ships could move in and out. The English youth is very fond of skating, and so are men of thirty years or over. Men of all classes have a passion for this sport. They would sometimes go three or four miles to reach a place where the ice was safe. Sheltered spots were flooded with men skaters, but I saw no women on the ice here.

January the 10th and 11th

Much Snow in America. The American Weekly Mercury N. 272. Philad. March 3, 1725. We hade such abundance of snow fall here yesterday, and last night, that it's near two foot deep. which has not been known here for some years past.

Earthquakes in North America. The American Weekly Mercury N. 237, Boston, June 15, 1724. On Thursday morning last some shocks of an Earthquake were felt here, by a considerable number of persons in different parts of the Town.

Ibid. N. 242. Philadelphia, Aug. 6, 1724. We had a small convulsion of the earth (or Earthquake) last night about the hours of 9 or 10 of the clock, which lasted about half a minute and was felt by many people in this city.

N. 409. Philad. Nov. 2, 1727. We have advice from New York, that on sunday night (:d. 29 Octob.) about ten a-clock they had a shock of an Earthquake, and about Two they felt a second shock which shook the pewter from off the shelves, and the China from off the cupboardsheads & chimney pieces, & set all the clocks a running down.

N. 412. Boston, Nov. 6, 1727. On the 20:th past, about 30 minutes past 10 a night, which was very calm & serene, and the sky full of stars, the town was on a suddain exceedinly surprised with the most violent shock of an Earthquake that ever was known. It began with a loud noise like thunder, the very earth reel'd and trembled to such a prodigious degree, that the houses rock'd and shook in so much that every body expected they should be buried in the ruins. Abundance of the inhabitants were wakened out of their sleep, with the utmost astonishment, & others so terribly afraighted, that they ran into the streets thinking themselves more safe there, but through the infinite goodness and mercy of God, the shock continued but about 2 minutes, and tho' some small damage was done in a few houses, yet by God's great blessing, we don't hear that any body received any hurt thereby. There was several times till the next morning heard some distant rumblings of it, but since then the earth has been quiet, tho' the minds of the people have still a great & just terror & dread upon them. On the next day prayers were offered in almost all the churches and the day was set apart as a public fast-day.

N. 420. Boston, Dec. 7, 1727. We hear from Newbury, that last week, *viz.* on wednesday and friday, they had there the repeated shocks of an Earthquake.

N. 422. Boston, Dec. 28, 1727. By Capt. Cooper late from Barbados, we have advice, that the Earthquake we had here October the 29:th, about half an hour past 10 in the evening was felt there the day before about noon; which is nigh 2000 miles from this place. The houses were in great convulsion, and the streets arose and fell like the waves of the sea, so that they were afraid the earth would sink under them, and they ran down to the wharves, to get into boats and vessels for their safety.

N. 428. Marblehead, Jan. 31, 1728. Yesterday between 1 & 2 a clock p. m. we had a terrible shock of an Earthquake, which began with a rumbling noise like the rolling a log over an hollow floor, & increased until it seemed like the discharging of several cannon at a distance; at which time the earth trembled so as to jar the pewter on the shelves in many houses; the whole shock lasted about 50 seconds. It's thought that had this shock been in the night in still weather, it would have appeared the greatest since the great shock on the 29 of October. This is the third shock we have had within these six days last past; and about the 30th since the 30th of October last.

EXTRACTS FROM THE MANUSCRIPT CONTINUATION OF KALM'S TRAVELS IN AMERICA

European News. A few moments ago I received the newspapers that were published to-day. In these I learned, from a dispatch of September 30, 1749, that the minister of the Russian Czarina[1] in Stockholm had delivered a memorial to the Royal Swedish Court requesting the permission to station troops in Finland, since she was a guarantee of Swedish freedom.[2] As soon as I had read but half of the Czarina's insolent, damnable and super-immoral demand, I became so angry, I must confess, and my blood circulation so violent, that every limb in my body shook for an hour as if I had had the ague, and I could not read or write a word. It would be better if the devil removed the whole gang of sympathisers with the Russian Czarina in Sweden, where she commands a few demons, or that they were all hanged in a gallows, rather than let them work their own will. I venture to believe that many a satan has heartily rejoiced at this damned shamelessness on the part of the Czarina. But there must be some honorable Swedish men left. God bless and keep His Royal Highness [Frederick av Hesse—Cassel], the apple and delight of all righteous Swedish eyes, Her Royal Highness [Ulrica Eleonora], and the whole Royal House! May our Lord also further His Excellency Count [C. G.] Tessin's useful projects for Sweden![3] May God induce many others to follow his footsteps! (February 3, 1750, N. S.)

The language of the Indians is difficult to speak, as you can see from the following *Nummatchekodtantamoonganunnonash*, which

[1] Elizabeth.

[2] The last war between Sweden and Russia had ended in 1743 with the treaty of Åbo. Kalm represents the turbulent feelings of a Swedish Finn toward Russia in 1750.

[3] Count Tessin was the contemporary champion of Swedish national independence and of the economic, scientific and artistic development of his country. He sponsored especially the friendship of France and Sweden at the expense of Russia, a policy which had a strong appeal to Kalm.

signifies *our lusts. Kummogdadonattoottummooc iteaon gannunnonash* means *our question.* Mr. [John] Eliot [apostle to the Indians], the first English clergyman in New England, learned this language from an Indian man-servant and later wrote a grammar of the dialect. He began his mission in October, 1646. (February 14, 1750).

Swedish seamen came often to Philadelphia on the English ships, either as boatswains, ordinary seamen, carpenters, gunners, or second mates. For the most part they performed the duties of a gunner. I was told that during the war between France and England, which terminated with the peace of 1748 [Aix-la-Chapelle], there were seventeen English privateers in North American waters and that the gunners on all of them were Swedes. All the English sea captains assured me that there was a large number of Swedes on the English merchant vessels, but that the number on the Dutch ships was still larger. They could earn quite a good sum here in a year, but many of them understood but little of economizing, and as soon as they landed directed their steps to saloons and gambling dens, where most of their pay was spent. Those who served on board the English ships especially did not become rich; for the British sailors were accustomed to live well and have no care for the morrow, and the Swedish seamen imitated them. On the other hand, those in the Dutch employ usually became well-to-do; because Holland is a parsimonious nation that knows the value of money, and gradually the Swedes working for it acquired that same quality. A large proportion of the Swedish sailors on foreign vessels were such as belonged to the Swedish navy or admiralty, and who had either obtained leave to enter service elsewhere for outside training in order to serve their own country skillfully at some later date, or had simply left without official permission. Almost all the Swedish sailors whom I met in America, and particularly those that belonged to the Swedish naval department, had neither thought nor desire of ever returning to Sweden. When it was pointed out to them that it was their duty to serve their fatherland in time of need, they always answered, as if with one voice, that they felt no obligation to do so, since their native land had been and continued to be ingrateful for their services. No matter how faithfully they had served or would serve they were never promoted to any higher position, and wholly ignorant and untrained youths were always given the preference instead of them. In fact, [the officials] had generally promoted such as hardly knew the name of a single rope or anything else that belonged to a ship, and when these commanded they had been obliged to keep a paper fastened on their arm on which were written the words of command. Nor did these green officers dare to take their eyes away from the paper. Yet they had treated their subordinates severely, not much better than if they had been dogs, and given them only spoiled food, although these subordinates could easily have been the teachers of the others so far as seamanship was concerned. (April 28 and May 17, 1750).

In one place we passed a house built of clay. It had been constructed according to a half-timber method with a cross-work of wood on which the roof had been laid. Between the wood the spaces had been filled with clay, which then had been allowed to dry. (May 5, 1750).

Swedish vs. English. In the morning we continued our journey from near Maurice River down to Cape May. We had a Swedish guide along who was probably born of Swedish parents, and was married to a Swedish woman but who could not, himself, speak Swedish. There are many such here of both sexes; for since English is the principal language in the land all people gradually get to speak that, and they become ashamed to talk in their own tongue, because they fear they may not in such a case be real English. Consequently many Swedish women are married to Englishmen, and although they can speak Swedish very well it is impossible to make them do so, and when they are spoken to in Swedish they always answer in English. The same condition obtains among the men; so that it is easy to see that the Swedish language is doomed to extinction in America; and in fifty or sixty years' time there will not be many left who can understand Swedish, and still less of those who can converse in it (May 7, 1750).

"Irish Bull." An Irishman had recently written a letter to a fellow countryman, in which he had also enclosed a copy of the same letter, with the postscript that he was sending both, for fear that in these troublous times one might be lost, and if the original didn't arrive the copy might. (May 17, 1750).

The Indian Catechism. The Rev. Mr. Lars Nyberg[4] told me that he had taken a copy of the Indian Catechism, which had been sent over from Sweden, and tested it on the savage natives of Virginia to see whether or not they could understand it, since it was said to be translated into their dialect.[5] But they could not comprehend a word of it, whereas it was found upon examining the Indians on the Delaware River that they understood a good deal. The late Mr. Peter Kock asserted several times that it was stupid to have spent so much money in the printing of the Red Man's Catechism, since it had not been translated into the right Indian dialect (May 21, 1750).

Smilax herbacea. The flower of this plant has the most disagreeable odor in the world, for it smells like a dead snake, and when you have once smelled of it the terrible odor will never leave your nose (June 7, 1750).

Honorary Titles. The following resolution by his Royal [Swedish] Majesty in regard to a petition by Bishop [Jesper] Svedberg [father of Emanuel Swedenborg] was graciously granted in Stockholm on April 1, 1721: the title of Magister,[6] in accordance with the request in the petition, was bestowed upon Dean Eric Björk of Falun and Dean Anders Sandel in Hedemora [both formerly pastors of New Sweden], with the further provision that they henceforth be excused from praesidium duties at ministerial conferences. After due consideration, too, of all the reasons set forth, His Royal Majesty likewise conferred the same honorary title on the preachers Abr. Lidenius and Samuel Hesselius, who now have charge of Swedish congregations in America (June 10, 1750).

Skin Gifts. It appears from Bishop Svedberg's letter that some people in America had sent skins both to the King of Sweden and to the Bishop, for which the Bishop at first expressed his thanks. But when the colonists continued sending the skin gifts he took offence, and wrote that he was attending to his duties from love for

---

[4] Later a member of the Moravian Church.

[5] Johan Companius's "Vocabularium Barbaro . . . Virgineorum" had appeared 1696 in Stockholm at Royal expense, in an exquisite monographed binding, together with a translation of Luther's Catechism into the Lenape dialect.

[6] The nearest American equivalent here is Master of Arts, but occasionally it corresponded more to our honorary doctor's title.

the Swedes and not for love of the skins, which he did not want (*ibid.*).

More about the Indian Catechism. In Christina, or the Swedish Church in Wilmington, there are something over a hundred copies of those catechisms which had been published in Sweden in the Swedish and Indian tongues and later sent over here. They are bound in a French binding and printed on excellent paper, which shows what an act of vanity it was. I was told that Dean Anders Hesselius had written home to Bishop Svedberg informing him that he had converted several heathen savages, when as a matter of fact, only one single Indian had through this means (*taliter qualiter*) been converted (*ibid.*).

The doors in the Ephrata Protestant Convent, about thirteen or fourteen miles from Lancaster [Pa.], are so narrow that only one person can pass through at a time, and if he is fat he cannot get in at all. Our Royal Councillor Cedercreutz[7] would therefore have to stay out. The doors are made of a single board of the *Liriodendron tulipifera* or tulip tree (June 13, 1750).

This same convent had its own printing press. Among the books that had been printed there was a German translation of the Gospel of Nicodemus (June 14, 1750).

In the city of Lancaster the Town Hall is located almost in the center of the town and round about is the marketplace, about the same as in Fredrikshamn, [Finland] (June 16, 1750).

King Charles XII Sends Books to New Sweden. In 1707 King Charles XII sent over [to New Sweden] a large number of books, Bibles, psalmbooks (hymnal, missal and prayerbook combined), A B C-books, and works of religious meditation. Half of these were sent to the Christina Congregation in Wilmington. Payments for them were to be made to a widow of a former clergyman (Anders Rudman?) (June 20, 1750).

Skins. The following year something over thirty skins, most of them mink, were dispatched to the Royal Secretary Peringer Liljeblad for his trouble with the packing and sending of the books, as decided in the parish meeting of May 13 (*ibid.*).

In a letter dated at Brunnsbo the 24th of November, 1714, Bishop Svedberg again acknowledges gratefully the receipt of the skins

[7] Herman Cedercreutz (1684-1754), count and diplomat, was noted for his obesity.

sent by the congregations, and specifies that the skins sent to the King are still in his custody (*ibid*).[8]

The same year, at the departure of Mr. Eric Björk [for Sweden], the Swedes sent 115 skins with him including pelts of the bear, mink, cat, fox, muskrat, wolf, otter, panther and wild cat. Some were sent to the King in Sweden, and others to Bishop Svedberg. Those belonging to the King were forwarded to Pomerania, where they were confiscated by the [German] enemy at the surrender of Stralsund [1715] (*ibid*).[9]

On the fifth of December, 1719, Bishop Svedberg again expresses his thanks for skins sent (*ibid.*).

About ten years later, on November 19, 1729, in his letter of recall to Rev. Jonas Lidman, Bishop Svedberg writes: "Bring back something nice with you, especially pelts, as Magister Björck did." Later, in a letter to the congregation of October 15, 1731, he thanks them for the skin gifts which they had sent him, but which he had not yet received (*ibid.*).

Phlox maculata had the peculiarity that when its seed vessels were thoroughly dried they would burst with a crack, leave their paper enclosure and scatter all over the room (July 3, 1750).

Origin of the Indians. Concerning their origin the Indians had the following legend: A large turtle floated on the water. Around it gathered more and more slime and other material that fastened itself to it, so that it finally became all America. The first savage was sent down from heaven, and rested on the turtle. When he encountered a log he kicked it, and behold, people were formed from it. In every city (of the Red Men) there is ordinarily one family which takes the name of "Turtle" (July 30, 1750).

Impatiens (*noli tangere*) was called "the crowing cock" by the Indians, because of the form of the flower (August 8, 1750).

The Dog. An old Indian said that when God had created the world and its people, he took a stick, cast it on the ground, and spoke unto man, saying, "Here thou shalt have an animal which will be of great service to thee, and which will follow thee wherever

---

[8] Charles XII was at this time, as always, busy with his numerous wars, though he returned to Sweden from Turkey in 1714.

[9] Frederick William I of Prussia declared war against Sweden in 1715 and in December captured Stralsund in Swedish Pomerania.

thou goest," and in that same moment the stick turned into a dog (August 11, 1750).

In number 1136 of the *Pennsylvania Gazette*, for September 20, 1750, Mr. [Benjamin] Franklin published under my name my whole article on Niagara Falls (October 3, 1750).[10]

Herr Adolph Benzel (later Benzelstjerna), the son of the late Archbishop Dr. Eric Benzelius, arrived yesterday in Philadelphia on a ship from Holland. He had been absent from Sweden about six years, and had spent most of that time in the French service. He had now come hither to examine this land, and to settle here if he could find suitable employment and living conditions (September 29, 1750).[11]

The Catechism which the followers of Count Zinzendorf had translated into Swedish and published is of historical consequence, because it is the first and only Swedish book printed in America (October 23, 1750).

A brief list of expressions is given below illustrating the present (1748-1750) language and method of speaking in New Sweden, which shows how far one has already deviated from the Swedish that is used in the Old Country. Soon it will be a new tongue and we can see that Swedish will in a short time die out.[12] [The illustrations follow. The English words or stems mixed in with the Swedish are reproduced in *italics*. For the benefit of those who may not read Swedish, a translation is given in parenthesis. Note the custom of placing Swedish endings on English roots, a practice well known of Swedish-Americans of to-day].

A. Denna hästen *amblar* braf. (This horse is a good stepper).

C. Jag vill *considerera* det innan jag *concluderar* något. (I shall consider it before I make a decision).

[10] This was the first description in English of Niagara Falls based on first-hand information. See bibliography of Kalm's writings on America at end of this volume, item 12. Cf. also J. L. Odhelius, *Åminnelsetal öfver Kalm*, p. 24.

[11] Benzelstjerna stayed in America and became a fortification officer and an intendant of forests in the English service. He married the daughter of a Swedish pastor at Raccoon.

[12] Kalm did not dream that a century later there would commence an immigration tide from Scandinavia that was ultimately to bring hundreds of thousands of Swedes to America, and perpetuate the Swedish language in North America for several generations at least.

D. Ni är *desperat* flitig. (You are "desperately" diligent).

G. Jag vill gå och öppna *gäten.* (I shall go and open the gate).

H. Om det skulla *happna* så. (If it should happen thus).

I. Var *isi* och *trubla* er intet. (Be at ease and don't trouble yourself).
Han tog in något, som *isade* honom mycket. (He took something [medicine] which eased [relieved] him very much).

K. Jag *kärar* det intet. (I have no care about it, i. e. pay no attention to it).
Rum voro *kipare* fordom än nu. (Formerly rooms were cheaper than now).
Det är en *klöfver* karl. (That is a clever fellow).
En å (a river) was always called a *kil* by the Swedes, as in *Schuylkill,* and corresponded to *creek* among the English.

L. Jag *lusa* min hustru på sjön. (I lost my wife on the ocean), said the Rev. Sam Hesselius when he came back home to Sweden. The Swedes in this locality used the word *lusa* for *mista* (lose) a great deal.[13]

P. Vill ni *plisa* sitt ner och äta sådant som vi ha? (Will you please sit down and eat of what we have)?
Jag har varit mycket *pårli* denna vinter. (I have been very poorly this winter).
Det är denna tiden ej så mycket vatten *pannorna* som i min barndom. (There is not so much water in the ponds nowadays as there was in my youth).

S. Ni kommer körandes så *smärtli.* (You are out driving in a dashing style, i. e. smartly).—Vill ni *småka* tobak? (You wish to smoke)?—Han *simar* vara en braf karl. (He seems to be a fine fellow).—Jag har legat och *slipat* så länge. (I have been lying down sleeping for such a long time). De begynna *skeda ut*[14] att tala svenska. (They are beginning to give up talking Swedish).—*Stenhäst* (stone horse) was always used for a stallion.

---

[13] The modern Swedish *lusa* means to louse, to free from lice, or delouse, hence the tragic-comic element in Hesselius's statement as written. Kalm evidently represented the sound of *o* in *lose* with a *u.* The modern careless Swedish-American would spell and pronounce it *losa,* reproducing the vowel sound of the English infinitive.

[14] Probably from the Dutch *uitscheiden,* to cease, to stop.

T. Hästen vill *trötta.* (The horse wants to trot).—*Trubla* er intet därom. (Don't trouble yourself about it).—All Swedes called the kitchen garden a *tina,* which is said to have come from the Dutch [*tuin,* garden].

V. Jag *vantar* intet. (I don't want, or need, anything).—Jag vill *väta på* er. (I shall wait upon, or come to see, you). Madame Tranberg used this expression in speaking to the Rev. Mr. Acrelius.—Vill ni komma och ta en liten *vak* med mig? (Will you take a walk with me)?—Mäster Professor, han rider efter och *väter på* frun och magistern. (He will ride behind and wait on your wife and yourself, Professor).—Hästen *väcker.* (The horse shies).[15]

Anglicism. Vill ni *komma våra vägar.* (Do you want, or intend to come our way, i. e. to visit us)?—Så göra de *våra vägar.* (That's what they do up our way, i. e. among our people, at our homes).—Jag *fattas veta* om han kommer, hvad det kostar. (I should like to know what it costs, if he comes).—Han bor just *öfver vägen.* (He lives right across the street, lit. over the way).—The word *karl(en)*[16] is always used for "this man", no matter how distinguished he may be, or whether he be present or not, for instance, Hvarifrån är denna *karlen*? (Where does this man come from)?

Swedish Songs. In Wicaco I received some Swedish songs that had been printed there in the year 1701. There were eight of them, all composed by the Rev. [Andreas] Rudman. They are of two kinds, or on two types of paper, and maybe of two different years, for on one part the title is given before the (two) songs, and the date printed, while the other six are undated. All are printed in octavo, with Latin script, and are said to be not only the first items printed in Swedish in this vicinity (i. e. in Philadelphia), but one of the first writings in any language to be printed here, since before that time there was no printing establishment in Philadelphia and everything was published in New York. (January 18, 1751).

The [*Royal Scientific*] *Society* in London has degenerated con-

[15] *Väcker* is possibly from the stem in the Dutch word *ontwijken,* to shun; Swedish *undvika,* to avoid; German, *weichen*; or simply the Swedish *viker,* sing., pres. ind., of *vika,* to turn or move suddenly, to turn to one side.

[16] German *Kerl,* English *churl.* To-day there is nothing really derogatory about the word in Swedish, but it is not used in polite society about and in the presence of a man from the upper classes. It is used in the sense of "fellow." *Herrn* should be used instead of *karlen.*

siderably since it was founded, and does not have the standing it enjoyed for several years thereafter. In the beginning there were members thoroughly expert in all sciences, and everything published at that time in the *Transactions of the Royal Society* was learned, useful, informative, and quite well prepared. That period may be said to extend from 1670 to 1720, or a little longer, and embraced England's great savants, such as Boyle, Rajus, Tyson, Whiston, Wallis, Halley, Flamstead, Newton, Lister, Sloane, and many others. It is true that many of them lived on after 1720, but very little by them is found in the *Transactions* subsequent to that date. Almost everything printed there during the period noted will be read with pleasure, because it is so well prepared; but later the society seemed to weaken more and more, to tire, to decline, so that that it became difficult to find anything of value in its publication. One had to hunt for it, and occasionally one perused almost a whole quarter of the journal without finding anything worth reading. Among the present members of the Society there are very few of outstanding merit. Sloane is still living, but is so old that he has hardly been able to talk for several years, to say nothing of writing.[17] Much the same condition obtains in the case of Dr. Mead Collinson, who has a scholar's inquisitiveness and is an ardent promoter of the natural sciences, but does not publish anything of his own. Miller [author of the Botanical Dictionary] publishes whatever is of value in his work in separate books, and only sends in to the *Transactions* of the Society his minor articles. The same is true of most of the other members that amount to anything. For this reason the Society has fallen into disrepute, even here in England,[18] so that even the more sensible people speak of it with mockery. Recently, too, the well-known Dr. [John] Hill (c. 1716-1775) published a small pamphlet which is specifically directed against the [Royal] Society, and in which he critically reviews a number of their treatises, showing by numerous examples how wretchedly their articles had been written, and how carelessly they had been printed. Dr. Hill's particular brochure, however, appears to have been written with too great heat of passion.[19] (April 16, 1751).

---

[17] Hans Sloane was at this time 91 years old. He died in 1753.

[18] Kalm was now in England on his return journey.

[19] Hill apparently attacked the Royal Society in more than one article. The *Dic-*

In Gothenburg they are said to have discovered an efficient exterminator of bedbugs, namely fresh coriander seeds stuffed into the holes and cracks where the vermin is found. Others claimed that the leaves of the *Lepidium rudevale* were best for that purpose. (May 5, 1751).

The [Swedish] commercial adviser [Magnus L.] Lagerström (1691-1759) asserted that he had discovered a method of planting and transplanting spruce trees, without running the risk of having them die in the process. It consisted in covering the bottom of the hole where the spruce is to be planted with small stones, placing the tree directly upon the stones, and then pressing the dirt around the roots in the usual way. If the spruce is placed right on the soil without any stones it usually perishes from the transplanting. (May 11, 1751).

The Beginning of Swedish Manufacturing. The councillor of commerce, [the noted Jonas] Alström (later Alströmer) told me that he first went to England in 1707, when he was 22 years old; that he was born in Alingsås; that he about the year 1714 first entertained the plan of introducing factories in his native land, since, travelling about in England a great deal, he had been employed by Swedish merchants in sending a large amount of English-made goods to Sweden. He therefore visited the British manufacturing establishments, to which he had free access, and studied them diligently. In those days he was allowed to inspect everything because they [the factory directors] were not then jealously apprehensive [of competition]. Thereupon he returned to Sweden when King Charles XII came back from Turkey (1714), thinking that there would immediately be peace, and that thoughts would be turned to industry. But when he came home he found that the war was waged more violently than ever, and that the subject least considered was factories. Consequently, a little while afterwards he

---

tionary of National Biography says: "Failing to obtain the requisite number of names for his nomination to the Royal Society, he attacked the society in several satirical pamphlets, specially vituperating Folkes and Baker, his former patrons, and in 1751 published "A Review of the Works of the Royal Society", holding up to ridicule the "Philosophical Transactions," to which he had himself contributed two papers a few years previously."

Hill was a noted author, extremely prolific, and conducted the *British Magazine,* 1746-1750. In 1759-1775 he published *The Vegetable System,* for which he obtained the Royal Order of Vasa of Sweden.

started off again; was taken prisoner by the Danes; but was set free again after seven weeks, under the pretext that he was an Englishman. After peace had been proclaimed he returned, but travelled first through a part of France and Holland, hired there a number of Frenchmen who were specialists in their respective manufacturing lines, whom he sent to Sweden at his own expense. At the Riksdag of 1723 he obtained several privileges for the industries. At the session of 1727 he caused the middle estate to send a deputation to the nobles, demanding the erection of factories, the total prohibition of certain imported goods, and a tariff of 5 per cent on others. It did not look very encouraging to him at first in this Riksdag, because various merchants of the middle class sought to persuade their fellow tradesmen to overthrow the proposal and destroy the recently constructed shops. And the latter, from all appearances, would have succeeded in their schemes if the burgomaster of Malmö, Stobée,[20] had not championed the cause. The latter was a very enterprising, aggressive man. (May 15, 1751).

[20] Possibly Lorentz Christoffer Stobée (1676-1756), army officer and civil administrator, who later became governor of "Göteborgs och Bohus Län."

A DESCRIPTION OF NIAGARA WATERFALL IN NORTH AMERICA

[THE following description of Niagara Falls is translated from a contemporary letter by Kalm to the librarian Carl Christoffer Gjörwell of Stockholm. It was originally the author's intention that this account should conclude the promised fourth part of his *Travels*. It is more detailed than the description in English sent to Benjamin Franklin. Cf. the *Förord* in Elfving's *Tilläggsband*].

A DESCRIPTION OF NIAGARA WATERFALL IN NORTH AMERICA

Waterfalls are found in many places in the world, where water in some river, creek or brook hurls itself down a rocky precipice from a considerable height. There are few countries which cannot boast of some such phenomenon. In Sweden, for instance, we have Trollhättan. The Ammä waterfall at Kajana [Finland] is not to be despised either. The well-known fall at Woxen [Voxna, Sweden] is large. But there are probably few waterfalls in the world that can be compared to Niagara in North America, when one considers its height and the amount of water which passes over it. Therefore we must without contradiction reckon it as among the largest on the globe.

In order that the truth of this statement may appear more clearly, I must mention the following before I describe the Falls.

In North America are five fresh-water lakes, each one large enough in size to be more like an ocean than a lake. These are Lake Superior, Lake Huron, Lake Michigan, Lake Erie and Lake Ontario. The Upper Lake (Lake Superior) is calculated to be 200 French miles long from east to west, and in several places 80 French miles broad, from north to south.[1] It is pretty big, the largest of

[1] Lake Superior is 400 miles long, hence about 140 French miles.

them all. Lake Huron is about 100 French miles long and between 30 and 40 miles broad. Lakes Michigan and Erie are about of the same size, namely 100 miles in length and 30 in width. I shall speak of Lake Ontario later. Into each of these bodies of water a large number of brooks and rivers empty. The nature of the outlet from these inland lakes is this: the water from Lake Superior flows through a narrow sound into Lake Huron. A little south of this sound Lake Michigan sends its water also into the same lake. The water of Lake Huron together with what it has received from the two above-mentioned lakes seeks an outlet in a long, narrow strait which is called Le Detroit and runs into Lake Erie. All the water thus collected in the four inland seas mentioned and in the numerous streams running into them makes its way first through a short, narrow sound, and then over Niagara's lofty falls, whereupon it forms a broad river of about six French miles in length and empties into Lake Ontario, and from there flows with the St. Lawrence River through the most thickly settled parts of Canada and to the ocean.

In order to learn more of the nature of the interior of America, its natural products, the customs of the inhabitants, etc. I undertook during the summer of 1750 a journey through the land of the Iroquois. These Indians or savages have for a long time had the reputation of being cannibals, because they were occasionally wont to roast their prisoners of war and eat their flesh, so as to inspire a greater terror among their enemies. This custom has now for the most part been given up. A traveller who visits their villages or cities must not be terrified if he sometimes should find the outside of the gable walls of their houses covered with human skulls. These are trophies of war and serve as victory proofs of the number of enemies slain. Nevertheless, however cruel they may be in warfare, when it is in no way advisable to meet them, they are, on the other hand, very friendly and hospitable when at home in times of peace, especially if they are sober, for then they exhibit greater hospitality than most of the Christians. A stranger has scarcely time to enter their dwelling before the Indian mistress offers food, while her husband tries in his way to entertain the visitor. Since Niagara is located near their land I felt it worth while to see it, as one of the most remarkable sights in nature. I knew that no

Swede before that time had ever had the opportunity to behold it.[2]

Well, after a very difficult and quite adventurous journey on horseback through the territory of the Iroquois, I finally arrived on the 13th of August, 1750, new style, at Fort Oswego belonging to the English and located on the great Lake Ontario, which more resembles a sea than a lake. This lake is situated between the 42nd and 44th degrees north latitude; its length from east to west is about 80 French miles; and its breadth is about half of that distance. In this lake there are only a few small islands to be found, and these near the shores; there are none further out. The water is as fresh as spring water, clear, and in some places over 60 fathoms deep. It never freezes over in winter, and ice forms only near the shores. From Oswego I made my way along the coasts of this lake in a flatbottomed boat or battoe as far as Fort Niagara, then belonging to the French, and where I arrived after six days of rowing on August 23. Fort Niagara is situated on the west shore of Lake Ontario, right near the place where Niagara River flows into it. The river and the lake have already washed away a part of the land nearest the fort, so that to prevent further damage it has become necessary to construct retaining walls there. The location of the place is pleasant: one has a view over Lake Ontario towards the north, and of Niagara River in the west and south. Woods of beautiful hardwood trees appear on the mainland. Here I was received with much courtesy by the French officers.

Early in the morning of the 24th of August, accompanied by the French officers and three soldiers, I started off for the famous Niagara waterfall, the distance from Fort Niagara to the Falls being six French miles, of which half is travelled on water and half on land. We first ascended Niagara River in a birch boat. The width of the stream here was said to be about twelve arpents, but this varied. The banks of the river consisted of high precipitous hills of red sandstone, in layers. The water in the river flowed very slowly at first; but the further up we came, the faster the current, so that after rowing a French mile we had great difficulty in work-

[2] This may be true, but it is not certain. Several Scandinavians had come to New York State in the seventeenth century, and among them may well have been a Swede who penetrated as far as Niagara Falls. But if he did, he did not write up his observations.

ing our way upstream, although we rowed close to the shore. Shortly after we had left Fort Niagara we saw the vapor of the Falls rising high toward the sky like a thick cloud, and this cloud could be seen during our whole journey, gradually increasing in size as we approached the falls.

After three French (about nine English) miles of hard rowing we stepped ashore to continue our trip on foot. It is difficult to come nearer with a boat, because the number of steep rapids encountered. First we had to climb up the high, steep river banks, then proceed three French miles by land, which has two high and tolerably steep hills to be crossed. On this road we met a great number of Indians of both sexes, who were engaged in carrying their skins and other goods to Quebec. These goods had either been purchased originally from the Indians by the French and were being sent on, or the Indians were taking them to Quebec on their own initiative. Several natives had their own horses which carried such wares in return for pay. In as much as one cannot row a boat from Lake Erie to Lake Ontario because of Niagara Falls, boats have to be carried over land this distance (of nine English miles). Of course only birch canoes as boats can thus be conveniently transported. To-day I saw four men carrying a birch canoe that was five and a half fathoms long and about five and a half feet wide in the middle. Finally about half past ten in the forenoon we reached the Falls ourselves. The air was clear and the wind southwest, which was the best we could have wished for, for it drove away the vapors, rising like dense smoke, from the side where we sat, and we could see the falls much more distinctly. Had the wind been contrary we could not have seen many feet through the fog. The temperature was rather high, too; at three in the afternoon the thermometer registered 26½ degrees C. I also tried to determine the temperature of the water in Niagara River. In that part of the River over which we had travelled in the morning the thermometer showed a uniform temperature of 22°. But the water close to the awe-inspiring falls showed a constant temperature of 24½°.

Now I shall, without any alterations, reproduce that description of the Falls which I made with pen and ink as I sat on the utmost brink of them, at hardly a fathom's distance from the place where an enormous mass of water hurls itself perpendicularly down from a height of 135 French feet, i. e. $147^{690}/_{1000}$ Swedish feet, or about $34\frac{2}{3}$ fathoms.[3]

The river at the falls runs from S.S.E. to N.N.W., and the falls themselves from southwest to northeast, not in a straight line but in the shape of a horseshoe or semicircle, since the island in the middle of the falls lies higher up than the ends of the semicircle. Above the falls and about in the middle of the river is an island, extending from S.S.E. to N.N.W., which is said to be about eight arpents long [about 1450 feet]. (The length of a French arpent here was $\frac{1}{84}$ of a French mile [and the latter is 2.9 of an English mile]). The island tapered in width toward its ends, being in the middle about a quarter of its length, with its lower extremity extending right up to the falls, so that no water tumbles down where the island touches the falls. The width of the latter, in its curved line, was said to be six arpents [about 1094 English feet]. The island lies exactly in the middle, so that the falls are split in two. The breadth of the end of the island in the falls is about $\frac{1}{3}$ of an arpent [61 ft.], and the length of the island is, as I have just said, estimated by all to be eight arpents.

On the two sides of this island flows all the water collected in the great and small sea-like lakes of Superior, Michigan, Huron and Erie and the many streams, brooks and rivers emptying into them. Before arriving at the island the water does not flow so rapidly, but as soon as it reaches the former, and divides into two branches, it begins to rush with such a fearful velocity that in many places it turns as white as the strongest rapids, and shoots up into the air. If a boat should ever reach that spot, no matter how good it might be, or how brave and strong the occupants were, it would still be impossible for them to reach land. They would be compelled to follow along down the terrific falls. In fact, they would not be able to keep their boat right side up for even the shortest time before it would be capsized by the violent current and the rocks in it. On the west side of the island the current is stronger, swifter and more snow-white, and there flows a greater quantity of water.

When a person stands down near the spot where the water begins to cast itself perpendicularly down the rock and looks back

[3] 135 French feet = 24 1/4 fathoms. Note by Elfving.

up the river, it is easy to see that the stream runs at a considerable incline: it seems as sloping as the side of a fairly steep hill. It is so inclined that the forest a short distance above and beyond it cannot be seen. From this it is easily perceived what forceful speed the water must have, even before it reaches the angle of the falls proper. And when it does reach that point, it immediately [of course] hurls itself downward. It is enough to make the hair stand on end on any observer who may be sitting or standing close by, and who attentively watches such a large amount of water falling vertically over a ledge from such a height. The effect is awful, tremendous!

During my stay in Montreal, 1749, I was anxious to learn the exact vertical height of the Falls from M. [Etienne Rocbert] de La Morandière [1701-1762] who was royal [army officer and] engineer and who a few years before had been commissioned by the governor-general of Canada, Monseigneur Beauharnois[4] to make a careful measurement of the same. He told me that he had measured them very accurately at three different occasions, almost every time he had happened to pass by, and found them at every measurement to be 135 French feet.[5] When I arrived at Niagara I asked all the gentlemen who were there, and who had often seen the Falls, how high they believed them to be. They answered almost unanimously that they had found their height to be exactly the same as that given by M. de La Morandière. When I to-day beheld the Falls themselves, where I met M. [Daniel] Joncaire [1716-71] who had spent ten full years here as an officer and had had charge of the goods that had almost daily been shipped between the Lakes of Erie and Ontario and therefore was better acquainted with the location than anyone else, I inquired of him how high he thought they were. He answered that he himself had never measured them, but that a few years ago a Jesuit had been there who had measured the Falls with a line or cord and that he [Joncaire] had helped him in his undertaking. He had there found the falls to be 150 French

---

[4] See page 440, note. He was governor-general from 1726-1747.

[5] The actual height of Niagara Falls on the Canadian side is 158 feet and on the American side 167 feet. The French estimate of 135 of their feet would be the equivalent of about 144 English feet, which is too short, if the length of the French foot as given in reference works is correct, *viz.* 1.066 English feet.

feet.[6] M. Joncaire assured me that they were certainly not less than 135 feet, and surely not over 150, but somewhere in between. He felt that M. de La Morandière's measurements must be nearer the true distance since it was a difficult matter to measure the height with a line; the line would be crooked because of the ledge and water. Since others had so accurately determined the height, I did not feel it was necessary to measure it again with a cord, especially since my time did not allow it, and some places were so slippery that the attempt would have been quite hazardous.

When the water rushes down the ledge and strikes the bottom it jumps up again in some places to a considerable distance. Elsewhere it boils and seethes with a snow-white foam, behaving just like the water does in a glass into which more is poured from above. The falling waters cause a loud roar, as can easily be imagined from the amount of water running over from such a great height. The noise of the falls can sometimes be heard at Rivière à la Boeuf, which is said to be located about fifteen French miles to the south; and at Fort Niagara, which is six French miles away, one can hear it very plainly in calm weather. One would be able to hear it there at almost any time, were it not for Lake Ontario, on which the water is seldom still, [drowning out the noise from Niagara] through its own din and roar in dashing against its shores.

M. Joncaire and all others declared that the falls occasionally made a louder racket than at other times. When its roar was unusually strong in any direction, it was said infallibly to indicate rain and bad weather. This was the reason, it was claimed, why the Indians of the neighborhood were able to prophecy the weather so accurately. All residents of Fort Niagara asserted that when the falls were heard clearly, i. e. as far as the fort, it signified invariably a northeast wind, and they thought this peculiar because the falls were southwest from the fort and one would have rather expected a southwest wind to hear it stronger.

Several who have spoken of these falls have declared that the

---

[6] This is about 159.7 English feet, which is reasonably accurate. Obviously the Jesuit with his string came much nearer to the actual height of the falls than did the supposedly more scientific calculations of the French engineer officer. Or, were the Falls lower in Kalm's day?

roaring noise is so deafening that people standing near them cannot hear each other speak unless they yell loudly close to the ears; but I did not find it so. We were close to the falls on all sides, but one could well hear what another spoke, provided he talked a little louder than usual; it was not necessary at all to shout. When the locusts or grasshoppers, which are described in the *Transactions* of the Royal Swedish Academy of Sciences for the year 1756,[7] made their piercing shrieks in the neighboring trees they easily drowned out the noise of the falls.

A large mass of vapors rise from the bottom of the falls that resemble a thick smoke rising high toward the sky. Caused by the violence of water contact these vapors, if the weather is calm, rise straight up to a great height and look like the heaviest cloud. But if there is a wind they are blown about as in a driving storm, and anybody enveloped by them will get as wet as if he had been dragged out of the sea. A couple of the Frenchmen who accompanied me climbed down a short distance below the falls, to examine the spot. The wind drove the mist at them so turbulently that they stood as in an impenetrable fog and thought they would suffocate. They left at once, and when they came up they were so drenched they were forced to take off almost all their clothes and dry them in the sun, in the interim walking about half naked.

If a spectator takes up his position anywhere near the falls whether it be in the woods or elsewhere, and the direction of the wind is from the falls, he will soon be so wet from the mist, even in the most glaring sunshine, that he will be glad to make his escape as soon as possible.

Everyone contended that from the other, east side of Lake Ontario, where only a small portion of this section can be seen, one could on a still morning behold something like a heavy mist rising above the woods in the vicinity of the falls, and that it looked like the smoke from a forest fire, although it was but the vapor from Niagara. The same phenomenon is noticed several miles from the falls, on Lake Erie, if one looks in the right direction. The morning seems to be particularly favorable to the rising of a formidable-looking fog from the falls, so much so that it is under those cir-

[7] This clause was of course inserted when Kalm later prepared his notes for the press.

cumstances hard to see the falls at all; but if a wind comes up the mist is dissipated considerably.

Almost daily, and especially in the spring and autumn at the migration time of the birds, a large number of sea fowl are found dead beneath the falls. There was said to be two reasons for this, 1. when the birds swim in the water a little above the island located in the middle of the falls, they like to let the current carry them slowly onward. This current approaching the precipice becomes swifter and swifter, and finally near the beginning of the falls becomes so rapid that when the birds hope to rise from the water they are no longer able to do so because of this swiftness, and are obliged to follow the current over the falls and perish. Long-time settlers of this vicinity declared unanimously that they had often seen a whole flock of sea birds sail over the falls. 2. When the mist in calm weather rises high and is very thick it happens that birds fly into it and never return. Either their wings get too soaked from the fog—because it drenches faster and more thoroughly than rain or other water—or the thundering noise from the falls frightens them so when they pass over into the heavy vapors that they try to alight, [are lost and killed]. All present agreed that they had often seen acquatic birds fly into the heavy mist and that they had found them dead beneath the falls a short time thereafter. As a further corroboration of the second reason it was pointed out that not only water fowl but land birds had been found killed at the bottom of the falls. The death of these cannot be ascribed to swimming too long on the waters above until they are borne to their destruction; they must have been lost in the fog. M. Joncaire, who has lived here for many years and visited the falls almost daily, assured me that he had an innumerable number of times seen whole flocks of swimming sea birds go over the falls, and that he had also perceived repeatedly how birds flew into the thick smoke and were later found killed below the falls. He could not remember, however, having seen any fauna from the land or forest suffer this fate.

In the autumn of 1749, among other birds, a swan sailed over the falls; but it was not seriously hurt, as it happened, and remained alive. It swam about beneath the falls for a month, but was not able to fly. Many tried to shoot it, but it was clever enough to move either to the other side or nearer the falls when a hunter appeared, so that it was impossible to get near it. Finally the swan disappeared, nobody knows whereto. Either it was eventually able to use its wings and fly off or some Indian killed it.

It is especially in the autumn, when birds fly south in large flocks, that a large number of them are lost in the falls, and that is the season therefore when they are collected. Both Indians and resident French soldiers from Fort Niagara are said to appear here daily then to gather the [killed] supply of sea fowls. The commandant of Fort Niagara, M. Beaujeu [8] assured me that the soldiers there lived in the autumn a long time principally from the birds found dead beneath the falls, which they prepared for food in various ways. The birds were said to make good food if they had not been dead too long.

During my sojourn in the English colonies the Englishmen claimed they had been told that in the autumn one could gather bags full of down near these falls from the birds killed and rotted. I asked those present if this story was true, and they answered that while it was not literally so, it was true that much down was obtained by plucking the birds found killed beneath the falls.

Besides birds a number of animals like deer are frequently found killed at the bottom of the cataract. They attempt to swim across the river above the falls to the island in the middle, are pulled into the current, and pass over the brink. They are usually dashed to pieces in the descent. Bears, seeing deer on the island, occasionally try to visit them, but are with much growling compelled to change their course and go over the falls. Later they have been discovered crushed to death at the bottom. Sometimes, in fact almost daily, fishes suffer the same fate. These are said to be good to eat.

Often it happened, too, according to a unanimous testimony, that drunken savages, who had propelled their birch canoes a considerable distance above the cataract, had gradually been drawn into the current and disappeared over the watery precipice. Only an arm was found of some of them.—Several told me that they had purposely thrown large trees into the stream above in order to

---

[8] Either Daniel-Hyacinthe-Marie Beaujeu (1711-1755) or his brother Louis Liénard de Beaujeu (1716-1802), both military men noted in Canadian history. See *Dictionnaire Générale du Canada.*

see how they would behave when they passed over the falls, but that they were completely lost in the fall; wherefore it was concluded that there existed a bottomless pit below. But although the trees might have stuck to the bottom at first, they might well have come loose afterwards unnoticed.

One cannot help being amazed and awed when standing above the cataract gazing down. It felt as if one were looking down from the highest church steeple. The falls that I had seen previously—Cohoes and Montmorency in North America, and Trollhättan in Sweden seemed but child's play in comparison. One could not gaze and contemplate without feelings of wonder and astonishment. The water seemed to flow gently and well-nigh lazily over the cliff at first; but the farther down it went the greater the speed. Even now there was an immense amount of water hurling itself out into space; but in the spring there is of course more, the most of any season. In its descent, near the bottom, the water strikes the rock in some places and there it foams and froths and roars in gigantic leaps, so that it gives the impression of a continuous firing of cannon, with a heavy, persistent and hastily appearing mist dashing forth; for the liquid is creamy white and looks like a thick vapor. When falling, the water at the top first appears green; but further down it takes on a snow-white tinge as in the strongest waterfall in the world, which this in fact really is. When the water reaches the bottom it sets all in motion beneath it, so that it swirls about in circles and whirls like the liquid in a seething cauldron. About two musketshots below the falls the current is not stronger but that one could row a good boat across there without danger; but a little below we meet again some strong and steep rapids, where the water jumps forward with such great violence that it turns white with froth. It would not be advisable to run these cascades in a boat, for it would be capsized and dashed to pieces.

Between the island right above the falls and the land on the east side was a distance of about two short arpents [365 ft.], it was said. To me it seemed but two good stone's throws. This islet is covered with tall trees, and is sometimes full of deer. These in attempting to swim across the river above the falls have been caught by the current and pulled along, a few being fortunate enough to strike this holm and thus escape the falls.

This islet had always been considered entirely inaccessible and no human being was thought ever to have visited it; but about 1739 a method of reaching it was discovered by the following adventure. Two natives of the Seneca Nation, who lived north of Lake Erie, were travelling on business down to Fort Niagara. Here they procured a quantity of French brandy, whereupon they started back over land, past the Niagara Falls, to Lake Erie. Here they stepped into their birch canoe, intending to visit an island in the lake a good distance above the falls to hunt deer. They had already imbibed so much of the French liquor that they knew but little of what they were doing, and during their paddling they continued to refresh themselves from time to time. Finally the brandy made them so heavy and sleepy that they lay down to sleep, letting the boat drift on the lake. The water which here moves gradually toward the falls carried the boat further and further down, until it finally reached the rapids a short distance above the upper end of the islet that lies in the middle of the river near the falls. Here one of the Indians, hearing the roar of the waterfall, woke up. He was horribly frightened, became sober immediately, and shouted with all his might to his companion that they were lost and in the twinkling of an eye would be right in the precipitous cataract. The companion woke up at once, both seized their oars (paddles?) with all the energy at their command, and since the boat had drifted quite close to the island, they sought desperately to reach it, and were lucky enough to land there. Now they were happy at first, to be sure, that they had so far saved their lives; but when they had had time to consider the situation, they felt that they had not gained much, and that it might have been just as well to have let the current carry them over the falls once and for all, for then the agony would have been brief. Here they would either have to starve to death, gradually, or else eventually hurl themselves into the waterfall anyway to end their days quickly. For how they should ever be able to return to their families again, alive, became a new problem, which both to themselves and others had always been deemed impossible to solve. But necessity urged them to try everything. On the holm grew the American linden tree; they peeled off its bark, and of its bast made a ladder as long as the height of the falls. One end of this ladder was fastened

From early print based on Kalm's description

Niagara Falls

to large trees that grew on the brink of the perpendicular ledge which constituted the nether extremity of the islet between the falls, and the other was let down to the water below the cataract. Then they climbed down the ladder to the rock at the bottom, where they threw themselves into the water, thinking that they might be able to swim to land, especially if they could reach the more quiet water below the falls before the place where the abovementioned rapids commence. As I have noted previously, one could safely row a boat on this intervening space. But the descending waters on the two sides of the island make strong surging waves at the bottom, which with great force were hurled back against the rock directly below the islet. Consequently as soon as the Indians had jumped into the water, these huge waves drove them so forcibly back on the rock that when they had tried a few times to swim they had been so bruised and buffeted about that most of their skin had been torn off their bodies. They realized then that it was impossible for them to reach the shore, and were therefore compelled to climb back up the ladder to the island. Here they were obliged to remain for nine days in all, without any food except what they had brought with them in the boat and the few deer that they had been able to catch. But of water to slack their thirst they had plenty.

Fortunately for them, some savages who happened to pass by on the mainland, caught sight of the prisoners and heard them call. The former hurried down to Fort Niagara, told the commandant of the plight of their brothers, and exhorted him to bring about their rescue, if possible. Immediately he called his officers and men into consultation and questioned the Indians, who were well acquainted with the current on both sides of the holm where the two others were marooned. They remembered that the water at the upper end of said holm was not very deep, nor was the current there very strong. They had occasionally observed six Indian boys wading far out into the water there. So they adopted the following plan: the commander had a few poles shod at one end with a ferrule-like spike, so that they became quite sharp. Thereupon he urged the visiting Indians to do their best to save their brothers, and showed them how they should go about it. He and his officers went along [with them] to see the plan put into execution. A couple of young

Indians, each one provided with four iron-shod poles and some food in a bag with which to refresh the wretches on the island, set out on the venture. They made their way along the shore at the east side of the river to a point a little beyond the extreme end of the island, and from there began to wade diagonally across to that spot, driving their iron-shod poles into the bottom at every step in order to support themselves against the violence of the current. At that time the water in the deepest places went just up above the knees. Consequently they reached the island safely, fed the wretches a little, and giving each one of the latter a pair of shod poles conducted them back to land. After the trail in this way had been blazed, the Indians continued at later times to visit the island without danger. That which attracted the natives to the place was, as noted above, the number of deer, elks and other animals that tried to cross the stream but which were carried to this islet, if they escaped passing over the falls. When the natives from the mainland saw a sufficiently large number of animals on the island they waded across and killed them. The two first Indians, who were so luckily rescued from death, were said to be still alive, when I visited the falls.

The bottom of the river above the falls, between the island and the mainland on both sides, consists of ledges or rock of a gray, compact limestone. The cliff which forms the falls, is of the same material. Now and then a stone of considerable size is found in the rapids above the falls, and also below where the water strikes the bottom. These stones are granite. The limestone of the falls lies stratum super stratum. The banks of the river below the falls are also of the same gray, massive lime rock, and are exceedingly high and steep, almost vertical. It is however, possible to climb up in some places; but it is certainly dangerous and the attempt should be made with great caution. On the west side of the river the bank is less precipitous.

The land about the falls is stony, and here and there a large bit of gneiss or granite is found.

Several people told me that one could pass below the falls right between the rocky wall and the mass of falling water, since the latter does not fall perpendicularly close to the rock but describes a curve. I asked M. Joncaire if this were so, and he answered in a decided negative, for, he said, the water in its descent hugged the wall very closely.[9] On the west side of the island there had formerly been a cliff projecting out farther than the rest of the wall, and when the water ran over this protruding part it had been possible to pass beneath the falls below that point; but this jutting rock had fallen down some time before, so that it was now impossible to go between the wall and the water.

The Indians make small bark boats which they use in fishing below the falls. Among other fish they catch a large number of small eels of nine or twelve inches in length, and all the dexterity needed for their capture is to go below the cataract and feel around with the fingers in the cracks, holes and crevices of the wet rock, find them and grab them. A large quantity is gotten this way. Some small native boys tried the method to-day for my benefit, and returned very soon with a large heap of them. The Indian lads were real daredevils; they walked right out to the very edge of the cataract or river and looked down, where it was not only vertical but where the rock besides had been worn away by the water. They waded a long distance into the water right above the cataract, and then proceeded to approach the falls themselves so close that there was not more than a foot to the outer ledge of the cliff where the water spills over from its terrifying height. There stood those rascals, gazing down! I was chilled inside, when I saw it, and called to them; but they only smiled, and still stood a while on the outermost brink. A single false step would have cost them their lives, had they had a thousand of them.—I suspect that the officers who accompanied me had commanded these Indian boys to show me their skill and daring.

When the wind blows the most furious hurricane seems to reign below the cataract, because of the mist being blown in all directions.

On a certain time of day when the sun is shining, we can always see a rainbow against the mist. I arrived a little after eleven in the forenoon, and we had clear, sunshiny weather all day, so that we could see almost constantly a beautiful rainbow on the east side of the river, near the bottom of the cataract, about six fathoms below it. When we stood on the edge or bank of the river the rainbow appeared beneath us. I saw it at eleven, twelve, one, and half

[9] M. Joncaire's reply was wrong, of course, as every visitor to Niagara knows. Obviously, after this denial, Kalm did not proceed to test the facts.

past one o'clock, but after I had been away to dinner and returned at a quarter past three no more rainbow was to be observed, presumably because the sun was lower in the sky. While it lasted the following phenomena about it were noticed. If the vapors were thick, two rainbows were seen, one outside the other; but the latter was then very faint. When the fog was thin, only one rainbow appeared. It had quite a large arc, and was on the outside of a flaming red color, the tinge of a flame mixed with smoke; but the inner part of the arc was greenish-yellow, and the central portion of a heavenly blue, though very light, the flame-colored and greenish-yellow parts made up the arc or bow, and the light blue filled the remainder. Because the mist was driven over the river with the wind, the rainbow looked like a floating stream, for although it stood still the vapors moved forward. The thicker the mist, the plainer the rainbow; too much fog made a double one. Occasionally when the mist was driven away, the rainbow disappeared almost entirely, and only pieces of it were visible. As one moved from one place to another the rainbow seemed to move also, so that one could constantly see a new one. Some have claimed that a rainbow may be seen here only at eleven in the morning; others that it can be seen all day. My own observations are recorded as above. The external rainbow was almost always of a green color, narrow, thin and faint.

The basin or the space which the water immediately below the falls occupies is near them as broad as the falls themselves, but a short distance below it becomes gradually narrower.

The boat landing above Niagara, when one wished to sail up Lake Erie was said to be about 15 arpents from the falls [about a third of a mile]. I thought it was but a little less than ⅛ of a Swedish mile [¾ of an English mile]. It is not considered safe to approach the falls nearer with a boat, lest one suffers the same fate as the two Indians, or something worse.

In the rocky sides of the river below the falls are a large number of holes, since it is stratified limestone. It was claimed that rattlesnakes hibernated there. In summer time they are scattered about in the woods and on the hills. To-day we saw three of them near the road, which were killed at once. In the beginning of autumn when it begins to grow colder, they gradually gather here from all directions and creep into the aforementioned holes; and when in spring it gets warmer they come out again from their hiding places, when one can sometimes see them by the hundreds. It was asserted that about six hundred of these hideous beasts had been killed this spring (1750) in this neighborhood, when they first came out of these winter quarters and were still torpid. It is well known that sometimes [—Kalm might have said frequently or generally—] there is no cure for the bite of this dangerous serpent; it is so poisonous or deadly.

ADDENDA TO THE DIARY

[THE following material, on New Sweden, is not all new of course. In fact, a large proportion of it, in its original form, has been used before, and notably by Israel Acrelius in his well-known work on New Sweden (1759).[1] But Kalm's account in its simplicity gives us many new slants or viewpoints, much interesting and illuminating historical gossip, and some details omitted by Acrelius. Besides, it serves as a useful corroboration of many facts of history connected with New Sweden. It represents first-hand sources in any event, and as such is some contribution to knowledge. However, a number of pages, which Kalm copied from the Wicaco Church Records, have been omitted here as being primarily of local interest.]

THE ARRIVAL OF THE FIRST SWEDES

I have talked with several of the oldest Swedes and sought to learn in what year the first Swedes arrived; but there was none of them who could give me definite information about it.[2] Their sentiments and stories I have already introduced [in these volumes].

SEPTEMBER THE 18, 1748

From Mr. Kock and Jacob Bengtsson I learned that after people had been sent over from Sweden to settle and cultivate this land, several years passed before the mother country cared to make any inquiries about the countrymen it had sent hither, or to learn how they were, and if they were living or dead. The Swedes who then came over settled near bays and rivers, where they had good opportunities for fishing. Most of them had no horses or beasts of burden, so that when they needed salt, which could not be procured

[1] This has been translated into English in part by Nicholas Collin (1841) and in full by William M. Reynolds (1874). The latter is called *A History of New Sweden or the Settlements on the River Delaware*, printed in the *Memoirs of the Historical Society of Pennsylvania*.

[2] The first Swedish settlers landed on the Delaware in 1638.

elsewhere than in New York, then belonging to the Dutch and called New Amsterdam, they went thither after it, riding in part on oxen and in part on cows, bought salt there and whatever they needed and brought it back home on the backs of the just-mentioned animals. And since they had no other people to associate with than the native Indians, they soon began to differ more and more in their actions and manners from the Europeans and old Swedes and began to resemble the Indians. At the arrival of the English, therefore, the Swedes to a large extent were not much better than savages. One of the first reasons for sending Swedes over here was connected with a man by the name of Printz.[1] He, together with some others, had been in prison in Sweden for some misdeed, and suggested to the Swedish government that if he were released he would leave the land, go to America and seek out such places as could be settled by Swedes and be of such advantage to Sweden as other colonies had been to other countries. His request was granted; a ship [*Fama*] was fitted out; and Printz, who was an experienced sailor, made captain of it. Several criminals who had been serving sentences in prison with him were released and made his companions on condition that they would settle here, start to till the land, and hunt for gold, silver and similar things. He sailed hither on the ship just mentioned, let the newcomers settle wherever they pleased, stayed here a while to govern them, and then after a time returned to Sweden. When the Swedes saw that they could not get what they needed from Sweden they surrendered to the Dutch[2] and became their subjects, enjoying their former privileges.

The nonagenarian, *Nils Göstafson* [Gustafson], told me that when the Dutch came and took the land the Swedes did not accept any favors from them; nor did the Dutch harm them in any way. Then when the English came and seized this territory there were a great many Dutchmen living among the Swedes; but a large proportion of them left for Surinam, which the Hollanders had received from the English in return for the American settlement.

---

[1] Johan Printz (1592-1663), governor of New Sweden. His misdemeanor, mentioned in the next sentence, consisted in a breach of discipline while serving in the Thirty Years' War. It was not in Printz's case regarded very seriously.

[2] The Swedes surrendered to the Dutch involuntarily in 1655 because of the sheer superiority in numbers of the latter, but undoubtedly there were some who did not care who the ruling power was.

Zachris Peterson told about his father, who died seven years ago at the age of ninety-five. He had emigrated from Uppland in Sweden, and related how the Swedes had formerly gone to New York to purchase some cows, but that on the return journey, when the English fleet had been at anchor in the Delaware River, the English had taken away a number of those same cows and left only one cow to two families.

Governor Printz is said to have been a big, very tall man, who inspired such a terror among the Indians that although they had conspired to annihilate the Swedes they became so frightened that they dared not do anything but agree to everything he proposed. In the beginning the danger from the Indians was so great that when the Swedes were plowing, someone had to walk behind the plowman with a gun in his hand to defend him if the savages should appear.[1]

Mr. Peter Rambo, who lived down in Raccoon, N. J., related to me on the 30th of January, 1749, the following concerning the first coming of the Swedes, which he had heard from his father who had died six years ago and had at the time of his demise been very old. His grandfather's name was Peter Rambo [also]; he had been born in Stockholm, had with others been hired to come here, and had gotten his freedom after three years to return to Sweden if he had so desired. Several of these so hired had returned to their homeland, but he had remained. But those who had been sent here because of some misdeed had not been allowed to return. The first Peter Rambo landed here when the original settlers had been here four years (1642). He was then unmarried, and when he had been here for a short while he married and had several sons of whom this Peter Rambo's father was the youngest. He was born 1661 and was 12 years younger than his oldest brother. The original Peter Rambo, when he emigrated, had brought apple seeds and several other tree and garden seeds with him in a box. He had also taken some rye and barley along. Later when the Englishmen came he had often told them that his hands had been the very first to sow seed in the settlement, thereby announcing that the first Swedes had not brought these seeds with them, and that conse-

[1] The relations between the settlers of New Sweden and the Indians are generally supposed to have been exceptionally friendly.

quently no European seed had been sown here before he upon his arrival had made a beginning. His grandfather had prospered, so that Governor Penn had often lodged at his house; and when the English first came here it had been rather difficult for some of them, so that Rambo not only helped them as much as he could but for ten years gave to everyone that came to him free food and lodging. The old man was very kind, but liked to drink a bit at times.—All the Swedes who first came had built cow barns, as the living Peter Rambo affirmed, and they still had them when he was a boy; but afterwards when they saw that the English did not use them they had abolished them.[1]

The Finnish Ship. Among the ships sent here from Sweden was one loaded almost entirely with Finnish passengers, who had been sent here to settle the land; but when they came near the American continent the vessel sprung a leak so that they were unable to pump out all the water that came in. They kept pumping however for three days, though the water finally got the upper hand and the damage was irreparable. Besides the crew, there were three hundred people on board. When the sailors saw that all hope of saving the ship was gone, they jumped into the [life]-boat under the pretext of investigating the leak, but in reality to save their own lives. But when one of the Finns, by the name of Lickoven (otherwise known as Jacob Eit) noticed it, he jumped into the boat also. The ship sank with all its passengers, but some of those in the lifeboat reached the shores of New England; yet no one came here, and information was received only through rumors. This ship was called *Det Finska Skeppet* (the Finnish ship). The ship doctor had silently been exhorted by the captain to board the lifeboat but he had not wished to leave the people, had stayed on the vessel and perished with the rest.

There seems to be no evidence that the Swedes here ever used dampers in their rooms or huts [as Kalm has mentioned before], either when they first came or later, because in addition to the fact that the winters were not so severe or long here as in Sweden they had a vast amount of forests of big trees right near their home [so that no saving of fuel was needed], and, besides, they had not brought any dampers with them from Sweden, and none were ob-

[1] Kalm has mentioned this fact before.

tainable here. I asked many old settlers if they had not heard them say that their ancestors had some at first, but they all declared they had never heard of any.

Some cows, a horse and a mare were brought over on the ship on which the first Rambo sailed to America, and the same evening they sighted land the mare foaled. It also snowed quite hard that night.—When the Swedes had been in this country for a while the English of New England learned that the former did not have all the cattle they needed. Therefore they drove down a large herd of cattle from the north and sold them to the Swedes, receiving good pay in skins and other goods that the Swedes had previously bought from the Indians.

The First Cause of Enmity between the Dutch and the Swedes. The Dutch and Swedes had always lived on good, friendly terms with one another until a number of Dutchmen came and settled in New-Castle, and until a ship arrived from Sweden that carried a captain by the name of [Sven] Scute, who had begun to shoot at the Hollanders to drive them away: this was the origin of the dissension between the two.

When the Swedes first arrived they lacked all kinds of tools, both for agriculture and other purposes.

Swedes and Indians. All the old Swedes told me with one voice that in former times the Indians had on several occasions banded together to kill the Swedish colonists, but through God's providence some old Indian man or woman had always secretly run to the Swedes and warned them about what their fellow-Indians had in mind. Sometimes the Swedes wanted to pay these secret messengers, but they would not accept anything, and hastily returned to their people. The Swedes then collected, and when the natives saw them prepared, they dared not attack, and a new peace treaty was drawn up between the two parties. The Indians always liked the Swedes better than the English, and the English better than the Dutch, whom they still hate a good deal.

Undated

Although the Swedes constituted only a handful of people in comparison with the Red Men, these were quite afraid of the former, and never ventured to attack them. Mr. Jacob Bengtsson told me, from assertions by his father and grandfather, that sometimes when the Indians stole a pig from the Swedes, the latter not only went and brought it back but gave the former a sound beating, without the Indians daring to strike back, to say nothing of a more serious revenge, and would return to the Swedes soon thereafter and be very humble and friendly.

November the 22nd, 1748

Erich Rännilson had an old paper about his family and its immigration which read as follows: "Erich Mulleen aged 46 year, from Helsingland in Swedland, arrived in Dellaware River in the Ship Örn the 26th day of May, 1652. His wife Ingeri Philips aged 36 year from Wermland in Swedland, arrived in Delaware River in the Ship Mercurius March 1654. This family as follow" listed a daughter Anna, 16 years; a son, Anders, 14; Olle, 11; Erich, 9; Johan, 7; a daughter, Eleonora, 4; and one Catharina, 1 year old. N. B. These ships were the last that brought people from Sweden.

Church Records on Early Swedish Settlements. I could obtain no information from the church records about [the time of the first settlement], for there were no such records before the Rev. Mr. Rudman and Rev. Björk came here. Everything that happened before is enveloped in the darkness of forgetfulness. However, according to their story, the first Swedes arrived about the time given by [Thomas] Campanius [Holm] in his description of New Sweden,[1] namely 1630 or soon thereafter [1638]. I shall not here repeat what I have already said on the matter, but only add what others have later related. There have formerly been several records extant which have dealt with the coming of the earliest Swedish colonists, their number and names, and the locations in Sweden from which they came, and an account of their first activities, not to mention several deeds concerning their properties; but after the English took possession, William Penn in particular has sought to get at all these documents, and they are now said to be among his collections.

The old man Gustafson, 91, whom I have mentioned before, told me that several ships with people, had come from Sweden; that

[1] *Kort Beskrifning om Provincien Nya Swerige uti Amerika*, 1702.

Printz was one of the first who arrived; and that he returned after he had been here a while, nobody after that hearing anything more from him; that the name of one of the ships which had brought settlers was *Örnen* (The Eagle), and another *Mercurius.*[1] Otherwise he could not give the year of the first settlement; only when he was a boy [about 1665] he had heard it said that the Swedes had then been a long while in the country, so that he felt they must have been here forty or fifty years before his birth.[2] In his youth the Swedes lived mostly near the kills, rivers, and the bays found in them, and only seldom in the interior where dwelt thousands of Indians, who had daily relations with the Swedes.

Cooperative Farming. The wife of the old man Måns Keen said she had heard from her father and grandfather that when the first Swedes arrived, every family, besides the grain brought for food, had carried a keg, pail or stoup of every kind of seed to sow when they came. They had also arranged it between themselves for one family to bring a cow, another an ox, etc. So that after arrival, when they were to plow for instance, it happened often that they [pooled their resources and] hitched a cow and an ox together to form a team.

Kidnaping and Scalping. Formerly the Indians used to steal a Swedish child now and then, which was never returned.—Once they killed a few Swedes and scalped them. They also scalped completely a little girl, and would have slain her if they had not seen a Swedish boat appear, when they fled. The girl was later cured, though she never got any more hair on her head. She was married, had several children, and lived long thereafter.

Of old *Åke Helm* I learned the following facts: His father came to this country with Governor Printz, who brought him along [as a servant or companion?] for his son. He was then twelve years old. He lived to be 75. This year, 1749, it is forty-five or forty-six years since he passed away, which agrees entirely with the record given by Campanius that Gov. Printz arrived in 1642 [1643].

---

[1] These are now well known facts among historians and need no elaboration. What we learn from Kalm here is the incredible ignorance of the Swedish colonists of 1750 about the details of their ancestors' settlements.

[2] Since this conversation took place in 1748, and Gustafson was 91 years old at the time, he must have been born in 1657, when the first Swedes had been in America only nineteen years.

The First Swedish church was located on Penako (Tinicum) Island, between Philadelphia and Chester, where Governor Printz had his residence, and where the Swedes also possessed a fort or redoubt.[1] Now the island is inhabited mostly by Quakers, though there is said to be one Swede left there. There is no sign left of the Swedish temple, and the cemetery is washed away by water; because the Delaware [River] has eroded the soil there so that the graves are now under water.

Old Mr. Helm reiterated that the Swedes had in part brought cattle with them, but that the largest number had been procured from New York. The cattle of his younger days had been larger than now, so that there was no comparison—they were getting smaller and smaller. They also gave more milk formerly than now. But in his youth settlers had not possessed very many of them; he who owned a cow was held to be rich and prosperous.

Finnish Settlers. Finns have also settled here. They have never had clergymen of their own, but have always had themselves served by the Swedish. They have always spoken Finnish among themselves. Most of them settled in Penn's Neck, where people have been found who until very recently spoke Finnish. But now most of them are dead, and their descendants changed into Englishmen. Helm believed that the copy of the Finnish Psalmbook, which had been presented to me by Zachris Peterson, was not only the oldest of all Finnish and Swedish books available here at this time, and which the Finns had brought with them, but that it was the only Finnish book procurable here; for Mr. Helm said he had often seen the copy, and had also had it as a loan, without ever seeing any more in that language. Most of the Finns came over on the ship *Örnen*.

Mr. Jacob Bengtson in Philadelphia imparted the information obtained from both his grandfathers, who were among the first Swedish settlers, that the majority of the early Swedes came via Gothenburg, and that they hailed from Västergötland. His own paternal grandparents came from that province. It was also possible to tell from the manner of speech of the oldest colonists that

[1] This may have been the second church of the Swedish Colony, being erected in 1646. The Rev. Reorus Torkillus "apparently" built the first edifice in 1641. See Amandus Johnson, "History of the Swedes in the Eastern States from the Earliest Times until 1782" in *The Swedish Element in America*, II, 21.

they had originated there. But a number came from Uppland and a large number from Finland. He knew of no other Swedish provinces from which early settlers had come.

Extracts from the Records of Wicaco Church

On the eighth of June, 1750, I borrowed from the Rev. Gabriel Näsman the Wicaco Church records to see whether I might find anything that could throw light on the history of this land, especially as the late Anders Rudman had exerted himself more to obtain and search the old transactions than anyone of the other congregations had done. The title of the book in translation is "A καὶ Ω. Church Records of the Parish Wicaco from the First Coming of the Swedes to America and its Part New Sweden, afterwards called New Netherlands and now Pennsylvania, so far as They have been Compiled from Friends and a few Writings, by Anders Rudman, Pastor."

Ever since Providence through the wise Christopher Columbus in the year 1492 discovered this new part of the world (although in itself as old as other parts), and Americus Vespucius had the good fortune to have the whole country named after him, the European nations have from time to time sought to colonize this land and become masters of its wealth. Among other nations our old Goths and Swedes have not lagged behind. I shall not dwell on their ancient and many commendable deeds, but limit myself to [an account of] their settlement in this country, and relate the events connected with it, such as I have learned them from many oral sources, and particularly from Capt. Israel Helm.

Before the Swedes' arrival on the [Delaware] River a few Dutchmen had a fort on the other side, on a place which the Indians called Hermaomissing, now Glo[uce]ster; but the Dutch named it Fort Nassau. The commander's name was Menewe [Peter Minuit or Minnewit], who could not get along well with the people and therefore returned to Holland, where after certain charges had been advanced he was removed.

Knowing the land and its condition he went to Sweden and informed the distinguished gentlemen there that the Dutch had settled on the east side of the [Delaware] River, but that the whole continent was still a wilderness, and that although there were so many heathens there that it was a bit adventurous to settle there, if they, as gentlemen of means and quality, would like to venture a settlement he would be willing to be its leader. The chancellor, Count Axel Oxenstjerna, the chief of the others, spoke to Queen Christina, who was so pleased by the proposal that through her order and permission a vessel was fitted out and sent from Gothenburg, called *Calmar Nyckel* (The Key of Kalmar), with a number of people who safely arrived in [America] and negotiated with the Indians for land extending all the way from the mouth of the [Delaware] to the [Trenton] falls. There they drove a few stakes into the ground, which the old folks of my early days in America told me they had often seen; and if one investigated the matter it was believed that the stakes could still be located, on both sides of the River, all the land of which was purchased of the heathens and the sale thereof recorded in deed and letter with the signature or marks of the Indians attached. The document was sent back to Sweden, and when I was in Stockholm it lay in the archives there. It is to be regretted that there is no copy here. Such a one was really made, to be sure, but it disappeared during the many troublous vicissitudes of the day. When [Governor] Rising and [Per] Lindeström[1] came to New Sweden they brought with them the names of the signers of the agreement, which all the Indians who were still living recognized, but they did not wish to hear the names again, since it is a custom of theirs not to speak of or name a dead person. The people settled at a place called Christina, erected a fort there which they named Fort Christina, after the Queen, and the locality has that name to this day. Måns Kling was the surveyor who mapped the whole settlement and reproduced the whole River with all its kills and streams on a chart that was sent to Sweden and deposited in its archives, which I saw the day before my departure and which Mr. Aurén copied in miniature and brought here. Minnewit returned; [his] ship came back laden with goods; but he died and Peter Holländare, a native Swede, took his place, stayed a year and a half and returned. Capt. Israel Helm said he had seen him in 1655 on Skeppsholmen, and that he was a major there.

[1] An engineer who in 1691 completed his *Geographia Americæ*, which was not published, however, in Swedish until 1923, and in English, in a translation by Amandus Johnson, 1925.

In the year [1643] Johan Printz arrived with two war vessels, the *Swan* and the *Charitas*, with full powers to become Governor over the settlement, which he really was, for ten years, during which time only two ships arrived namely *Svarta Kattan* (The Black Cat), on which there were no settlers or passengers, and later the *Swan*, which brought the Rev. Johan Campanius, Capt. Fisk, etc. Since the time passed but slowly for Mr. Printz, and no more ships came over because of the war between Sweden and the Emperor [Ferdinand II in the Thirty Years' War], and because Governor Printz had made himself detested among his own people by his ultra-severity, and especially after the Dutch had begun to come too close, having established themselves five miles below Christina—for those and other reasons he returned to Sweden.[1] His influence and command of man-power had not been sufficient to prevent the Dutch from making that settlement. In the meantime the ship *Örnen* had left Sweden with Director Johan Rising, the chief surveyor and engineer Peter [Per] Lindeström, and several officers and other men, who arrived safe to be sure, but unfortunately got into a quarrel with the Dutch on Sandhuken (New Castle), capturing their fort on their first arrival, Trinity Sunday, 1654, which caused all the later trouble. To understand this rightly we must go back to the very beginning: before the coming of the Swedes to the [Delaware] River, the Dutch had settled on the Hudson River, where they had begun to build a town, calling it New Amsterdam. Then when they started to prosper, some came across to this River, and as a protection erected a fort on the New Jersey side, where Gloucester now is located, calling it Fort Nassau. Their Governor did not get along well with his people, and therefore Minnewit returned to Holland, and from there went up to Sweden with a history of the settlement. Immediately thereupon the influential Swedish gentlemen with the sanction of the higher authorities came together and decided to establish a colony here, and so sent him back with a number of settlers, who upon their arrival negotiated

[1] Kalm's source is not wholly accurate here. The *Charitas* had been in New Sweden in 1641. Printz and Rev. Campanius sailed from Sweden together in the fall of 1642, on the *Fama* and the *Swan*, arriving in New Sweden, 1643. In 1646 the *Gyllene Haj* arrived, and in 1648 the *Swan*, which in the interim had returned to Sweden, arrived again in New Sweden with a large cargo for the Colonists.

Governor Printz was probably not as black, i.e. as severe, as Kalm's source has painted him. Cf. Amandus Johnson, *The Swedes on the Delaware*, 1927, pp. 142 ff.

with the Indians for land on the River, and built Fort Christina, allowing the Dutch to dwell peacefully on the other side. Later Printz appeared with full powers as governor. During his time the Dutch commenced to interfere with the trade between the Indians and the Swedes, for which reason Printz held them in subjection, and also had Fort Elseborg [Elfsborg] constructed, where all their ships had to pay toll. Then they moved from Nassau—since there was no longer any great danger from the natives—down to this side, on Sandhuken, and erected a fortification there. Mr. Printz forbade it, but since his power was too weak to prevent it, the Dutch went on with their plan. This irritated some good Swedes, who would not suffer the intrusion of strangers, who against the wishes of the real owners flauntingly settled and encroached upon their purchased territory. Consequently some crossed over to Sweden to tell the tale, and among them Mr. Scute, whereupon the ship *Örnen* sailed hither with the command to set about with gentle means to recover Sandhuken and place the Dutch in obedience. But the Germans [Dutch] would have none of that, and after Sandhuken had been taken, Governor Hugwesand [Peter Stuyvesant] of New Amsterdam arrived with a troop of soldiers and captured the whole River on behalf of the West India Company in Amsterdam. Stuyvesant regretted it, for which reason he became ready at once to return the Fort, but Mr. Rising, expecting the means of a greater revenge from Sweden, would not accept the offer without orders from abroad. One can judge the condition in the land from the following pass made out by Mr. Rising to old Nils Matson:

Be it known through the Governor of New Sweden and the most humble servant of His Royal Majesty and King that since the bearer, the honest and intelligent Nils Matson, a freeman from Sillesön, at this inopportune time in New Sweden, when we Swedes have so unexpectedly been attacked by the Dutch with hostile intent, is unable to move away so hastily, and must for the sake of his belongings remain here until a more convenient time, and since he has on that account requested a testimonial letter from me, which I have not for the sake of justice been able to refuse him, I hereby testify that he, during the whole time that he has stayed in the country has acted like a reliable and faithful servant of the Crown, that he has willingly aided in the repair of the Fort and buildings and in the performance of other public service, that he volunteered in the last struggle by repairing to Fort Trefaldighet and there participating in the defense of the land, and that he on the way was made a prisoner and taken on a ship where he for about three weeks had to endure much mockery. In the meantime his home was robbed by the enemy, and his wife divested of everything, having constantly maintained her loyalty as a good subject during these troublous times. In testimony of which, I attach my seal and signature. Actum Fort Christina. September 24, 1655.

L. S. Johan Rising.

Now the settlement was under Dutch rule; Sweden was involved in war; the gentlemen who formerly had constitued a "company" were incapable of action, and cared but little about any colony, since the venture gave but small or no profit. The king and government, who had received title to the settlement lands from the [New Sweden] Company were busy with other big affairs, and so the situation had to remain as it was, and the land was now under Dutch and now under English rule, since these two peoples were at war. But the fact that this River belonged to the Swedish crown is attested among other proofs by a letter in the hand of Queen Christina, dated at Stockholm, August 20, 1653, and given Captain Hans Amundsson Bask. So that the Dutch try in vain to veil their actions by declaring that they took the land from a managing company and not from the Swedish government. It is true that in the beginning it was a company which under the protection of Queen Christina carried on its commerce here, but the real power lay vested in the queen, all the ships were sent by her, and all costs were defrayed by her generous hand in that prosperous time, all of which is duly attested by this document literally translated into English by M. Carl Springer.

Wee Christina by the Grace of God, Queen of Swedland, Gothen & Wenden, Great Princess of Finland, Dutchess of Estland etc.

Be it known, that we of favour & because of the true & trusty service, which is done unto us & the Crown of our true & trusty servant Capt. Hans Amundsson Bask, for which service he has done

& further is obliged to do so long as he yet shall live, so have we granted and given unto him freely, as the virtue of this our open Letter is, & doth Shew & Specify, that is, we have freely given & granted unto him, his wife and heirs, that is heirs after heires, one certain piece & tract of land, being & lying in New Sweadland, Marcus Hook by name, which doth reach up to and upwards to Upland Creek & that, whit all the priviledges, appartenances and conveniences thereunto belonging, both in wett and dry, what soever name or names they have and may be called none excepted of them, that is, which hath belonged unto the afores:d tract of land of age, & also by law and judgement may be claimed unto it, & he & his heirs to have & to hold it unmolested for ever for their Lawfull possession & inheritance, so that all which will unlawfully lay any claim there unto, they may regulate themselves hereafter, so that they may not lay any further claim or pretence unto the aforsaid tract of land for ever hereafter. Now for the true confirmation here of have we this with our own hand underwritten and also manifested with our seale in Stockholm the 20 of august in the year of our Lord 1653.

Christina
L. S. Niels Tungell.
Secretary.

After the settlement had so innocently declined, the Swedes underwent many sufferings, which if all were to be enumerated would make a reader resentful. I shall only mention two, from which an opinion may be formed. 1. In the year 1656 the *ship Mercurius* arrived with several Swedes, among them Anders Bengtsson, who is still living and to-day, April 5, 1703, gave a vivid narrative of the event. He told how the Dutch forbade the vessel to pass up the River, and how they ingloriously would have sent it back, had it not been for the heathens who liked the Swedes and who collected, boarded the ship and in defiance brought it up past the fort. 2. The daughter of Governor Printz, Madame Armgott, sold Tennakong and neglected to make a reservation for the church and the bell. The buyer of the land laid claim to both, and the Swedes were compelled to repurchase their church bell for two days . . . since Madame Armgott had gone away, who had promised them their bell, but who had not kept her promise. [Then follows a copy of her written obligation, in Dutch].[1]

Just as New Amsterdam, now New York, a few times passed under Dutch, now English, rule, so this [Delaware] River which by the Germans is called South River to distinguish it from East and North River at New York—passed in its entirety under English jurisdiction, the Hollanders getting possession of Surinam again. Afterwards Penn received it as a property reward for the large services rendered the English government by his father. . . .

Mr. William Penn arrived with a group of his followers, namely Quakers, whom England was glad to get rid of, and began settling the country. He founded a city at Wicaco, called Philadelphia, to which the landowners objected for a long time, but who were finally won over by kind words and other means. Philadelphia had a modest beginning, but could within twenty years take pride in its own splendor and strength. Penn, as proprietor and governor absolute, returned home, and put Mr. William Markham in his place. The Swedes were offered the opportunity of banding together and living after their own laws, but they submitted to the English, which they and their children bitterly but uselessly regret.

Thereupon it happened that an impostor appeared, who claimed the name of Königsmarck.[2] He started a revolt; got many followers, especially among the Finns; was captured, branded and exiled. His confederates lost their land and suffered great harm; and besides, many innocent colonists were brought into ill repute with W. Penn, who otherwise was very kind to them, presenting them with a small box of catechisms and books as well as a copy of a Bible in folio for their church.

[Then follows several pages (193-221) of local church history up

---

[1] Laus Deo den 24 Maius A:o 1673.

Iik ondergeschrewen Armegot Printz bekenne gelewert te habben aen de gement hier Oowen von de Augsburgsche Confessie, de Klock di aen Tennakom geweesen sijnde, dat sij dar meede daen wat haer belifwen sar en belowe haer te bevrijden voon alle naemanige dit gedaen sijnde, voor ondergeschrewen getuygen. Datun als oowen.

Het merk van
P.K.
Peter Kock
Het merk van
X.
Jonas Nilson

Armegast Prins.

[2] His real name was Marcus Jacobson. See below, pp. 732 ff.

to about the year 1735, giving the plight of the colonists because of the lack of pastors, their urgent correspondence with the Swedish authorities to procure clergymen, and their final success in these negotiations which resulted in the arrival in New Sweden (1697) of Anders Rudman, Eric Björk and Jonas Aurén. A list of parishioners and statistical details of births, burials and weddings are given, as are also the wages of the laborers who helped build the new church. A description of the unique administration and troubles of Jacob Fabritius, German pastor of a Swedish congregation, is briefly sketched, and an enumeration made of the religious tracts, catechisms and other books which were sent to New Sweden with Rudman[1] *et. al.* by the Swedish king and government. Since most of these local details, however, are already reasonably well known to the readers most interested, and have been pretty thoroughly exploited by previous writers they are here omitted].

Remarks on [William] Penn. Mr. Lewis Evans told me the following about Penn: there is good reason to believe that W. Penn was not a real, good Quaker at heart, for he was in reasonable favor with King James II of England, who as we know, was a strong Catholic. It is true that W. Penn in a letter to the archbishop in London assured him that he was not a Catholic; but as a politician he might well have been clever enough to make that assertion; it would not have been wise in those days to let anyone know that he was of the Catholic faith. However, all Quakers wish to believe that he was a member of their Society; but it is known that often Catholics conceal themselves under that name among the English.

It is believed by most people that the form of government which Penn furnished to his province, and in accordance with which it was to be ruled, was drawn up by Penn himself, since it bears his name; but there is strong evidence that it had first been outlined and formulated by Sir William Jones [1631-1682] in England, who was one of the best men that England at that time possessed in the field of law. He is said to have been attorney-general of England [from 1675-1679].

Thomas Penn [1702-1775], who is now [hereditary] proprietor, is one of the meanest and stingiest of men. [For example] he tries

[1] Rudman and his party spent ten weeks on the ocean between England and America. The MS copy of his diary of the trip is preserved in the Yale University Library.

to obtain a personal examination of all former deeds on the land which his father gave away, and when he under some pretext succeeds in procuring them he often does not return them at all but has the property in question resurveyed, and then he lops off a piece, maintaining that the party has more than he is entitled to. Many residents in Philadelphia have lost land in this way, and it makes no difference of what nationality they may be, if he only can find some excuse to get the land. It is known that the land on which Philadelphia is built formerly belonged to a Swede,[1] and that William Penn made a trade with him, giving him twice the amount of land in another locality, four English miles from Philadelphia. But the present proprietor thought it was too large a grant, so he had the land surveyed three or four times, and each time he cut off a slice from it.

A New Academy Established. When this year, 1749, a new academy was established for the education of youth, the trustees of the same wrote to the proprietor Thomas Penn, in London, thinking the news would give him a singular pleasure. He answered he wished them luck in their undertaking, but he felt that the country was still too young to found academies;[2] he feared the public would not be able to stand the expense. The residents of the city, who heard this, thought he would have done better if he had donated a few acres of land for said academy, for he himself would in the future have profited the most from the gift. Whenever they had wanted to build lodging houses (dormitories) for the children (students), he would have received a good income in rent if these had been erected on his land.

Houses and Buildings in New Sweden (Addenda). Since it has frequently happened that a disastrous fire has broken out because of having the [large] oven in the cabin or dwelling house, the cus-

[1] It belonged to the Svensson (or Swensson) brothers, as Kalm has previously pointed out.

[2] The University of Pennsylvania had been founded in 1740. The article on Thomas Penn in *The National Cyclopædia of American Biography*, lists him nevertheless as a "benefactor if not a founder of the College of Philadelphia (afterward merged in the University of Pennsylvania), as well as of the hospital and library there," but adds that "these gifts were well within his power, as he was the chief proprietor of one of the largest feudal estates in the world." The proprietary estates were exempt from taxation, which later led to serious complaints and disputes. Benjamin Franklin was in 1757 sent to England to present the Colonial views to Thomas Penn, who had returned to London in 1747.

tom has now been abandoned entirely. So I have not seen an oven in a cottage anywhere. It is now built separately in the yard, a short distance from the houses, and is generally covered by a little roof of boards, to protect it from rain, snow and storms. Usually it is elevated a few inches above the ground so that chickens and other small animals can stand under it when it rains. The reason why some colonists still construct both fireplaces and ovens of nothing but clay is that in many places here it is impossible to find a stone as big as a fist, not to mention larger ones; that but little experience has been had in brick-making over a fire; and that, besides, a first settler who erects a house on a plantation cannot afford to buy brick.

From his maternal grandfather Rambo, who had been one of the earliest settlers, Mr. Jacob Bengtson had learned that the board ceilings in the first colonial houses had been covered with earth to prevent the heat from escaping through the top. No dirt is used now, only thin boards, and not any too many of these. Probably there was no real damper in those days, and when it was very cold and the fuel in the fireplace had all been burned, someone in the house very likely crawled up on the roof and shut in the heat by placing a board wrapped in cloth over the chimney. There were no stone houses before the English came. The Swedes knew nothing of brick-making or burning lime; the first settlers of New Sweden were ignorant folk.

The Swedes. From an old man, Mårtin Gäret [Martin Garret], seventy-five, I learned the following observations: he said he had never seen or heard of any Swedes who used dampers here, but he had heard them say that they used them in Sweden.[1] However, he himself had made a cover which he placed over the chimney on cold nights, thereby retaining much more heat than usual. But it was a lot of trouble to climb up on the roof of the house every night and morning.

The fireplaces made at the time were built in a corner of the dwelling room. He still had such a hearth in his room, which resembled those we build in our guest chambers or bedrooms. They are now called Swedish fireplaces here, and are said to be quite

[1] Kalm was extremly interested in the economy of heat, as was natural from one who came from the Scandinavian North. Time and again he comes back to the subject of dampers.

rare. The most common ones now are English, which are as large as our kitchen hearth, though the bottom of them is not higher than the floor of the room.

The Swedes raised many peach trees in his youth, and a sufficient number of apple trees, but no cider was made then. They made a malt beverage at that time.

Formerly the houses of the Swedes were all of wood, with clay smeared between the logs, like those now built here by the Irish. They have no glass in their windows, only small loopholes with sliding shutters before them, just like our Finnish cabin windows.

The weather was much the same as it had been in years past.

He did not know whether the people formerly had better teeth than now; he still had most of his own.[1] He had never liked tea;[2] according to Rev. Mr. Jonas Lidman,[3] the eating of hot bread was responsible for the bad teeth.

Fever and ague had formerly been rare, and when it did come it was not so severe as it is now; that virulent form of fever was not known.

Jacob Bengtson was now a man of about sixty years of age. His father was Anders Bengtson, who had emigrated on the ship *Mercurius* that arrived in America right after the Dutch had taken possession of it. This Anders Bengtson is the same man mentioned so often [in the records] as being among those who wrote to Sweden after clergymen immediately before Mr. Rudman's day (before 1697). He was also the one who [on Sundays when there was no preacher] read to the Swedish congregation out of a book of religious reflections and sermons [postilla], after the days of Fabritius, and before Rudman and Björk had arrived. From this Mr. Bengtson I learned the following:

Rev. Jacob Fabritius was blind for several [9] years. He was of German nationality but born in Poland. He could not preach in Swedish—he did this in German—which most of the Swedes of that day understood. When he was to preach he had someone read the sermon to him and then he would expound it.

---

[1] This is opinion or information supplementary to what Kalm had said about the subject before.

[2] It will be remembered that Kalm thought the excessive tea-drinking in the Colonies might have something to do with the prevalence of poor teeth.

[3] Pastor of the Wicaco Church from 1719-1730.

Photograph by Philip B. Wallace

Old Swedes' Church, Wilmington, Delaware

The bell which now hangs in the Wicaco Church [Gloria Dei] belonged formerly to the church on Tinicum, and was brought hither from Sweden when the settlement belonged to the Swedish government. When the Tinicum Church was abandoned the Swedes of Wicaco had it brought hither; but when Mr. Rudman and Mr. Björk came, and two separate congregations were formed, Wicaco and Christina [Old Swedes' at Wilmington], a dispute arose as to who should have it. The members of Christina felt it belonged to them as much as it did to Wicaco, and wanted to claim it, while those of Wicaco refused to give it up. The quarrel lasted until a Swedish captain by the name of Jacob Trent arrived and settled it. He pointed out that while the bell really belonged to both parties, it could not very well be cut in two, and that therefore it had better remain where it was. He had on his vessel two ship-bells, he said, and was willing to present one of them to Christina Church. This he did, and the quarrel was settled.

This Captain Jacob Trent was said to be a Scotchman by birth, though born of a Swedish mother. He arrived in New Sweden the year after Mr. Rudman (i. e. in 1698) with a large ship, well-manned, for he had more than ninety men on board, and armed with big cannon. Most of the officers were Swedish, and among them was also a nobleman by the name of Oxhufvud. Mr. Bengtson thought it was a Swedish ship. It brought a large number of Scotchmen who came to settle here, and whom the captain sold to the Englishmen, who payed for their passage. He spent a whole winter in this country, near Chester. The English did not know what to think of him. At times they thought of making a prisoner of him, but there were too few people in the locality, and his ship too formidable, so that they did not dare attack him. His crew were for the most part Swedes and it was believed that he had been sent by the King of Sweden to inspect the country. Many Swedes who were carried here on this vessel remained in America. From here Trent went to the Barbados, where he was deprived of most of his men, because he seemed too strong and dangerous on the high seas. He was allowed only as big a crew as was absolutely necessary to the navigation of his ship. Some time afterwards he died, and only a few of his men returned to Sweden, among whom, however, was the above-mentioned Oxhufvud. Captain Trent had a brother in town who was a wealthy merchant, and from whom Trenton is said to have received its name, because he first settled in that locality.[1]

Mr. Jacob Bengtson's maternal grandfather was Peter Gunnarson Rambo. He is the man mentioned in Postmaster [Johan] Thelin's letter to these [Swedish] congregations, in which it is mentioned that he has a sister in Stockholm, etc.[2] His name first was Peter Gunnarson Ramberg, but changed the latter to Rambo. He came to this country from Västergötland with the Swedish Governor [Printz], and was then a bachelor, and was still living when Mr. Rudman arrived, though he died soon thereafter. He is the ancestor of all the many Rambos who now inhabit this land or have ever dwelt here. His oldest son's name was Gunnar.

Mr. Isaac Bengtson's father was Anders Bengtson, also from Västergötland.—I asked the former whether he knew of any Swedish clergymen here from Lock (enius)'s time.[3] He answered that he could not name any, but that he heard that two or three lay buried in Traanhuken, a place just below Christina-kill, where one of the first Swedish churches was built, but that is all he knew.

The Swedes whose names are Johnson are also from Västergötland.

Those who bear the name of Keen were said to be of Dutch extraction.[4] The ancestor's name was Kusk (?) which was later changed to Keen, pronounced "Kijn."

Kock, or as it is now written, Cock, represents a large family here. The ancestor of that family, who came here from Sweden, was Pehr Johnson, but since he was cook on the ship he was given the name Cock, and kept it ever since.[5] He was born in Uppland or in that vicinity, and died before Rudman's days.

---

[1] Trenton was named in 1719 after William Trent, Speaker of the House of Assembly, probably the brother of Jacob, as Kalm relates.

[2] Johan Thelin, postmaster at Gothenburg, served as effectual intermediary between the Swedes of America and his own government, when they needed more clergymen. The letter in question is dated Nov. 16, 1692, and is printed in an atrocious English translation by Carl Springer in the omitted part of the Kalm-Elfving *Tilläggsband*, pp. 192-201.

[3] Rev. Lars Carlsson Lockenius arrived in New Sweden in January, 1648.

[4] This may be true, but the original (Swedish) name was Kyn. See Gregory B. Keen, *The Descendants of Jöran Kyn of New Sweden*. Jöran Keen was a Swedish-born soldier, who sailed from Stockholm with Johan Printz in 1642.

[5] Cock arrived 1641 on the ship *Charitas*. Amandus Johnson gives his original name as Pehr *Larsson* Cock. See *Swedish Settlements*, 712.

Most (?) of the people who settled at Christina or Traanhuken were from Finland, such as Mr. Björk's mother-in-law, and others. The ancestor of the Holsten (Holstein) family came from Germany, according to the testimony transmitted by the first Holsten settler. Tolsa, Mullika, and Likonen may be recognized as Finnish names. Toy was said to have come from Holland, but old Garret claimed the family of that name came from Finland.

[As mentioned before], the name of the family who owned the land where Philadelphia is located is Svensson (Swensson). The original settler was Sven Gunnarson. He had three sons, Sven, Olof and Anders; of these the first was born in Sweden, Olof on the way hither, and Anders in this country, one of the first Swedes born in America. There had only been a couple of children born here before him. From these three brothers Penn purchased the land on which the city was built. . . . North of Wicaco Church there stands yet an ancient wooden structure, which belongs to this family, and which is now rented to some Germans. This house is older than Philadelphia, but now it is rapidly approaching its final doom.

The wife of Lars Lock was the first female child born of Swedish parents in the colonies.

Isaac Bengtson's mother had often told him that when she was a girl she had seen a man carrying the mail between Boston and Virginia on an ox which he rode. He had visited her father both on the down as well as on the return trip, and had lodged there over night. He went away in August and returned the following May. There were no horses here at that time.

Usually one or two merchant boats appeared in the Delaware every year, laden with salt, linen, etc. The Swedish bought goods from them; but sometimes it happened that no "yacht" came during the entire summer. In that predicament the colonists had to obtain their materials in New York, and to get there either had to walk or ride on oxen or cows. If they went on foot there were generally several together in a party, and then they carried their purchases on their backs. Now and then Indians were hired to do the lugging. They had no money and paid for their goods either with skins or wampum, which they got from the Indians.

The Swedes whose name is Iockum (Jokum or Yokum) came originally from Germany.—Anders Whiler (Wheeler) who is counted among the Swedes, was an Englishman on his father's side but a Swede on his mother's.—The first Stille to appear in America was Olof Stille.—Bure is believed to be Finnish,[1]—Wallrave is a common name in Christina, but the Swedish ancestral place of the family is unknown.

From the above it may be seen that not all who here pass for Swedes have come from Sweden; but some have come from Germany, Holland, England, and other places, either because one of the parents was Swedish or they were married to Swedes.[2]

Among the living old Swedes here is a man by the name of Garret. His older sister is still living and is over seventy years old. Her husband, Johan Scute, is dead, and was the son of the [Sven] Scute who was vice governor of New Sweden when that belonged to the Swedish crown. There are no Scutes left on the male side.

In Mr. Rudman's observations quoted above I mentioned an impostor by the name of Königsmarck. Mr. Isaac Bengtson told the following tale about him.—This rascal, of Swedish birth, had been transplanted from England to Maryland to serve a number of years as a servant or slave. He was duly sold, but escaped and joined his fellow-countrymen in New Sweden, pretending to be of a distinguished family in Sweden whose name was Königsmarck. He said the Swedish fleet lay outside in the Bay and would at the first opportunity capture the land from the English; that he had been sent to encourage the Swedes who dwelt here to cast off the foreign yoke; and exhorted them to join in slaying the English as soon as they heard the Swedish fleet approach. A large proportion of the Swedish colonists let themselves be persuaded, and concealed the alleged Königsmarck in the Colony for a long time, that no one might learn about his presence. They carried the best food and drink they had to him, so that he lived exceedingly well, and what is more, they went to Philadelphia and bought powder, bullets, lead, etc. to be ready at the first signal. He had the Swedes called together to a supper, and after the drinks had been passed he ex-

[1] That is, Swedish-Finnish, in this case. Probably a large proportion of the so-called Finns who came to America at the time were of Swedish origin, though there were some, as we have seen with pure Finnish names, in which case the father at least was of the original Finnish blood.

[2] It should be remembered, too, that up to the eighteenth century, Sweden had possessions in Northern Germany, whose inhabitants, like the Finns, were (politically) Swedish nationals.

horted them to throw off the old rule, reminded them of what they had suffered, and finally asked them whether they sympathized with the King of Sweden or the King of England. A few declared themselves at once for the Swedish ruler, but Peter Kock pointed out that since the land was English and the settlement had been duly ceded to the English crown he ought to support the English sovereign. Thereupon he ran out, slammed the door, and braced himself in front of it so that the alleged Königsmarck could not get away, and called for help to arrest him. The impostor tried to force open the door, and Kock stabbed his hand with a knife; though the swindler got away [temporarily]. But Kock reported the matter to the English, who set out and made the alleged Königsmarck a prisoner. Captain Kock then demanded his real name, for, he said, "We can see that you are not of noble blood." He then admitted that his name was Marcus Jacobson. He was so ignorant that he could neither read nor write. After being branded he was sold in the Barbados as a slave.[1] The Swedes who had sided with him lost half of what they owned—land, cattle, clothes and other goods.

Mr. Bengtson could not remember the Swedish church on Tinicum, for it had fallen into decay before his day, but he recalled clearly the old wooden church at Wicaco which had remained standing many years after the present stone [brick] structure had been erected. It was located immediately south of the present temple, and so close to it that one could barely squeeze into the church between the two. Rev. Fabritius is buried in the choir of the old church, or now right in front of the southern church door.—Iongh [Jongh or Young] who lived during the days of Fabritius is said to have been a schoolmaster in this place.

Mr. Jacob Bengtson related the following story about Governor Printz: when he arrived in his ship the Indians came by the thousands to see him, as soon as they learned that he had come. He then entertained a few of them and had such trifles distributed among them as were considered very precious by them. The shores were full of their boats. They had heard the roar of the cannon and marveled mightily over it, calling them Manito or God. The Governor asked them if they would like to have such a Manito, since they

---

[1] James Kirke Paulding, The American author, wrote a hasty novel of New Sweden. He called it *Koningsmarcke, the Long Finne,* the only striking name he could remember, apparently.

had none before. At first they did not know what to answer, but finally they approved of it and were happy at the prospect of possessing such a gift. The Governor had the cannon loaded heavily with balls and brought on land; a long thick cable was attached to it, and the natives were told to pull it away. All took hold of the cable and pulled; the cannon was heavy and sank down in the mud so that they could scarcely jerk it from the spot. The Governor told them they must all walk in a straight line in front of it, otherwise it would not follow them. They did so, but without much success. On the shore lay some embers from the many fires they had first built on the shores; so the Governor suggested to them to take a firebrand and exhort the Manito to action by burning its tail, when it would move faster. An Indian took a burning ember, smiled and laughed, and applied it to the fuse, which he took to be the tail. The cannon went off and killed all the Indians in the cable-team in front. The others were so terrified at this, believing Manito had become wroth over their action that they dashed headlong for their boats and began to row with all their might toward the other side of the river. Governor Printz then had the cannon loaded with chain shot and fired at the fleeing Indians. This not only mowed down many more natives but the trees where they were trying to land. This increased the terror of the Indians, who felt they must be safe on the opposite shore, when Manito was on the one behind. They abandoned their boats and anything else they had, fled to the woods, and did not stop until they had arrived at the sea near Cherbour (?), which was over six or seven Swedish miles distant, and dared not return for a long while.[1]

Otherwise Governor Printz exercised rather good control over the natives, who were afraid of him. One of their chiefs had the honor of being admitted to his table whenever he visited, and the Governor favored him in several other respects. Once when this sachem had been intoxicated among his own people he had ventured to assert that next time he dined with the Governor he would

---

[1] So far as the editor can learn there is no foundation whatsoever for this stupid tale. Governor Printz did not like or trust the Indians, but did not go out of his way to provoke them; he was too prudent to do otherwise, and had been instructed to keep on friendly terms with the natives. This was not always possible, but the killings on either side were relatively few. The above legend shows what proportions a story may assume during the course of a century.

cut his throat. The Governor learned about it. A short time later the chief came on his visit to the Governor and expected as usual to be entertained; but the latter kicked him out, and reminding him of the former hospitality told him that thereafter he would be forced to eat with his dogs since he [the Governor] had no more respect for him.

When the Swedes had dwelt in this country for a while the Indians on the New Jersey side conspired to kill all the Swedes on a certain night. In these days the colonists were not so scattered as they are to-day, and had concentrated in just three or four locations, with several living together in each place in a common dwelling place like a blockhouse. The Indians had planned to start the massacre at one end [of the settlement] down by the River, and then to proceed inland during the night up to the other sections. The plan was to slay before the victims could learn of the fates of those already killed. But during the night or the day before [the massacre was to take place] the sachem's squaw or queen appeared and warned the Swedes. They made preparations for defense, and also warned their neighbors so that they might be prepared. When the savages arrived and perceived that the Swedes stood ready to receive them, they realized that they had been betrayed, returned hastily, and sought to ferret out the traitor. After a thorough investigation they learned that it had been their own chief's wife. Thereupon they transfixed a stake lengthwise through her body, and thus let her miserably perish. The next day they came and made a new treaty with the Swedes.—Thus Jacob Bengtson.

Extracts of Pennsylvania Laws. Law concerning liberty of conscience. Whosoever believeth in God the Father, His Son and Holy Ghost, and acknowledges the Holy Scriptures and wishes to live in peace, shall have religious freedom.

The governor alone has the right to purchase land from the Indians. [N. B. One observes through the whole book of statutes] that a large number of laws had formerly been enacted, but that the majority of them had subsequently been repealed.[1]

Swearing is punishable with a fine of five shillings for the first offense, or five days in jail; second offense, six shillings; third, 10

[1] This was 186 years ago. It is clear that we have not made much progress in the method of lawmaking since 1750.

shillings or as many days in prison.—The blasphemer has to give 10£ to the poor and spend three months in the penitentiary at hard labor.

Justices have the right to lay out roads. If anyone fills them in [i. e. blocks them by obstructions of any kind] he is fined 5£.

Fences must be five feet high.

A ship carrying sick people may not approach within a mile of Philadelphia or any other city. The breaking of this ordinance is punished by a fine of 100£.

If someone cashes a check on an English bank and it is returned with a protest, he must pay a fine of 20£ and restore the money.

The produce of the land, such as wheat, rye, corn, barley, oats, pork, meat and tobacco shall in trade be valid as money, unless the contract has definitely stipulated payment in silver.

Fire in a chimney entails a fine of 40 shillings. Every house must have a water barrel and a leather bucket; failure to comply with this provision brings a fine of 10 shillings. Smoking on the street costs 1 shilling and the fine is used for purchasing leather buckets, pumps, etc. for the public welfare.

There is a bounty of three pence per dozen on blackbirds, and threepence apiece on crows. The heads should be shown to a town official.

Anyone selling rum to the Indians, secretly or otherwise, must pay a fine of 10£.

He who works on Sunday is fined 20 shillings.

An adulterer must suffer twenty-one lashes on his bare back and spend a year in prison at hard labor or pay a fine of 50£. Besides, the wife who has suffered shall have the right to divorce the guilty from bed and board. If he be guilty a second time, he shall receive the same number of lashes and seven years of hard labor in prison, or pay a fine of 100£, and a like amount for every subsequent offense. Fornication is punishable with a fine of 10£ or twenty-one lashes on his bare body at a whipping-post. If a woman gives birth to an illegitimate child and at the birth or in court accuses someone, he shall be father to the child. Anyone proven father shall support the child.

Bigamy is punished by thirty-nine lashes on the bare back, life imprisonment at hard labor, and declaration of invalidity of the second marriage.

Pigs must have a yoke, and rings in their snout. Swine without these may be killed with impunity within fourteen miles of the Delaware or any other navigable river. (Afterwards this law was extended to include all Pennsylvania.) No pigs may run loose in Philadelphia, Chester or Bristol. Offenders against this ordinance lose the pigs thus running wild.

All saloons and taverns must be licensed.

To prevent degeneration of horses small stallions, eighteen months old or more are forbidden to run loose. They must have a height of thirteen hands, counting four inches to a hand. Such a stallion may be gelded by anyone at the risk of the owner, who should besides pay a fine of 10 shillings.

A bounty of fifteen shillings was first paid for wolves, but this was later raised to 20 shillings. For a young wolf cub the bounty at first was 7 shillings sixpence, which was raised later to 10 shillings. An old red fox brought 2 shillings' reward, a young red fox cub, 1 shilling.[1]

---

[1] In addition to the material of the diary notes, which ends here, readers of Swedish will find some supplementary observations on America in Kalm's private correspondence with his teacher Linné and others, and especially in his letters from America, of course. This correspondence, to which we have already referred, has been edited and published by J. M. Hulth in series of *Bref och skirfvelser af och till Carl von Linné* (under the general editorship of Th. M. Fries and J. M. Hulth), Första Afdelningen, Del VIII, Uppsala, 1922, pp. 1-118. A few other references to Kalm's travels in America are found in other volumes of this large collection of Linné letters, the publication of which began in 1907, but all of the essential material of general interest to America is found in his diary as here published. Yet we refer to the above correspondence for some personal, intimate matters.

METEOROLOGICAL OBSERVATIONS

In the first column of these tables, the Reader will find the days of the month; in the second, the time or hour of the day when the observations were made: in the third, the rising and falling of the thermometer; in the fourth, the wind: and in the fifth, the weather in general, such as rainy, fair, cloudy, etc.

The thermometer which I have made use of is that of Mr. Celsius, or the Swedish thermometer so called [the Centigrade], as I have already pointed out in the preface.[1] To distinguish the degrees above freezing point from those below it, I have expressed the freezing point itself by 00, and prefixed 0 to every degree below it. The numbers therefore which have no 0 before them, signify the upper degrees. Some examples will make this still more intelligible. On the 17th of December it is remarked, that the thermometer, at eight o'clock in the morning was at 02.5. It was therefore at 2 degrees and 5/10 below the freezing point: but at two in the afternoon it was at 00.0, or exactly upon the freezing point. If it had been 00.3, it would have signified that the thermometer had fallen 3/10 of a degree below the freezing point: but 0.3 would signify that it had risen 3/10 of a degree above the freezing point. Thus likewise 03.0 is three degrees below the freezing point; and 4.0, four degrees above it.

The numbers in the columns of the winds signify as follows: 0, is a calm; 1, a gentle breeze; 2, a fresh gale; 3, a strong gale; and 4, a violent storm or hurricane. When, in some of the last tables, the winds are only marked once a day, it signifies that they have not changed that day. Thus, on the 21st of December, stands N.0 fair. This shows that the weathercocks have turned to the north all day; but that no wind has been felt, and the sky has been clear all the day long.

Before I went to Canada in summer 1749, I desired Mr. John Bartram to make some meteorological observations in Pennsyl-

[1] This preface has not been translated literally in this work but the substance of it has been incorporated in the Introduction.

vania, during my absence, in order to ascertain the summer heat of that province. For that purpose, I left him a thermometer, and instructed him in the proper use of it; and he was so kind as to write down his observations at his farm, about four English miles to the south of Philadelphia. He is very excusable for not putting down the hour, the degree of wind, etc., for being employed in business of greater consequence, that of cultivating his grounds, he could not allow much time for this. What he has done, is however sufficient to give an idea of the Pennsylvanian summer.

AUGUST 1748

D.	H.		Ther.	Wind	The Weather in General
1	5	m	20.0	E S E 2	Fair
	2	a	24.5	E 2	
2	5	m	22.0	E 2	
	2	a	24.5	E 2	
3	5	m	22.0	E 1	
	2	a	25.5	S S W 1	Cloudy with some rain.
4	5	m	22.0	S 1	Alternately fair, cloudy and rainy all day.
	1	a	21.0	S 1	
5	5	m	17.0	S S W 1	Chiefly rainy.
6	7	m	17.0	S 2	Cloudy.
	2	a	19.0	S 2	Somewhat cloudy, but chiefly fair.
7	5	m	15.5	S S W 2	Alternately fair and cloudy.
8	5	m	18.0	S S W 0	Fair all day.
	3	a	19.0	S S W 0	
9	6	m	17.5	W N W 0	
	4	a	21.0	W N W 1	
10	6	m	18.5	E 1	Fair.
	3	a	20.5	E 1	
11	6	m	47.0	E N E 1	Somewhat cloudy.
	12:30	a	18.5	S W 1	Fair.
	4		22.0	S W 1	
	6		22.00	W 3	
12	6	m	16.0	N W 1	Cloudy with some drizzl. rain at ten.
	4	a	19.0	N W 1	Cloudy, fair, some drizzl. rain altern.
13	6	m	17.0	W N W 2	Cloudy with some rain; foggy; sometimes
	2	a	18.5	W N W 2	fair.
14	5	m	18.0	W S W 0	Somewhat cloudy; fair from 11m. to 3a.
	4	a	20.0	W S W 0	Cloudy.
15	5	m	18.0	W S W 0	Cloudy; sometimes fair; at ten o'clock fell
	2	a	19.5	N E 2	a thin fog.

D.	H.		Ther.	Wind	The Weather in General
16	6	m	18.3	N N E 2	Somewhat cloudy; sometimes fair.
	2	a	18.5		Dark; rainy at night.
17	6	m	18.5	E N E 2	Dark; with some drizzling rain.
	2	a	19.5		Drizzling rain all the afternoon.
18	6	m	19.0	E 2	Drizzling rain all the day.
	2	a	20.5		
19	6	m	19.5		Cloudy.
	2	a	20.0		Scattered clouds.
20	6	m	19.5		Fair.
	2	a	21.5		Scattered clouds: sometimes rain.
21	6	m	20.8	E 1	Somewhat cloudy, fair at nine.
	2	a	21.3		Thin clouds.
22	5	m	21.0		Fair: about twelve it became cloudy.
	1	a	23.5	E S E 1	Cloudy.
23	5	m	22.2		Scattered clouds.
	7			S E 2	
	2	a	24.2		Scattered clouds, dark towards eve.
24	5	m	23.5	W S W 2	Violent rain.
	6			W 2	
	7			W N W 1	About seven it cleared up.
	9			N W 1	
	2	a			Scattered clouds.
25	6	m	24.5	W 1	Scattered clouds.
	10			W N W 3	
	2	a	23.5		
26	6	m	24.0	W 2	Fair. At night a great halo appeared round the sun.
	2	a	24.5	S W 2	Dark. A strong redness at sunset.
				W S W 1	Cloudy. At ten it began to rain, and it rained all day.
27	6	m	24.5	S E 2	
	11			E 3	
	1	a		N E 4	Rain.
	4		21.5	N 1	Scattered clouds.
28	7	m	23.0		
	2	a	23.5	S W 1	
29	6	m		S W 3	Towards evening drizzling rain and lightning.
	2	a	25.5	N W 2	Scattered clouds; air very cool.
30	6	m	23.5		
	2	a	21.5	S W 1	Fair: in the morning it began to grow cloudy; at night lightning, hard rain, and some thunder.
31	6	m	22.2		

SEPTEMBER 1748

D.	H.		Ther.	Wind	The Weather in General
1	7	m	20.0	N W 2	Scattered clouds.
	2	a	21.5		Clouds passing by. Rain and strong winds all the afternoon.
2	6	m	19.0	N W 1	Scattered clouds all day.
	2	a	20.5	N W 0	At night a great halo round the moon.
3	6	m	21.5	W S W 0	Scattered clouds.
	2	a	23.0	S 1	It became more cloudy. In the evening appeared a great halo round the sun.
4	6	m	23.3	E 1	Scattered clouds.
	12	n	27.5	E S E 1	
	2	a	24.0		
5	6	m	24.5	S E 3	Scattered clouds.
	12	n	26.5		
6	6	m	27.0	S E 2	Scattered clouds.
	1	a	28.5		At night a great halo round the moon, and the sky very red.
7	6	m	27.5	E 3	Dark sometimes. The sun shone through the clouds.
8	12	n	28.5	N E 2	Scattered clouds.
8	6	m	26.0	N N E 2	Scattered clouds all day.
	1	a	26.5		
9	6	m	24.5	N 1	Scattered clouds all day.
	1	a	24.5		
10	5	m	24.0	N N W 1	Fair.
	1	a	24.5		
11	6	m	23.2	W N W 1	Fair.
	2	a	25.0		At night a halo round the moon.
12	6	m	24.0	A Calm	Fair, and very hot.
	12:30	a	26.0		
13	5	m	25.5	S E 1	Fair.
	1	a	26.5		
14	6	m	25.5	S E 1	Fair; but a cool wind all the morning.
	1	a	26.5		
15	5	m	23.0	S E 1	Scattered clouds.
	1	a	27.5		It grew more cloudy. In the evening and ensuing night, violent rain and winds.
16	5	m	21.5	N N E 1	It rained hard all day.
	2	a	21.5		
17	5	m	25.5	N W 1	Cloudy.
	1	a	21.0		Scattered clouds.

D.	H.		Ther.	Wind	The Weather in General
18	6	m	13.0	Calm	Fair.
19	1	a	24.5	NNE 1	Fair all day.
20	6	m	14.0	NE 1	Scattered clouds.
21	6	m	11.0	NE 0	Scattered clouds.
	1	a	23.0		
22	7	m	10.5	NE 1	Fair.
	1	a	25.0		
23	6	m	11.0	NNE 1	Fair.
	2	a	28.0		
24	6	m	14.0	NE 1	Fair
	2	a	28.0		It grew dark. At night came rain, which continued late.
25	6	m	18.0	NW 1	Dark. At 8, scattered clouds.
	2	a	28.0	NE 1	Scattered clouds.
26	6	m	15.5	NNE 1	Fair.
	2	a	27.5		
27	6	m	17.0	NE 1	Cloudy. Fair at 8, and all the morning.
	2	a	27.0		Cloudy.
28	6	m	14.0	NE 1	Fair and cloudy alternately.
	2	a	20.0		
29	7	m	15.5	NE 1	Cloudy.
	2	a	20.5		Fine, drizzling rain.
30	7	m	16.0	NE 0	Alternately fair and cloudy.

October 1748

D.	H.		Ther.	Wind	The Weather in General
1	6	m	19.0	W 1	Fair. Scattered clouds at 8.
	2	a	18.5		Scattered clouds. Dark towards night.
2	6	m	18.5	SW 0	Cloudy.
3	6	m	15.0	NW 1	Cloudy.
	1	a	18.0		Scattered clouds. Late at night a great halo round the moon.
4	7	m	6.0	NW 1	Fair.
	1	a	16.0		
5	7	m	2.0	N 1	Fair.
6	7	m	2.0	NE 1	Fair.
	1	a	18.0		At night a great halo round the moon.
7	6	m	7.0	ENE 1	Cloudy. Fair at 9, and all day.
8	6	m	14.0	ENE 1	Cloudy. Scattered clouds at 8.
9	6	m	18.0	SSE 1	Rain all the morning.
	3	a	23.0		Cloudy.
10	6	m	20.0	SW 0	Fog, and a drizzling rain.
	2	a	23.0		Fair.

D.	H.		Ther.	Wind	The Weather in General
11	7	m	20.0	S W 1	Fog, which fell down. Fair at 8.
	2	a	26.0		Fair.
12	6	m	8.0	W N W 1	Fair all day.
	8			W 1	
	2	a	20.0	W S W 1	
13	6	m	2.0	W N W 1	In the morning, hoary frost on the plants.
	2	a	17.0	W S W 0	Fair all day.
14	6	m	5.0	S S W 0	Fair.
	2	a	21.0		
15	6	m	4.5	S S E 0	Fair.
	2	a	24.0		
16	6	m	11.0	E N E 0	Cloudy.
17	6	m	8.0	N E 1	Cloudy.
	2	a	18.0		Cloudy. Violent rain all night.
18	6	m	12.0	N W 0	Cloudy.
	5	a	4.0	S W 0	
19	6	m	00.0	W S W 1	Scattered clouds.
	2	a	9.0		
20	5	m	01.0	W N W 1	Fair.
	2	a	9.0		
21	7	m	00.0	W 0	In the morning ice on standing water, white hoary frost on the ground; fair all day.
	1	a	15.0		
22	6	m	00.0	W 0	Fair.
23	6	m	4.5	N N E 1	Fair.
	1	a	16.0		
24	6	m	4.5	N 0	Fair.
	2	a	18.0		
25	6	m	4.5	S W 1	Fair. Air very much condensed in the afternoon.
26	6	m	4.0	S W 0	Fair.
	3	a	19.0		
27	6	m	1.0	S W 0	Fair.
	3	a	17.0		
28	6	m	9.0	E 2	Heavy rain all day.
29	6	m	14.0	W 1	Fair.
	1	a	20.0		At night I saw a meteor, commonly called the shooting of a star, going far from N. W. to S. E.
30	6	m	3.0	N W 1	Fair.
31	7	m	4.0	W 1	Fair.
	1	a			

NOVEMBER 1748

D.	H.		THER.	WIND	THE WEATHER IN GENERAL
1	7	m	3.0	S 1	Fair.
2	6	m	4.0	N 0	Fair.
	3	a	18.0		
3	7	m	7.0	N W 1	Fair.
	1	a	14.0	S E 0	
4	7	m	1.0	S W 0	In the morning the fields were covered with white frost.
	12	n	19.0		A fair day.
5	7	m	4.0	S W 1	Fair.
	1	a	17.0		
6	7	m	4.5	N E 1	Fair.
	1	a	12.0		Towards evening somewhat cloudy.
7	7	m	7.0	E N E 1	Cloudy.
	4	a	11.5		
8	7	m	11.5	E N E 2	Drizzling rain.
	12:30	a	18.0	E S E 3	Heavy rain.
9	7	m	17.0	S E 1	Drizzling rain.
	9		15.0	S S W 1	At eight it cleared up.
	1	a	17.0		Scattered clouds.
10	7	m	6.0	S S W 2	Fair.
	12:30	a	13.0	W N W 2	
11	7	m	4.0	W S W 1	Cloudy.
	12:30	a	12.0		Scattered clouds.
12	6	m	03.0	S W 1	Fair.
	2	a	11.5	N W 2	Cloudy.
	4		5.0		
13	7	m	00.0	N N E 1	This morning ice on the water.
	2	a	5.5		Fair.
14	7	m	0.5	N 3	Fair.
	1	a	8.0	N 2	
15	7	m	3.0	S 2	A strong red aurora.
	1	a	8.0		Cloudy and continual drizzling rain.
16	7	m	4.5	W 1	Fair.
17	7	m	01.0	W 1	Fair and cloudy alternately.
	1	a	8.0		Sometimes drizzling rain.
18	7	m	4.0	S 1	Fair.
	3	a	6.5	N W 2	
19	7	m	03.0	W 0	Fair.
	2	a	11.5		
20	7	m	01.0	N N E 1	Fair.
	2	a		S 1	

D.	H.		Ther.	Wind	The Weather in General
21	7	m	15.0	SW 2	Fair.
	1	a	19.0		
22	7	m	20.0	E 1	Rain all day.
	2	a	10.0		
23	8	m	16.0	S 1	Cloudy, foggy, and rain now and then.
	8	a		SW 4	
24	7	m	00.0	WNW 3	Fair.
25	7	m		NW 0	It was very cold last night, and fair today.
26				NW 0	Alternately fair and somewhat cloudy, and always pretty cold.
27					Fair; scattered clouds: pretty warm in the air.
28					Cloudy, foggy, and quite calm.
29					Somewhat cloudy.
30				N 1	Fair, and a little cold.

December 1748

D.	H.		Ther.	Wind	The Weather in General
1				N 1	Fair.
2				WSW 1	Fair, and cold; a great halo round the moon at night.
3				WSW 1	A pretty red aurora, however a fair day.
4	7	m	6.0	SSW 0	Fair.
	3	a	18.0		
5	7	m	5.5	NNE 1	
	4	a	9.5		
6	7	m	6.5	SSW 1	Cloudy.
	3	a	14.0		Somewhat fairer: hard rain in the next night.
7	7	m	13.5	SW 1	Cloudy.
	2	a	19.0		Fair.
8	7	m	5.0	S 1	Cloudy.
	2	a	13.5		Rain and wind next night; thick, but scattered clouds.
9	7	m	12.0	SW 2	
	2	a	10.0	WNW 2	
10				WNW 2	Scattered clouds.
11	7	m	2.0	SSW 1	Fair.
	2	a	12.5		
12	7	m	0.5	NE 1	Cloudy, rain, and fog all day from nine o'clock.
	2	a	10.5		

D.	H.		Ther.	Wind	The Weather in General
13	8	m	7.5	S W o	Foggy, and cloudy.
	2	a	10.0		Next night a strong N. W. wind.
14	8	m	1.0	N W 2	Scattered clouds.
	2	a	2.0		
15	8	m	07.0	W N W 1	Fair and cloudy alternately.
	2	a	01.0		
16	8	m	01.0	W 1	Fair.
	2	a	1.5		
17	8	m	02.5	N W 1	Cloudy, some snow, the first this winter.
	2	a	00.0		
18	8	m	03.0	W 1	Fair.
	2	a	4.0		
19	8	m	1.0	W 1	Cloudy.
	2	a	8.0		Fair.
20	8	m	01.5	W S W 2	Scattered clouds: about six at night were
	2	a	7.5	W S W 1	quite red stripes in the sky, to the north.
21	8	m	07.0	N o	Fair.
	2	a	2.0		
22	8	m	04.5	S E o	Fair.
	2	a	13.0		It grew cloudy in the afternoon.
23	8	m	13.0	S S W o	Heavy rain.
	2	a	18.0		Foggy and cloudy.
24	8	m	13.0	W S W o	Thick fog.
	2	a	17.0	S W 1	Fair; but late in the evening a hard shower of rain.
25	8	m	18.0	S 3	Last night was a storm, rain, thunder, and lightning.
	2	a	18.5	S S E 2	Heavy rain all day.
26	8	m	3.0	W 3	Last night a violent storm from W. and S. and heavy rain. The morning was cloudy, and some snow fell.
	2	a	3.5	W N W 3	Clears up.
27	8	m	04.0	W N W 3	Fair.
28	8	m	07.0	W o	Fair.
	2	a	8.0		
29	8	m	3.0	N N E 1	Somewhat cloudy, and intermittent showers.
	2	a	13.0	— — — o	
30	8	m	8.0	N N E 1	Cloudy and foggy all day.
	2	a	10.0	— — — o	
31	8	m	6.0	W 3	Fair.
	2	a	4.0	N W 1	At night a halo round the moon.

JANUARY 1749

D.	H.		THER.	WIND	THE WEATHER IN GENERAL
1	7:30	m	07.0	N W 0	Fair.
	2	a	4.0	— — 0	
2	7:30	m	04.5	W N W 1	Alternately fair and cloudy.
	2	a	5.5	— — — 1	
3	7:30	m	2.0	N W 1	Cloudy.
	2	a	2.0	— — 1	
4	7:30	m	02.0	W 1	Fair.
	2	a	11.0	— 1	
5	7:30	m	03.0	W 0	Fair.
6	7:30	m	03.0	W 0	Fair, but darkened towards night, with
	2	a	14.5	— 0	some snow.
	5	a	14.5	N W 3	
7	7:30	m	01.0	W N W 1	Somewhat cloudy.
	2	a	3.0	— — — 1	
8	7:30	m	04.0	W N W 1	Fair.
	2	a	8.0	— — — 1	
9	7:30	m	03.0	W N W 1	Aurora, cloudy, heavy rains at night.
	2	a	8.0	— — — 1	
10	7:30	m	15.0	S 2	Cloudy, and showers, some snow at night;
	2	a	2.0	W 4	at 9 m. W.S.W. 3; at 11 m. S.W. 4; at
	4	a			2 a W. 4.
11	7:30	m	03.0	W N W 3	Cloudy.
	2	a	04.0	— — — 3	
12	7:15	m	04.0	W N W 3	Fair.
	2	a	01.5	N N W 2	
13	7:15	m	07.5	W N W 2	Fair.
	1	a	03.0	— — — 2	Cloudy.
14	7:15	m	05.5	W N W 1	Cloudy, and snowed all day; it lay more
	1	a	02.0	— — — 1	than two inches deep.
15	7	m	07.0	W N W 0	Fair.
	2	a	3.0	— — — 0	
16	7	m	08.9	N W 3	All last night W N W 4.
	8	m	09.0		Fair all day.
	2	a	08.0	— — 1	
17	7	m	011.0	N N E 0	Cloudy, snowed all day, and ensuing
	7	a	09.0	— — — 0	night.
18	7	m	012.0	N W 1	Cloudy, and snowed in the morning, fair
	10	m	011.0	— — 1	all the afternoon, and the ther. at 011.0:
					snow lay 5 inches deep.
19	7	m	015.5	W 1	Fair.
	1	m	010.5	— 1	

D.	H.		Ther.	Wind	The Weather in General
20	7	m	012.5	W 1	Fair.
	2	a	07.0		
21	7	m	022.0	W N W 0	Fair.
	2	a	03.0	W 1	
22	7	m	05.0	W 1	Fair.
	2	a	01.0	W 1	Cloudy.
23	7	m	010.0	W N W 1	Fair; a great halo round the moon at night.
	7	a	3.0		
24	7	m	01.0	N N E 0	Cloudy, snowed all day.
	2	a	4.0	N E 0	
25	7	m	00.0	W N W 0	Fair.
	2	a	4.0	W 0	
26	7	m	013.0	W N W 1	Fair.
	2	a	1.0	— — — 1	Cloudy; at three in the afternoon it began to snow.
27	7	m	07.0	W 1	Fair; halo round the moon at night.
	2	a	00.0	— 1	
28	7	m	01.0	W N W 1	Cloudy; snowed almost all day.
	3	a	4.0	— — — 1	
29	7	m	05.0	N N E 1	Fair.
	3	a	03.0	— — — 1	
30	7	m	013.0	W N W 1	Fair; halo round the moon at night.
	3	a	4.0	— — — 1	
31	7	m	04.0	W N W 1	Fair; halo round the moon at night.
	3	a	8.0	— — — 1	

February 1749

D.	H.		Ther.	Wind	The Weather in General
1	7	m	03.0	W N W 1	Fair; a halo round the moon at night.
	1	a	11.0	W 1	
2	7	m	5.0	W N W 0	Fair.
	2	a	6.0	W 0	
3	7	m	00.0	W 0	Fair.
	2	a	19.5	— 0	
4	7	m	5.5	W 0	Cloudy; at ten at night wind N N E 3, snow.
	2	a	11.0	—	
	4	a		N N E 2	
5	7	m	06.0	N N W 2	Fair.
	1	a	03.0	N W 2	

D.	H.		Ther.	Wind	The Weather in General
6	7	m	010.5	N W 0	A cracking noise was heard in all houses the night before. Aurora.—Fair all day, — at 7 in morn. N.W. 0 — at 9, W.N.W. 1 — at 11, W 1 — at 2 in the afternoon, W.S.W. 1.
	2	a	3.0	W S W 1	
7	7	m	01.0	N N E 1	Cloudy — fair — at 7 in the morn. N. N. E. 1 — at 9, N 1 — at 10, W.N.W. 1 — at 12, N W 1.
	2	a	1.0	N W 1	
8	7	m	09.0	N W 0	Fair.
	2	a	7.0	W 1	
9	7	m	03.0	W 1	Fair.
	3	a	16.0	— 1	
10	7	m	7.0	W 1	Pretty clear; a violent storm with rain all the ensuing night.
	1	a	11.0	S S W 4	
11	7	m	9.0	S S W 2	Fair; rain towards night; at night a light similar to an Aurora Borealis in S. W.
	1	a	11.0		
12	7	m	4.0	S S W 3	Fair; about nine at night a faint Aurora Borealis in S.W.
	1	a	10.0		
13	7	m	2.0	W N W 2	Cloudy.
	3	a	5.0	N W 2	Fair.
14	7	m	06.0	N W 1	Fair.
	3	a	02.5	W N W 2	Flying clouds.
15	6:45	m	010.5	N W 1	Fair; at eight in the evening an Aurora Borealis.
	2	a	03.0	W N W 0	
16	6:45	m	013.0	W N W 0	Fair.
	2	a	00.0	N W 1	
17	6:30	m	02.0	W N W 1	Cloudy and snow; wind all the afternoon long.
	2	a	00.0	W 1	
18	6:30	m	2.0	W N W 1	Cloudy.
	2	a	00.0		
19	6:30	m	03.0	N N E 2	Cloudy; rain all day, mixed with snow and hail.
	2	a	01.0		
20	6:30	m	1.5	N W 1	Cloudy.
	2	a	4.5		
21	6:30	m	00.8	N W 0	Cloudy; at 5 in the morning we heard a waterfall near a mill, about a mile S S of us making a stronger noise than common, though the air was very calm—at 10 began a rain which continued the whole day.
	4	a	4.0	N N E 1	
22	6:30	m	3.0	W N W 2	Fair.
	2	a	3.5		

D.	H.		Ther.	Wind	The Weather in General
23	6:30	m	06.0	W 2	Fair.
	4	a	4.0		Some clouds gathered round the sun.
24	6:30	m	4.0	S S W 1	Cloudy.
	3	a	10.0	W 1	
25	6	m	3.0	W N W 0	Alternately fair and cloudy.
	2	a			
26	6	m	012.0	N N W 1	Fair; cloudy at night; at eight in the evening was a halo round the moon and the clouds in S. quite red.
	3	a	02.0		
27	6	m	04.0	N 2	Cloudy, and snow in the morning; but fair at 4 in the afternoon.
	3	a	01.0		
28	6	m	04.5	N W 4	Flying clouds.
	3	a	03.5	W N W 4	

March 1749

D.	H.		Ther.	Wind	The Weather in General
1	6	m	09.0	W N W 2	Fair. A great halo round the moon at night.
	3	a	01.5		
2	6	m	06.0	N W 2	Fair. A faint halo round the moon at night.
	4	a	2.5		
3	6	m	04.0	N W 1	Fair. Cloudy afternoon. About 8 at night the clouds in S W were quite red. At 9 it began to snow.
	2	m	6.5	S 1	
4	6	m	0.5	E S E 1	Cloudy. Heavy rain at night.
	2	a	7.0	S 1	
5	6	m	4.0	W 1	Alternately fair and cloudy. The next night calm.
	2	a	11.0	W 3	
6	6	m	4.0	W 2	Fair.
7	6	m	00.0	W S W 1	Alternately fair and cloudy in the morning. In the afternoon cloudy, with intermittent rain and thunder.
8	6	m	2.0	W N W 0	Fair. About 8 at night we saw what is called a snowfire to the S.W. — See Vol. I, p. 252.
	3	a	20.0	W S W 2	
9	6	m	5.0	N 1	Fair.
	3	a	13.5		Cloudy. Snowfire in S.W. about 8 at night.
10	6:30	m	5.0	S S E 1	Cloudy. Snow and rain all day, and next night.
	2	a	6.5	S E 1	
11	6	m	9.0	S S E 1	Cloudy and heavy rain in the morning. Cleared up in the afternoon.
	3	a	14.0	W 1	

D.	H.		Ther.	Wind	The Weather in General
12	6	m	9.0	N N W 0	Cloudy in the morning. Cleared up at 10.
	3	a	15.0	E N E 0	Towards night cloudy, with rain.
13	6	m	9.5	N N E 2	Cloudy, with heavy rain. Fair at 4 in the
	2	a	8.0	10 m. N 3	afternoon.
14	6	m	4.0	W N W 2	Fair.
	2	a	10.0		
15	3	m	00.0	W S W 0	Fair. Cloudy towards night.
	3	a	13.0	W 2	
16	6	m	2.5	N N E 3	Snow violently blown about all day.
	3	a	01.0		
17	6	m	01.0	N W 2	Cloudy. Cleared up at 8 in the morning.
	3	a	5.0		
18	6	m	02.0	W S W 0	Fair. The fields were now covered w'th
	3	a	4.0	W 2	snow.
19	6	m	02.0	W N W 1	Fair.
	3	a	6.0	N W 2	
20	6	m	05.5	W 0	Fair. Cloudy towards night.
	3	a	11.5	S W 1	Cloudy.
21	6:30	m	2.0	S S E 0	Cloudy. Intermittent showers.
	3	a	14.5		
22	6	m	10.0	S S E 0	Cloudy.
	3	a	19.5		
23	6	m	15.0	S S E 1	Heavy rain.
	3	a	19.0		
24	6	m	8.0	S W 1	Fair.
	3	a	15.0		
25	6:30	m	6.5	W N W 3	Fair.
	3	a	11.0		Flying clouds.
26	6	m	00.0	W N W 2	Fair.
	5	a	11.0	S W 2	Flying clouds. About 8 at night a snow-fire on the horizon in S.W.
27	6	m	3.0	W N W 1	Fair.
	3	a	9.0		
28	6:30	m	3.0	S 1	Rain all the day, and the next night.
	3	a	12.0		
	11	a		N N W 3	
29	6	m	1.0	N N W 2	Fair.
	2	a	6.0		
30	6	m	03.0	E 1	Fair. Cloudy at noon: began to snow,
	2	a	4.0	S E 1	which continued till night, when it turned into rain.
31	6:15	m	5.0	N 1	Cloudy.
	3	a	14.0		

April 1749

D.	H.		Ther.	Wind	The Weather in General
1	6	m	5.5	N N E 1	Rain in the morning,—aftern,—and in the night.
	3	a	3.5	E 1	Snow, with much thunder and lightning.
2	6	m	0.5	N N E 1	Snow almost the whole day.
	3	a	0.5		
3	6	m	02.0	N W 1	Fair.
	3	a	9.0		
4	6	m	02.0	W 1	
	3	a	16.0		
5	6	m	00.5	N 1	Fair.
	3	a	19.0	S W 1	Sun very red at setting.
6	6	m	4.0	S W 1	Fair.
	3	a	23.0		
7	6	m	13.0	S 2	Fair. Cloudy afternoon.
	3	a	24.0		About 7 in the evening it began to rain, and continued till late at night.
8	7	m	9.0	N W 3	Flying clouds.
	3	a	13.0		
9	6	m	1.0	N 1	Alternately fair and cloudy. Snowed in the evening, and at night.
	3	a	7.0		
10	7	m	2.5	N E 1	Cloudy. Began to rain at ten, and continued all day till night.
	3	a	6.5		
11	6	m	5.0	N E 1	Rain almost the whole day.
	3	a	9.0		
12	6	m	2.0	W N W 2	Fair. Afternoon cloudy, with hail and rain.
	2	a	13.0		
13	6	m		N W 2	Fair.
	2	a		S W 1	Cloudy.
14	6	m		E 1	Cloudy; fair at eight. Cloudy towards night.
	2	a			
15	6	m		E 1	Almost quite fair.
	2	a			
16	6	m	6.5	W N W 2	Fair.
	2	a	13.5	— — — 1	
17	6	m	7.0	S 1	Alternately fair and cloudy.
	3	a	16.0	S W 1	Rain.
18	7	m	6.0	N 0	Fair.
	3	a	18.0	N W 3	
19	5:30	m	2.0	N N W 0	Fair.
	3	a	20.0	W 2	

D.	H.		Ther.	Wind	The Weather in General
20	6	m	2.0	S W o	A hoar frost this morning. Fair and very hot all day.
	3	a			
21				S W 1	Fair; with hot vapors raised by the sun.
22	5	m	13.0	S o	Almost fair.
	3	a	23.0		
23	5:30	m	11.0	W 1	Fair.
	3		25.5		
24	6	m	12.0	S 1	Cloudy, intermittent drizzling showers.
	3	a	22.0		
25	6	m	18.0	S o	Rain the preceding night, and now and then this day. At night thunder and lightning.
	3	a	24.0		
26	6	m	28.0	W 1	Fair.
	3	a	30.0		
27	6	m	17.0	W 2	Fair.
	3	a	25.0		
28	6	m	7.0	W o	Fair.
	3	a	24.0		
29	6	m	7.0	N 2	Fair.
	3	a	17.0	E 2	
30	5	m	3.0	E 1	Flying clouds.
	3	a	15.5	S 1	

May 1749

D.	H.		Ther.	Wind	The Weather in General
1	4	m	01.5	S o	Hoar frost this morning,—fair.
	3	a	18.5	S W 1	
2	5	m	1.0	W 1	Fair.
	3	a	23.0		
3	5:30	m	4.0	W 1	Fair.
	3	a	27.5		
4	5	m	16.0	W 1	Fair.
5	5	m	13.0	S 3	Flying clouds.
	3	a	27.0		
6	5	m	14.5	N o	Fair.
7	5	m	13.0	N o	Somewhat cloudy.
8	5	m	4.0	N o	Fair.
9	6	m	14.0	S 1	Rain almost the whole day.
	3	a	14.0		
10	6	m	13.0	S S W o	Intermittent showers.
	3	a	16.0		

D.	H.		Ther.	Wind	The Weather in General
11	6	m	12.0	W S W o	Fair.
	3	a	28.0		
12	6	m	13.0	W N W 2	Fair.
	3	a	20.0		
13	5	m	9.0	N W 1	Fair.
	3	a	18.5		
14	5	m	00.5	N W o	Fair.
15	5	m	9.0	S S W 2	Cloudy.
	3	a	20.0		Rain.
16	5	m	17.0		Cloudy.
	4	a	23.0		
17	5	m	20.0	S 1	Rain intermittently all day; and lightning very frequent at night.
	3	a	24.0		
18	5	m	13.0		Fair.
19	5	m	17.0	W 2	Fair.
20	5	m	19.0	W 1	Fair.
	3	m	24.0		
21	6	m	20.0		Fair.
22				S W 1	Fair. Very hot.
23	5	m	17.0	S W 1	Fair.
	3	a	33.5		
24	12	m	32.0	S W 1	Fair.
25	8	m	23.0	S W 1	Fair, and very warm.
	2	a	28.0		
26	8	m	21.0	W N W 2	Flying clouds; at night thick clouds, with storm and rain.
	3	a	25.0		
27	7	m	17.0	W 2	Thick, scattered clouds.
	2	a	25.0		Pretty cool.
28	7	m	15.0	W 1	Flying clouds.
	2	a	25.0		
29	7	m	16.0	W 2	Flying clouds.
	2	a	25.0		
30	5	m	13.0	W N W 1	Fair.
	—	a	25.0	W 1	Cloudy.
31	5	m	13.0	S W 1	Somewhat cloudy.
	1	a	27.0		Fair.

June 1749

D.	H.		Ther.	Wind	The Weather in General
1	5	m	23.0	S W 1	Rain the preceding night.
2				S E 1	Morning cloudy,—cleared up at ten,—flying clouds.

D.	H.		Ther.	Wind	The Weather in General
3	7	m	24.0	S W 1	Flying clouds; afternoon, thunder clouds with rain from the N. W.
4	3	a	26.0	N W 1	Flying clouds.
5	5:30	m	15.5	S 1	Fair.
	3	a	22.0		
6	5	m	18.5	S W 1	Alternately fair and cloudy.
	3	a	23.0		
7	All d.		20.0		Cloudy and rainy.
8	6	m	15.5	N W 0	Cloudy.
	3	a	23.0	— — 1	Flying clouds.
9	5	m	13.0		Fair.
10	5	m	11.0	S W 1	Fair.
	3	a	22.5		
11	7	m	20.0	N 1	Flying clouds.
	2	a	33.0	S W 1	Thunder storm, with rain.
12	6	m	23.0	N 0	Fair.
	3	a	32.0	S 2	Somewhat cloudy.
13	5	m	19.0	S E 2	Almost fair.
	3	a	27.0		
14	6	m	26.0	S 1	Fair.
	3	a	25.0		Thunder clouds, with rain.
15	6	m	18.0	N 0	Fair.
	3	a	26.5		
16	6	m	20.0	N N E 1	Fair.
	2	a	28.0		
17	5:30	m	18.0	N 0	Fair.
	3	a	27.5		
18	5	m	21.0	E S E 1	Fair.
	3	a	32.0	N E 1	Thunder, with heavy showers.
19	6	m	20.0	N N W 1	Fair.
	3	a	27.0		
20	5	m	18.0	S 1	Fair.
	3	a	26.0		Cloudy.
21	5	m	23.0	S W 0	Cloudy, with some showers.
22	5	m	9.0	W 1	Fair.
23	6	m	17.0	S 1	Fair.
	—	a		N W 1	Cloudy.
24	6	m	20.5	S 1	Cloudy, afterwards fair.
	—	a		S W 1	Thunder and rain.
25	5	m	23.0	S 1	Fair.
	2	a	32.0		
26	5	m	14.0	N 1	Fair.

D.	H.		Ther.	Wind	The Weather in General
27	6	m	15.0		Fair.
28	6	m	18.0	S 1	Fair.
	1	a	35.0		
29	7	m	6.0		Fair.
30	5	m	11.0	S 1	Fair.
	3	a	31.0	W 1	

July 1749

D.	H.		Ther.	Wind	The Weather in General
1				N 3	Flying clouds.
2	5	m	7.5	N 2	Fair.
3	8	m	26.0	N 1	Fair.
	2	a	28.0	— 1	Thunder storm, and rain at night.
4	6	m	20.0	S 1	Cloudy; intermittent showers in the
	—	a		N 2	afternoon.
5				W 1	Fair.
	4	a	26.0	— 1	Cloudy; rain at night.
6	5:30	m	18.0	S W 1	Rain all the preceding night; fair in daytime.
7	4:30	m	17.0	N W 0	Fair.
8	6	m	16.0	N 0	Alternately fair and cloudy. A halo round the sun in the forenoon.
9	7	m	21.0	S W 0	Rain the preceding night. In daytime,
	3	a	22.0		cloudy with some showers.
10	4:45	m	18.0	S W 0	Fair; sometimes flying clouds and showers.
	3	a	24.5	— — 1	
11	5	m	17.0	S S E 1	Fair.
	2	a	26.0	— — — 1	
12	5	m	22.0	W 1	Fair.
13	6	m	20.0	S S W 1	Fair.
	3	a	33.0	— — — 1	
14	5	m	21.0	W S W 1	Fair.
	2	a	28.0	— — — 1	
15	5	m	26.0	N N E 1	Fair.
	3	a	28.0	— — — 1	
16	5	m	14.0	S 0	Fair; sometimes cloudy.
	10	m		S S E 1	
17	5	m	19.0	S 1	Fair.
	3	a	24.0	— 1	Cloudy.
18	5	m	15.0	N N E 0	Fair.
	2	a	25.0	— — — 0	

D.	H.		Ther.	Wind	The Weather in General
19	5	m	19.0	S S W 1	Cloudy; rain.
	—	a			Pretty fair.
20	5	m	19.0	S 1	Fair.
	3	a	24.0	— 1	Cloudy; some rain.
21				S 0	Fair.
	3	a	27.0	— 0	Flying clouds.
22	5	m	16.0	S W 2	Fair.
	3	a	27.0	S W 2	
23	6	m	19.0	S S W 1	Alternately fair and cloudy.
	3	a	28.5	— — — 1	
24	6	m	20.0	S W 1	Fair.
	3	a	29.0	— — 1	
25	5	m	20.0	W S W 0	Fair.
	3	a	29.5	— — — 0	
26	5	m	21.0	S 0	Fair.
	3	a	30.0	— 1	
27	5	m	22.0	W 1	Cloudy; intermittent showers.
	3	a	21.5	— 1	
28	6	m	17.0	W 1	Fair.
	3	a	27.0	— 1	
29	6	m	16.0	N W ½	Fair; flying clouds at night, and showers.
	2	a	24.0	— — 1	
30	6	m	14.0	W N W 1	Fair.
	2	a	26.0	— — — 1	
31	6	m	16.0	E 1	Cloudy; rain almost all day.
	3	a	22.0	— 1	

August 1749

D.	H.		Ther.	Wind	The Weather in General
1	6	m	22.0	N E 1	Cloudy. Some showers.
	3	a	28.0	— — 1	
2	4:30	m	16.0	N E 1	Fair.
		a		S E 1	Cloudy. Fair towards night.
3	5	m	13.0	S W 2	Fair.
4		m		N E 2	Cloudy. Some showers.
	2	a	21.0	— — 2	
5		m		N E 1	Fair.
		a		S W 1	
6	5	m	16.0	N E 3	Heavy rain all day.
	3	a	16.0	— — 3	Some thunder.

D.	H.		Ther.	Wind	The Weather in General
7	6	m	13.0	E S E 1	Cloudy. Frequent showers.
	3	a	16.0	— — — 1	
8	6	m	16.0	S W 1	Cloudy. Some showers.
	3	a	27.0	— — 1	
9	6	m	14.0	S W 1	Flying clouds.
	1	a	20.0	— — 1	Rain at night.
10	6	m	14.0	S W 1	Flying clouds.
	3	a	24.0	— — 1	
11	6	m	15.5	W 1	Cloudy.
12	6	m	14.0	W 1	Flying clouds.
	2	a	25.0	— 1	
13	7	m	15.5	N W 1	Fair.
	2	a	30.0	— — 1	
14	6	m	16.0	N E 2	Fair.
	2	a	26.0	— — 2	
15	6	m	14.0	N E 1	Fair.
	2	a	28.0	— — 1	
16	5	m	14.0	S E 1	Fair. At night thunder and rain.
	3	a	26.0	— — 1	
17	5	m	14.5	S 0	Flying clouds.
	3	a	27.0	— 0	
18	5	m	16.0	W 1	Thunder and rain in the morning. At ten
	3	a	29.0	— 1	in the morning flying clouds.
19	6	m	17.0	W 1	Fair.
	3	a	30.0	— 1	
20	5	m	16.5	S W 0	
	3	a	28.0	— — 0	
21	5	m	17.0	S W 1	Fair.
	2	a	29.0	— — 1	
	5	a	27.0	— — 1	
22	5	m	19.0	N E 2	Rain all day.
	3	a	17.5		
23	5	m	16.5	S W 3	Rain early in the morning. At 10 m.
	2	a	22.5	— — 3	flying clouds.
24	6	m	13.5	S W 2	Flying clouds.
	2	a	22.0	— — 2	
25	5	m	7.0	S W 2	Fair.
	4	a	20.5	— — 2	
26	5	m	13.0	N E 1	Alternately fair and cloudy.
	3	a	18.0	— — 1	Much rain this afternoon.
27	5	m	10.5	S W 1	Flying clouds.
	2	a	23.0	— — 1	

D.	H.		Ther.	Wind	The Weather in General
28	5	m	10.0	S W 1	Fair.
	2	a	20.0	— — 1	
29	5	m	13.0	N E 2	Fair.
30	5:30	m	11.0	N E 2	Fair.
31	6	m	13.6	S 1	Fair and cloudy alternately.
	3	a	18.5	— 1	Intermittent showers.

September 1749

D.	H.		Ther.	Wind	The Weather in General
1	5:30	m	14.5	N N W 1	Fair.
	3	a	30.0	— — — 1	
2	5:30	m	9.0	N 1	Fair.
	2	a	18.0	S S W 1	
3	5:30	m	7.5	S 1	Somewhat cloudy. Fair now and then.
	2	a	20.0	— 1	
4	6	m	14.0	S 1	Now and then a shower; and in the inter-
	2	a	17.5	— 1	vals fair.
5	6	m	14.0	N E 2	Fog. Rain all day. Now and then thunder.
6	10:30	m	15.0	N E 2	Fog, and drizzling rain all day.
	10:30	a	15.0	— — 2	
7	7	m	17.0	S W 1	Fog and rain.
	3	a	22.0	— — 1	Fair.
8	5:30	m	15.0	S S W 1	Fair.
	4	a	28.0	— — — 1	
9	5	m	17.5	E N E 2	Fair.
	3	a	25.0	— — — 2	
10	5:30	m	16.0	N E 2	Fair.
	3	a	26.0	— — 2	
11	5:30	m	15.0	E N E 0	Fair.
	3	a	25.0	— — — 0	
12	7	m	14.5	N N E 1	Fair.
		a		S W 1	
13	5:30	m	14.0	N E 1	Fair.
	1:30	a	24.5	— — 1	
14	5	m	15.0	N E 2	Fair.
	1	a	22.5	— — 2	
15	5:30	m	16.0	N N E 3	Fair. Forenoon, a halo round the sun.
	2	a	19.0	— — — 3	
16	5:30	m	8.5	N N E 1	Fair.
	3	a	20.5	— — — 1	
17	5	m	12.0	S W 0	Fair.

D.	H.		Ther.	Wind	The Weather in General
18	6	m	17.0	S W 1	Fair.
	3	a	27.0	— — 1	
19	6	m	14.0	S W 1	Fair.
	3	a	26.0	— — 1	
20	6	m	19.0	S W 1	Fair.
	3	a	26.0	— — 1	Cloudy. Rain towards night.
21	6	m	15.0		Fair.
	3	a	19.5		
22	6	m	13.0	E 0	Somewhat cloudy.
	3	a	22.0	— 0	
23	6	m	14.0	S W 0	Fair.
24	6	m	18.0	S W 2	Fair. Rain at noon.
	2	a	26.0	— — 2	Flying clouds in the afternoon.
25	7	m	16.0	W 1	Alternately clear and cloudy.
	2	a	17.0	— 1	
26	8	m	12.5	N E 1	Fair.
	3	a	11.5	— — 1	Cloudy and rainy.
27	6	m	9.3	N 1	Rain all day.
	3	a	14.0	— 1	
28	6	m	8.0	S W 1	Heavy rain all day.
	3	a	14.0	— — 1	
29	6	m	8.0	S 1	Fog.
	1	a	13.0	— 1	Flying clouds.
30	8	m	14.0	S W 2	Drizzling rain.
	2	a	18.0	— — 2	Somewhat clear.

October 1749

D.	H.		Ther.	Wind	The Weather in General
1	7:30	m	9.0	N W 1	Rain.
		n		— — 1	Somewhat fairer.
2	7	m	2.0	W 1	Hoarfrost this morning. Fair all day.
3	6	m	3.5	S W 1	Fair.
	1	a	12.0	— — 1	
4	6	m	11.0	S 1	Rain.
5	6	m	10.5	N E 1	Cloudy.
		a	11.0	— — 1	
6	6:30	m	10.0	E N E 1	Rain all day.
	3	a	12.0	— — — 1	
7	6:30	m	10.0	E N E 1	Flying clouds.
	2	a	14.0		
8	6:30	m	7.0	S 1	Fair.
	3	a	18.0	S 1	

METEOROLOGICAL OBSERVATIONS,

Made by Mr. John Bartram, near Philadelphia,
During my absence, in the Summer of the year 1749.

June 1749

D.	Ther. Morn.	Ther. Aft.	Wind	The Weather in General
1	22	25	W	Cloudy.
2	20	27	W	Cloudy.
3	23	28	W	Showers.
4	22	28	W	Fair.
5	18	25	W	Fair.
6	18	25	W	Cloudy.
7	22	22	N E	Cloudy.
8		21	N E	
9		21	N	
10	14	22	E	
11	22	23	E	
12	25	25	E	
13	23	25	E	
14	25	27	E 3	
15	24	28	E	Fair.
16	22	26	E	
17	23	27	E	
18	25	27	E	
19	23	24	N W	
20	17	26	W	
21	24	26	W	
22	18	27	W	
23	15	29	W	
24	22	30	W	
25	22	31	W	
26	23	30	N	
27	19	32	W	
28	24	36	W	
29	25	37	W	
30	25	36	N	

JULY 1749

D.	THER. MORN.	THER. AFT.	WIND	THE WEATHER IN GENERAL
1	21	30	W	
2	18	27	N W	
3	26	28	S W	Heavy showers.
4	24	36	N W	
5	22	32	W	
6	22	34	N W	Rain.
7	20	35	W	Hard showers.
8	20	35	N E	Rain.
9	20	29	N	Fair.
10	16	29	N	Fair.
11	17	33	N W	Fair.
12	20	35	W	Fair. Rain at night.
13	22	33	W	Fair.
14	26	30	W	Hard showers.
15	20	29	N	Fair.
16	21	30	E	Rain.
17	29	29	N E	Cloudy.
18	18	19	N E	Rain.
19	18	33	W	Fair.
20	19	33	W	Fair.
21	22	31	W	Fair.
22	23	23	W	Heavy showers.
23	23	25	W	Heavy showers.
24	20	36	W	Fair.
25	27	36	W	
26	28	32	W	
27	24	30	W	Fair.
28	19	27	W	Fair.
29	23	30	W	Rain.
30	30	34		
31	21	34		

AUGUST 1749

D.	Ther. Morn.	Ther. Aft.	Wind	The Weather in General
1				
2	18	32		
3	17	30		
4	18	33		
5	22	39	W	
6	18	37	N 2	
7	17	27	W	
8	14	25	N W	
9	12	24	N W	
10	13	24	N W	
11	11	25	N W	
12	14.5	30	N W	
13	18	31	N W	
14	18	30	W	
15	15	30	W	Rain.
16	23	33	N	
17	14	34	N W	
18	18	37	W	
19	18	25	S W	
20	20	26	N E	Rain.
21	20	25	N W	
22	23	34	N W	
23	17	34	W	
24	18	30	W	
25	20	32	N W by W	
26	10	24	N W	Fair.
27	12	20	N W	Fair.
28	13	23	N W	Fair.
29	22	24	W	Fair.
30	17	25	E	
31	20	29	E	

SEPTEMBER 1749

D.	THER. MORN.	THER. AFT.	WIND	THE WEATHER IN GENERAL
1	19	30	E	Hard showers.
2	18	20	E	Rain.
3	19	25	E	Rain.
4	22	25	E	Foggy.
5	23	21	N E	Cloudy.
6	23	37	N E	Cloudy.
7	24	34	N E	Cloudy.
8	24	32	N E	Cloudy.
9	23	33	N E	Rain.
10	23	32	W	Rain.
11	19	25	N E	
12	13	25	N E	
13	12	20	N E	
14	12	33	N E	
15	13	27	N E	
16	20	26	N E	
17	17	27	E	
18	16	34	S E	
19	12	30	S W	
20	17	26		
21	17	25	W	
22	15	30	E	
23	20	29	E	
24	21	29	W	
25	23	28	W 3	
26	20	15	E by N	Thunderstorm.
27	15	19	N W	
28	10	20	N W	
29				
30	6	26	W	

OCTOBER 1749

D.	THER. MORN.	THER. AFT.	WIND
1	13	25	W
2	14	29	N W
3	8	15	N
4	13	29	W
5	17	30	E
6	18	30	E
7	16	21	N W
8	11	22	N W

October 1749

D.	H.	Ther.	Wind	The Weather in General
6	6:30 m	8.0	E N E 1	Cloudy and rain.
	3 a	10.0		
7	6:30 m	8.0	E N E 1	Cloudy, nine o'clock clear; then clear and scattered clouds.
	2 a	12.0		
8	6:30 m	5.0	W 1	Fair.
	3 a	10.0		
9	6:30 m	5.0	S W 1	Fair and scattered clouds.
	3 a	15.0		
10	6:30 m	5.0	S W 2	Fair and scattered clouds.
11	6:30 m	13.5	S 1	Cloudy, occasional rain.
	2 a	16.5		
12	8 m	13.0	S S E 1	Cloudy and rainy; afternoon heavy rain.
13	11 m	4.0	W N W 1	Cloudy and drizzling.
	2 a	7.0		Rain stopped, toward evening clear.
14	6 m	02.5	W	Fair. Ground white with hoar frost.
15	6 m	1.0	W N W 2	Fair, later partly cloudy.
	2 a	11.0		Wind at 10 P. M., N 2.
16	6 m	6.0	S W 2	Fair and scattered clouds.
	2 a	12.0	E N E 2	
17	6 m	6.0	S 1	Cloudy.
	2 a	—	N E 3	
18	6 m	02.0	N E 3	Strong gale during night. Day, fair.
	2 a	4.0		
19	6:30 m	04.0	S W 2	Cloudy during morning. Ice one half inch thick on standing water.
20			N E 2	Cloudy with occasional sunshine. Air cold.
21	6:30 m	1.0	N E 1	Fair, cloudless.
	2 a	8.5	S W 1	Sunset without a cloud.
22	6:30 m	4.5	S W 3	Cloudy, rain occasionally.
	2 a	4.0		Mostly fair.
23	6:30 m	5.5	S W 2	At sunrise clear, otherwise at intervals cloudy with an occasional light shower.
	2 a	11.0		
	3:15 a	14.0		Mostly fair.
24	6 m	4.5	S W 1	Fair.
25	6:30 m	3.0	S W 1	Fair and scattered clouds.
	a			Somewhat cloudy.
26	6 m	05.0	S W 1	Last night N W 2 and cold; Morning, fair. Afternoon partly cloudy.
27	1 a	10.0		During night heavy rain and thunder; rain continued until 9 A. M., then fair.

D.	H.		Ther.	Wind	The Weather in General
28	6:30	m	06.0	N E 1	Cold at night. Generally fair, though partly cloudy in afternoon, Air chilly. Wind during night, N W 2.
29	1	a	0.5	S S E 1	Snow during preceding night and during most of day.
30	7	m	00.0	N N W 1	Fair.
	2	a	2.0		
31	7	m	09.5	N N E 1	Fair.
	2	a	0.5		

November 1749

D.	H.		Ther.	Wind	The Weather in General
1	7	m	05.0	N N E 0	Morning, fair; afternoon somewhat cloudy
	2	a	10.0	S 2	cloudy. Change of wind about noon.
2	7	m	6.0	S 1	Cloudy, rain, heavy atmosphere.
	3	a	8.3		
3	7	m	6.0	W N W 1	Cloudy and torrential rain all day.
	2	a	7.0		
4	7	m	5.0	N N W 2	Morning, heavy rain. Afternoon, cloudy, without rain.
5	7	m	3.0	W S W 1	Morning, somewhat cloudy. Afternoon,
	3:30	a	6.5		fair.
6	6:30	m	04.0	S 1	Fair, cloudless. Glorious weather for time
	2:30	a	11.5		of year.
7	7	m	2.5	S 1	Morning, fair until 9, increasing cloudi-
	2	a	11.0		ness. Quick shifting of wind as follows: At 12 noon, N 1; 6 P. M. N W 2; 10 P. M. N 4; 12 midnight, N E 4.
8					
9	8	m	02.0	S W 1	Cloudy.
	2	a	1.5	N 2	
10	7	m	0.5	N E	Fair. Glorious autumn weather.
	2	a	8.0		
11	7	m	4.0	W S W 1	Morning fair but atmosphere seemed
	2	a	6.0		heavy.
12	7	m	02.0	N W 1	Fair.
	3	a	5.0		
13	7	m	03.5	N W 1	Fair.
	4	a	5.0		
14	7	m	02.0	N E	Fair. In afternoon, patches of filmy
	3	a	6.0	S 1	clouds.

D.	H.		Ther.	Wind	The Weather in General
15	10	m	4.5	N E	Heavy mist until about 11 A.M. Fair for
		a		S 1	remainder of day. Evening, cloudy with a little rain.
16	8	m	7.5	S 1	Cloudy.
	2	a	8.5		
17	7	m	5.0	S W 1	Cloudy.
18	9:30	m	1.0	N W 2	Fair.
		a	6.0		
19	7:30	m	3.5	S 3	Fair until 10 A. M. Increasingly cloudiness until sunset, when it cleared.
20	8	m	4.3	S 2	Cloudy. Sun shone all day, but most of
	3:15	a	11.5		the time through clouds, like a red disk.
21				N N E 2	Cloudy, but sun shone through clouds all day like a red disk.
22	7	m	4.5	S W 0	Cloudy, with some sunshine.
23			14.0	S W 1	Thin clouds, sultry weather. In the evening, rain. Temperature given is estimated only.
24	12		n8–10.0	W 1	Fair.
25	5	a	5.0	W 2	Severe frost during preceding night; Day, fair; beautiful for this time of year.
26	7	m	03.0	N W 1	Early morning, cloudy. After 9 A. M.,
	2	a	5.0		Fair.
27	7:30	m	05.0	S S W 0	During the night, rain; day, cloudy.
	2	a	5.0		
28	7:30	m	00.0		
	2	a	2.5	N W 1	Fair, with scattered clouds in the morning; small flurries of snow.
29	7	m	03.0	N W 1	Cloudy with flurries of snow, especially
	3	a	3.5		toward evening.
30	8	m	03.0	W 1	Fair, later cloudy.
	3	a	1.0		

December 1749

D.	H.		Ther.	Wind	The Weather in General
1	7	m	1.5	W S S 2	Early morning, cloudy. After 10 A. M.
	2	a	4.0		Fair. Wind in the evening, N W 1.
2	7:15	m	04.0	W S W 1	Fair in the morning; in the afternoon,
	3	a	3.0	N W 2	cloudy and snow, occasionally becoming a storm. Late evening, clear.

D.	H.		Ther.	Wind	The Weather in General
3	7:15	m	010.0	N W 2	Fair.
	9	m	010.5		
	3	a	9.0		
4	7	m	013.0	S S W 2	Early morning, fair; 9 A.M. cloudy, with sun appearing at intervals.
	3	a	6.0	N W 1	
5	8	m	5.0	S W 2	
	3	a	3.0	W N W 3	Fair. At 10 A.M. wind shifted to W 2.
6	8	m	08.0	W 1	Fair. Glowing sunset; later, cloudy.
	2	a	5.0		
7	7:30	m	04.0	S W 0	Cloudy, quite foggy, thick atmosphere.
	2	a	2.0		
8	7	m	03.5	S W 0	Cloudy, with heavy, misty atmosphere.
	2	a	4.0		Rain at 4 P. M. continuing into night.
9	7	m	02.0	S W 1	Early morning, fair. 10 A.M., cloudy; Early afternoon, fair. At 4 P. M. a cloud from N W with snowstorm. Clear again at 6 P.M.
	2	a	3.0	N W 1	
10	7	m	010.0	N W 1	Fair.
	3	a	04.0		
11	7:30	m	011.0	N W 0	Fair, later cloudy; at 8 P. M., snow.
	3	a	3.0	E 0	At 1 P. M. the wind was S W 0.
12	7	m	1.0	S W	Cloudy.
13	7	m	1.0	W 2	Early morning, cloudy; at 9 A.M., fair.
14	8	m	02.0	W 1	Fair; in evening, cloudy.
	2	a	6.0		
15	8	m	5.0	E 2	Rain all day and far into night. Thunder and lightning late at night.
	2	a	4.0		
16	8	m	00.0	W N W 2	Cloudy.
	2	a	5.0		
17	8	m	5.0	W N W 2	Early morning, fair; after 9 A.M. scattered clouds with a little hail.
		a	4–5.0		
18	7	m	3.25	N W 2	Fair.
	2:30	a	1.5		
19	7	m	05.0	N W 2	Fair.
	2	a	01.0		
20	7	m	06.0	W N W 1	Fair.
21	7:15	m	01.25	N E	Early morning, rain; showers all day.
	2	a	00.0		
22	7:15	m	00.0	E	Cloudy at first, at 10 A. M. fair.
	2	a	4.0		
23	7:30	m	04.0	S W 1	Fair.

D.	H.		Ther.	Wind	The Weather in General
24	7	m	06.5	S W 1	Fair; at 4 P.M., cloudy. 7 P.M., fair.
		a		W N W 1	
25	7:15	m	08.0	S W	Fair.
	2	a	01.5	W N W	
26	7:30	m	08.0	E 0	Cloudy in morning. Afternoon, snow.
27	7:15	m	03.5	W N W	Fair, scattered clouds.
	2	a	01.5	S W 0	
28	7:30	m	08.0	N W 3	Fair, with large scattered clouds.
	3:30	a	04.5		Afternoon, fair.
29	7:15	m	012.0	W N W	Fair, moderating in evening.
	2		02.0	W S W 1	
30	7	m	04.0	S W 1	Cloudy.
	2	a	0.5		
31	7:30	m	09.0	N W	Scattered clouds; at 5 P. M. snow flurries.
	2	a	05.5	W 1	At 7 P.M., fair.
	10:30	a	011.0		

January 1750

D.	H.		Ther.	Wind	The Weather in General
1	7	m	014.5	N W	Fair.
	2	a	010.0		
2	7:15	m	017.0	N W 1	Fair.
	2	a	06.0		
3	7	m	08.0	N W 0	Cloudy. At 7 A.M. rain, turning into ice, making the ground slippery.
	2	a	03.0		
4	7	m	01.0	W N W 1	Cloudy in early morning; at 8 A.M., fair; at 12 noon, cloudy; at 4 P.M., fair.
	2	a	2.0		
5	7	m	04.0	W N W	Cloudy with flurries of snow.
	2	a	1.0		
6	7	m	06.5	N W 1	Fair.
	3	a	02.0		
7	7	m	010.5	W N W 1	Fair with scattered snow clouds.
	2	a	04.0		
8	7	m	011.5	N N W	Fair.
	2	a	02.0		
9	7	m	05.0	W N W 1	Fair, bright sunshine.
	2	a	2.0		
10	7	m	08.0	N W 1	Beautiful sunshine, with scattered clouds in the evening. Wind at 10 A.M., S W.
	3	a	5.0		
11	7	m	05.5	S W	Fair.
	2	a	7.0		

A BIBLIOGRAPHY OF
PETER KALM'S WRITINGS ON AMERICA

1. *En Resa til Norra America.* På Kong. Swenska Vetenskaps-academiens befallning, och Publici kostnad, I-III, Stockholm, 1753-1761.

   The second and third volumes deal with North America. Now rare.

2. *Pehr Kalms Resa till Norra Amerika,* utgiven av Fredr. Elfving och Georg Schauman. Skrifter utgivna av Svenska Litteratursällskapet i Finland. Första delen, 1904; andra delen, 1910; tredje delen, 1915; fjärde delen (Tilläggsband) sammanställt av Fredr. Elfving, 1929.

   A reprint published in Helsingfors. It has valuable prefaces in vols. I and IV, and the last volume contains the previously unpublished diary notes by Kalm, which were found by Georg Schauman in the university library at Helsingfors. The Preface to vol. I contains also a record of the more important reviews of Kalm's *Resa,* both in Sweden and abroad. I refer to this preface.

Translations of Kalm's Resa

3. *Reise nach der nordlichen America,* welche auf Befehl der Königlichen Schwedischen Akademie der Wissenschaften und auf allgemeine Kosten, von Peter Kalm, . . . ist verrichtet worden. Leipzig, G. Kiesewetter, 1754–1764. 3 Th.[1]

[1] Part I of the Leipzig edition was translated by Carl Ernst Klein, a Pomeranian Legation-preacher, who had settled in Stockholm; Parts I–II of the Göttingen version by J. P. Murray; and Part III by his brother J. A. Murray. Parts II and III are identical in the two German versions. The two Murrays were Swedish-Germans of Scotch ancestry—Johan Philip Murray (1726–1776), professor of philosophy at Göttingen, and the more famous Johan Andreas Murray (1740–1792), botanist, Linné pupil, professor of medicine and author of *Apparatus medicaminum,* who in *Biografiskt Lexicon* is credited with the German translation of Part III of Kalm's work. J. Andreas Murray was born in Stockholm, whither his father had moved in 1736 and become pastor of the German church. It should be noted, incidentally, that the German versions, the only ones to be made directly from the original, include that portion of the work relating to England, which is omitted in the English, Dutch, and French translations. (See next item).

4. Des Herren Peter Kalm . . . mitgliedes der Königlichen Schwedischen Akademie der wissenschaften, ***beschreibung der reise*** die er ***nach dem Nördlichen Amerika*** auf den befehl gedachter akademie und öffentliche kosten unternommen hat . . . Eine uebersetzung, Göttingen, Wittwe A. Vanderhoek, 1754-1764. 3 Th.

Appeared in the series of *Sammlung neuer und merkwürdiger Reisen zu Wasser und zu Lande.* IX–XIter Theil. Both German editions are quite rare, in America at least.

5. *Travels into North America;* containing its natural history, and a circumstantial Account of its Plantations and Agriculture in general, etc. By Peter Kalm, Professor of Economy in the University of Aobo in Swedish Finland, etc. Translated into English by John Reinhold Forster, F.A.S. [With map, cuts "for the illustration of Natural History," not found in the original, and notes.] Vols. I–III. Warrington, 1770–1771. Part dealing with Norway and England omitted.

It is obvious from a statement in the preface, xv, that the translation is made from the German version.[2] But it was a decided success and a second edition was issued almost immediately.

6. Same, second edition, abridged, London, T. Downdes, 1772. 2 vols.

[2] This English translation was the result of temporary financial distress on the part of the translator, who was invited to London from Germany for another plan that did not materialize, found himself stranded there, and so took up the Englishing of certain "Reisebeschreibungen" to get a living. So far as Kalm's work goes, most of the translating was in reality done by Forster's talented sixteen-year old son, Johann George Adam Forster (1754–1794). We may assume, however, that his father superintended the job and supplied the prefaces and notes. See the *Allgemeine Deutsche Biographie,* vii, 168, 173.

A preface by Forster to the third volume accuses Kalm of prejudice against the English in favor of the French. Maybe there is a little extra kindliness toward the French of Canada, who were exceedingly polite to Kalm during his visit there, but if so, it is an unconscious favoritism and in no way serious. Kalm is in the main a just, objective observer, scientific, and sometimes honest to the point of a childish naiveté. He constantly cites his authorities for phenomenon, real or alleged, that he has not observed with his own eyes, and leaves the final judgment to his reader. The usual criticism of Kalm is that he is hardly at all *subjective* and certainly not at all literary or sensational. Style, for example, has but little meaning to this Swede, but fact everything, and to record this in a simple, practical form is his great objective. In comparing English colonial women with those of French Canada Kalm gives the palm of glory to the Canadians, but in the budding troubles between England and her American colonies, for instance,—troubles which Kalm was one of the first to observe—Kalm's sympathy is with England. He had no prejudice against Englishmen. On the other hand, Forster, either because of conviction or circumstances, was or had to be favorable to the English, of course.

7. Same, reprinted in *A General Collection of the best and most interesting voyages and travels,* edited by J. Pinkerton, 1808–1814, vol. 13, 4to. London, 1812.[3]

8. *Kalm's Account of his Visit to England on his way to America in 1748.* Translated by Joseph Lucas. With two maps and several illustrations. London and New York, 1892.

This is a separate translation of the part of Kalm's *Resa* which deals with England only, and before omitted in any foreign version. It contains a life of the author, a translator's preface and a facsimile of the title-page of the original opposite its own title-page. It is a careful work, octavo, of 458 pages plus an index. Many words and sectional titles are given in both Swedish and English, viz., "Snake-Oil, Orm-olja." The character â is used instead of å.

9. *Reis door Noord Amerika,* gedaen door den Heer Pieter Kalm . . . Vercierd met kopern platen . . . Te Utrecht, 1772. 1–2.

This Dutch translation is based on Forster's English and Murray's German version. It is a handsome work in quarto, and, as noted before, the part relating to England is omitted. The name of the translator is unknown.

10. Jacques Philibert Rousselot de Surgy, *Histoire naturelle et politique de la Pensylvanie et de l'établissement des Quakers dans cette contrée.* Tr. de l'allemand. P.M. d.s. censeur royal. Précédée d'une carte géographique. Paris, Garneau, 1768.

The "compiler's chief sources were a German translation of P. Kalm's Resa till Norra America" and Gottlieb Mittelberger's *Reise nach Pensylvanien im Jahr 1750.*

11. *Voyage de Kalm en Amérique,* analysé et traduit par L. W. Marchand. Montréal, par T. Berthiaume, 1880. Two vols. Published in *Mémoires de la societé historique de Montréal.* Provided with notes and index.

The part dealing with the United States is condensed into an "analyse" of 151 pages with occasional citations of literal trans-

[3] In addition, according to the British Museum Catalogue, vol. 2 of John Hamilton Moore's *A new and complete Collection of Voyages and Travels,* London, 1778 (2nd ed. 1785?) contains some material based on Pehr Kalm.

lations. The emphasis here is, of course, on Canada, and the portion treating of it is carefully translated.[4]

Kalm did pioneer work in describing Niagara Falls, a task which was intended to conclude a finished fourth part of his *Resa,* but which never appeared in the original edition. A contemporaneous letter to the librarian Gjörwell on Niagara was recently published, however, in Elfving's above-mentioned *Tilläggsband* in 1929, pp. 162–180. But Kalm did more than that. On September 2, 1750, he addressed a letter on Niagara Falls "to his friend in Philadelphia," Benjamin Franklin. It was composed in English and was the earliest account of Niagara in that language.[5] It was printed as follows:

12. (a) In No. 1136 of the *Pennsylvania Gazette* for September 20, 1750.

(See Kalm's own testimony in Elfving's *Tilläggsband,* p. 157.)

I have not seen this first printing of the article. Dow does not mention it, but Professor A. J. Uppvall of the University of Pennsylvania has verified this printing.

(b) A letter from Mr. Kalm, a gentleman of Sweden, now on his travels in America, to his friend in Philadelphia, a particular account of the Great Fall of Niagara. *Gentleman's Magazine,* Jan., 1751, 21:15–19.

An engraved picture based in all probability on Kalm's description appeared the following month in the same magazine. It was the first view after Hennepin's (1697) to be founded on actual sight of the Falls. Various printed and pictorial reproductions of the Falls soon appeared which were based on Kalm's published letter.

(c) Same, reprinted in John Bartram's *Observations,* etc., London, 1751, where it is termed "a curious Account of the Cataract of Niagara." Bartram's *Observations* with Kalm's description of Niagara was reprinted at Rochester, New York, in 1895. Kalm's letter is found on pages 79–94.

This description is much less scientific and detailed than the one given in his letter to Gjörwell.

(d) Same, in Dodley's *Annual Register,* 4th ed., London, 1765, 2:388–394.

---

[4] In 1900, in the city of Lévis, Canada, near Quebec, there appeared a pamphlet, *Voyage de Kalm au Canada* by J. Edmond Roy, a member of the Royal Society of Canada. Though the brochure contains some new material it is of course based chiefly on Kalm's *Travels.*

[5] Charles Mason Dow, *Anthology and Bibliography of Niagara Falls,* I–II, Albany, 1921. I, 62–63. Dow refers to Kalm as an "eminent Swedish Botanist" and reproduces his account of Niagara, pp. 53–63 of vol. I.

(e) The Falls of Niagara, 1764. From a newspaper of the day. In *Mass. Mag.*, 1790, 2:592.

This is Kalm's account almost word for word. . . . It "reads like a careful revision of the earlier description." [6]

(f) Same, in Dow's *Anthology*. See note 5.

The *Enciclopedia Universal Ilustrada* [7] states in its article on "Pedro Kalm" that his letter to Franklin on Niagara Falls "fue traducida á muchos idiomas," and the *Biografiskt Lexicon*, its probable source, assures us that six editions of it appeared in one year in England and America, and that it was translated into both German and French. I have been unable to verify these statements, but assume them to be true. A condensation by an anonymous writer of the account of Niagara Falls appeared in *Uppfostringssälsk. Tidning*, 1782, nos. 45 and 46.

13. The passenger Pigeon . . . Accounts by Pehr Kalm (1759) [8] and John James Audubon (1831). Smithsonian Institution, *Annual Report* for 1911. Washington, 1912, 407–424.

Between the years 1749 and 1778 Kalm contributed seventeen articles on American subjects to Kong. Vetenskaps Academiens *Handlingar*, two of them running through two or three continuations. They deal with the climate, trees, insects, animals, plants, and other agricultural topics. We shall reproduce the titles of the whole list.

14. 1749. Anmerkningar om historia naturalis och Climatet af Pensylvanien. Pp. 70–79.—An abstract from a letter of October 14, 1748.

15. 1750. Lobelia, såsom et specificum mot Lues Venerea, 280–290.

Five species of this herb, discovered by Kalm, described as a cure for venereal diseases. Kalm possibly learned this from the Indians, who made a medicine from the lobelia plant. There are hundreds of species of it. See Webster's *Unabridged Dictionary*.

16. 1751. Huru socker göres i America af lönnens saft, 143–159.
A treatise on maple sugar.

17. 1751. Huru dricka göres i America af gran-ris, 190–196.
An account of how to prepare spruce beer.

---

[6] Dow, *Ibid.*, 1, 63.

[7] It is interesting to note in this connection that neither the *Encyclopedia Britannica* (14th ed.) nor *Der Grosse Brockhaus* devotes any special article to Pehr Kalm.

[8] Obviously a translation of Kalm's Swedish article on "Vilda Dufvor i Norra America" which has been printed in Svenska Vetenskaps Academiens *Handlingar* for 1759. See below, item no. 23.

18. 1751. (a) Om Amerikanska Maysen,[9] 305–319.
    1752. (b) Om maisens [9] skötsel och nytta, 24–43.

19. 1752. Några Nord-sken observerade i America, 145–155.

20. 1752. (a) Om Skaller-Ormen, 308–319.[10]
    1753. (b) Fortsättning af berättelsen om Skaller-Ormen, 52–67.[10]

    Kalm was immensely interested in rattlesnakes.

    1753. (c) Om Botemedlet emot Skaller-Ormens bett, 185–194. Kalm gives a bibliography on the subject.

21. 1754. Om de Amerikanske skogs-lössen [ticks], 19–31.

22. 1756. Beskrifning om et slags gräshopper i Norra America, 100–116.

23. 1759. Beskrifning på de vilda Dufvor i Norra America, 275-295.

    As always, Kalm furnishes a bibliography. See note 8.

24. 1764. Om maskar, som fördärfva skogarna i America, 124–139.

25. 1767. Om Norr-Americanska Svarta Valnöts-Trädets egenskaper, nytta och Plantering, 51–64.

26. 1769. Om Norr-Americanska hvita Valnöts-Trädets egenskaper och nytta, 119–127.

27. 1771. Thermometrika Rön på Hafs och Sjöars vattens värma, 52–59.

    Based largely on observations made in America.

28. 1773. Om Tuppsporre-Hagtorns nytta till lefvande Häckar, 343–349.

    On cockspur-hawthorn.

29. 1776. Beskrifning på Norr-Americanska Mulbärs-trädet, 143–163.

30. 1778. Om Americanska Valnöts-Trädet Hiccory, 262–283.

    Based in part on information received from Benjamin Franklin.

31. In 1751 there appeared, also, in Stockholm a separate anonymous pamphlet by Kalm with the title "Berättelse om naturliga stället, nyttan samt skötseln af några växter ifrån N. America."

---

[9] Note the slight difference in spelling.

[10] "Vide, Medical, & cases and experiments, translated from the Swedish, London, 1758, p. 282."—A propos of rattlesnakes, J. R. Forster in his English translation of Kalm's *Resa* gives this reference, 1, 116. Evidently Kalm's articles on the subject were already mentioned, and perhaps translated, in this London publication of 1758. I have not seen the book.

Besides the writings above, Kalm as frequent praeses at the University of Åbo was naturally the chief source of information or inspiration, or both, for a large number (146) of theses by his student respondents. Among these there are at least six items of *Americana* for which we can unhesitatingly give the presiding officer main credit of authorship. The six theses, none of them very long, have in the Yale library been assigned to the Rare Book Room. They are:

32. Anders Chydenius, Americanska näfwerbåtar. . . Åbo, J. Merckell. 1753.

    A master's thesis on Indian birch canoes.

33. Daniel Backman, Med Guds wälsignande nåd och wederbörandes tilstånd yttrade tankar om nyttan, som kunnat tillfalla wårt kjära fädernesland, af des nybygge i America, fordom Nya Swerige kalladt. Åbo, 1754.

34. Andreas Abraham Indrenius, Specimen Academicum de Esquimaux, gente americana, quod in regio Fennorum lycaeo . . . Åbo, 1756.

35. Georg A. Westman, Itinera priscorum Scandianorum in Americam. Dissertatione graduli.

    Publicly read in Åbo, 1747, but not printed there, it appears, until 1757.

36. Sven Gowinius, Enfaldiga tankar om nyttan som England kan hafva af sina nybyggen i Norra America. Åbo, 1763.

    An important pamphlet of 22 pages.

37. Esaias Hollberg, Norra americanska färge-örter. Åbo, 1763.[11]

[11] This bibliography was first published in *Scandinavian Studies and Notes,* May, 1933, pp. 89–98. It is here reproduced with the permission of the editor of that journal, Professor A. M. Sturtevant.

INDEX

This comprehensive index covers both volumes of the work. Volume I contains pages 1 through 401 and Volume II contains 402 through 776.

CATALOGUE OF DOVER BOOKS

Americana

THE EYES OF DISCOVERY, J. Bakeless. A vivid reconstruction of how unspoiled America appeared to the first white men. Authentic and enlightening accounts of Hudson's landing in New York, Coronado's trek through the Southwest; scores of explorers, settlers, trappers, soldiers. America's pristine flora, fauna, and Indians in every region and state in fresh and unusual new aspects. "A fascinating view of what the land was like before the first highway went through," Time. 68 contemporary illustrations, 39 newly added in this edition. Index. Bibliography. x + 500pp. 5⅜ x 8. T761 Paperbound **$2.00**

AUDUBON AND HIS JOURNALS, J. J. Audubon. A collection of fascinating accounts of Europe and America in the early 1800's through Audubon's own eyes. Includes the Missouri River Journals —an eventful trip through America's untouched heartland, the Labrador Journals, the European Journals, the famous "Episodes", and other rare Audubon material, including the descriptive chapters from the original letterpress edition of the "Ornithological Studies", omitted in all later editions. Indispensable for ornithologists, naturalists, and all lovers of Americana and adventure. 70-page biography by Audubon's granddaughter. 38 illustrations. Index. Total of 1106pp. 5⅜ x 8.
T675 Vol I Paperbound **$2.25**
T676 Vol II Paperbound **$2.25**
The set **$4.50**

TRAVELS OF WILLIAM BARTRAM, edited by Mark Van Doren. The first inexpensive illustrated edition of one of the 18th century's most delightful books is an excellent source of first-hand material on American geography, anthropology, and natural history. Many descriptions of early Indian tribes are our only source of information on them prior to the infiltration of the white man. "The mind of a scientist with the soul of a poet," John Livingston Lowes. 13 original illustrations and maps. Edited with an introduction by Mark Van Doren. 448pp. 5⅜ x 8.
T13 Paperbound **$2.00**

GARRETS AND PRETENDERS: A HISTORY OF BOHEMIANISM IN AMERICA, A. Parry. The colorful and fantastic history of American Bohemianism from Poe to Kerouac. This is the only complete record of hoboes, cranks, starving poets, and suicides. Here are Pfaff, Whitman, Crane, Bierce, Pound, and many others. New chapters by the author and by H. T. Moore bring this thorough and well-documented history down to the Beatniks. "An excellent account," N. Y. Times. Scores of cartoons, drawings, and caricatures. Bibliography. Index. xxviii + 421pp. 5⅝ x 8⅜. T708 Paperbound **$1.95**

THE EXPLORATION OF THE COLORADO RIVER AND ITS CANYONS, J. W. Powell. The thrilling first-hand account of the expedition that filled in the last white space on the map of the United States. Rapids, famine, hostile Indians, and mutiny are among the perils encountered as the unknown Colorado Valley reveals its secrets. This is the only uncut version of Major Powell's classic of exploration that has been printed in the last 60 years. Includes later reflections and subsequent expedition. 250 illustrations, new map. 400pp. 5⅝ x 8⅜.
T94 Paperbound **$2.25**

THE JOURNAL OF HENRY D. THOREAU, Edited by Bradford Torrey and Francis H. Allen. Henry Thoreau is not only one of the most important figures in American literature and social thought; his voluminous journals (from which his books emerged as selections and crystallizations) constitute both the longest, most sensitive record of personal internal development and a most penetrating description of a historical moment in American culture. This present set, which was first issued in fourteen volumes, contains Thoreau's entire journals from 1837 to 1862, with the exception of the lost years which were found only recently. We are reissuing it, complete and unabridged, with a new introduction by Walter Harding, Secretary of the Thoreau Society. Fourteen volumes reissued in two volumes. Foreword by Henry Seidel Canby. Total of 1888pp. 8⅜ x 12¼. T312-3 Two volume set, Clothbound **$20.00**

GAMES AND SONGS OF AMERICAN CHILDREN, collected by William Wells Newell. A remarkable collection of 190 games with songs that accompany many of them; cross references to show similarities, differences among them; variations; musical notation for 38 songs. Textual discussions show relations with folk-drama and other aspects of folk tradition. Grouped into categories for ready comparative study: Love-games, histories, playing at work, human life, bird and beast, mythology, guessing-games, etc. New introduction covers relations of songs and dances to timeless heritage of folklore, biographical sketch of Newell, other pertinent data. A good source of inspiration for those in charge of groups of children and a valuable reference for anthropologists, sociologists, psychiatrists. Introduction by Carl Withers. New indexes of first lines, games. 5⅜ x 8½. xii + 242pp. T354 Paperbound **$1.75**

GARDNER'S PHOTOGRAPHIC SKETCH BOOK OF THE CIVIL WAR, Alexander Gardner. The first published collection of Civil War photographs, by one of the two or three most famous photographers of the era, outstandingly reproduced from the original positives. Scenes of crucial battles: Appomattox, Manassas, Mechanicsville, Bull Run, Yorktown, Fredericksburg, etc. Gettysburg immediately after retirement of forces. Battle ruins at Richmond, Petersburg, Gaines'Mill. Prisons, arsenals, a slave pen, fortifications, headquarters, pontoon bridges, soldiers, a field hospital. A unique glimpse into the realities of one of the bloodiest wars in history, with an introductory text to each picture by Gardner himself. Until this edition, there were only five known copies in libraries, and fewer in private hands, one of which sold at auction in 1952 for $425. Introduction by E. F. Bleiler. 100 full page 7 x 10 photographs (original size). 224pp. 8½ x 10¾. T476 Clothbound **$6.00**

A BIBLIOGRAPHY OF NORTH AMERICAN FOLKLORE AND FOLKSONG, Charles Haywood, Ph.D. The only book that brings together bibliographic information on so wide a range of folklore material. Lists practically everything published about American folksongs, ballads, dances, folk beliefs and practices, popular music, tales, similar material—more than 35,000 titles of books, articles, periodicals, monographs, music publications, phonograph records. Each entry complete with author, title, date and place of publication, arranger and performer of particular examples of folk music, many with Dr. Haywood's valuable criticism, evaluation. Volume I, "The American People," is complete listing of general and regional studies, titles of tales and songs of Negro and non-English speaking groups and where to find them, Occupational Bibliography including sections listing sources of information, folk material on cowboys, riverboat men, 49ers, American characters like Mike Fink, Frankie and Johnnie, John Henry, many more. Volume II, "The American Indian," tells where to find information on dances, myths, songs, ritual of more than 250 tribes in U.S., Canada. A monumental product of 10 years' labor, carefully classified for easy use. "All students of this subject . . . will find themselves in debt to Professor Haywood," Stith Thompson, in American Anthropologist. ". . . a most useful and excellent work," Duncan Emrich, Chief Folklore Section, Library of Congress, in "Notes." Corrected, enlarged republication of 1951 edition. New Preface. New index of composers, arrangers, performers. General index of more than 15,000 items. Two volumes. Total of 1301pp. 6⅛ x 9¼. T797-798 Clothbound **$12.50**

INCIDENTS OF TRAVEL IN YUCATAN, John L. Stephens. One of first white men to penetrate interior of Yucatan tells the thrilling story of his discoveries of 44 cities, remains of once-powerful Maya civilization. Compelling text combines narrative power with historical significance as it takes you through heat, dust, storms of Yucatan; native festivals with brutal bull fights; great ruined temples atop man-made mounds. Countless idols, sculptures, tombs, examples of Mayan taste for rich ornamentation, from gateways to personal trinkets, accurately illustrated, discussed in text. Will appeal to those interested in ancient civilizations, and those who like stories of exploration, discovery, adventure. Republication of last (1843) edition. 124 illustrations by English artist, F. Catherwood. Appendix on Mayan architecture, chronology. Two volume set. Total of xxviii + 927pp.
Vol I T926 Paperbound **$2.00**
Vol II T927 Paperbound **$2.00**
The set **$4.00**

A GENIUS IN THE FAMILY, Hiram Percy Maxim. Sir Hiram Stevens Maxim was known to the public as the inventive genius who created the Maxim gun, automatic sprinkler, and a heavier-than-air plane that got off the ground in 1894. Here, his son reminisces—this is by no means a formal biography—about the exciting and often downright scandalous private life of his brilliant, eccentric father. A warm and winning portrait of a prankish, mischievous, impious personality, a genuine character. The style is fresh and direct, the effect is unadulterated pleasure. "A book of charm and lasting humor . . . belongs on the 'must read' list of all fathers," New York Times. "A truly gorgeous affair," New Statesman and Nation. 17 illustrations, 16 specially for this edition. viii + 108pp. 5⅜ x 8½.
T948 Paperbound **$1.00**

HORSELESS CARRIAGE DAYS, Hiram P. Maxim. The best account of an important technological revolution by one of its leading figures. The delightful and rewarding story of the author's experiments with the exact combustibility of gasoline, stopping and starting mechanisms, carriage design, and engines. Captures remarkably well the flavor of an age of scoffers and rival inventors not above sabotage; of noisy, uncontrollable gasoline vehicles and incredible mobile steam kettles. ". . . historic information and light humor are combined to furnish highly entertaining reading," New York Times. 56 photographs, 12 specially for this edition. xi + 175pp. 5⅜ x 8½. T964 Paperbound **$1.35**

BODY, BOOTS AND BRITCHES: FOLKTALES, BALLADS AND SPEECH FROM COUNTRY NEW YORK, Harold W. Thompson. A unique collection, discussion of songs, stories, anecdotes, proverbs handed down orally from Scotch-Irish grandfathers, German nurse-maids, Negro workmen, gathered from all over Upper New York State. Tall tales by and about lumbermen and pirates, canalers and injun-fighters, tragic and comic ballads, scores of sayings and proverbs all tied together by an informative, delightful narrative by former president of New York Historical Society. ". . . a sparkling homespun tapestry that every lover of Americana will want to have around the house," Carl Carmer, New York Times. Republication of 1939 edition. 20 line-drawings. Index. Appendix (Sources of material, bibliography). 530pp. 5⅜ x 8½. T411 Paperbound **$2.25**

Nature

AN INTRODUCTION TO BIRD LIFE FOR BIRD WATCHERS, Aretas A. Saunders. Fine, readable introduction to birdwatching. Includes a great deal of basic information on about 160 different varieties of wild birds—elementary facts not easily found elsewhere. Complete guide to identification procedures, methods of observation, important habits of birds, finding nests, food, etc. "Could make bird watchers of readers who never suspected they were vulnerable to that particular virus," CHICAGO SUNDAY TRIBUNE. Unabridged, corrected edition. Bibliography. Index. 22 line drawings by D. D'Ostilio. Formerly "The Lives of Wild Birds." 256pp. 5⅜ x 8½. T1139 Paperbound **$1.00**

LIFE HISTORIES OF NORTH AMERICAN BIRDS, Arthur Cleveland Bent. Bent's historic, all-encompassing series on North American birds, originally produced under the auspices of the Smithsonian Institution, now being republished in its entirety by Dover Publications. The twenty-volume collection forms the most comprehensive, most complete, most-used source of information in existence. Each study describes in detail the characteristics, range, distribution, habits, migratory patterns, courtship procedures, plumage, eggs, voice, enemies, etc. of the different species and subspecies of the birds that inhabit our continent, utilizing reports of hundreds of contemporary observers as well as the writings of the great naturalists of the past. Invaluable to the ornithologist, conservationist, amateur naturalist, and birdwatcher. All books in the series contain numerous photographs to provide handy guides for identification and study.

LIFE HISTORIES OF NORTH AMERICAN BIRDS OF PREY. Including hawks, eagles, falcons, buzzards, condors, owls, etc. Index. Bibliographies of 923 items. 197 full-page plates containing close to 400 photographs. Total of 907pp. 5⅜ x 8½.
Vol. I: T931 Paperbound **$2.50**
Vol. II: T932 Paperbound **$2.50**
The set Paperbound **$5.00**

LIFE HISTORIES OF NORTH AMERICAN SHORE BIRDS. Including 81 varieties of such birds as sandpipers, woodcocks, snipes, phalaropes, oyster catchers, and many others. Index for each volume. Bibliographies of 449 entries. 121 full-page plates including over 200 photographs. Total of 860 pp. 5⅜ x 8½.
Vol. I: T933 Paperbound **$2.35**
Vol. II: T934 Paperbound **$2.35**
The set Paperbound **$4.70**

LIFE HISTORIES OF NORTH AMERICAN WILD FOWL. Including 73 varieties of ducks, geese, mergansers, swans, etc. Index for each volume. Bibliographies of 268 items. 106 full-page plates containing close to 200 photographs. Total of 685pp. 5⅜ x 8½.
Vol. I: T285 Paperbound **$2.50**
Vol. II: T286 Paperbound **$2.50**
The set Paperbound **$5.00**

LIFE HISTORIES OF NORTH AMERICAN GULLS AND TERNS. 50 different varieties of gulls and terns. Index. Bibliography. 93 plates including 149 photographs. xii + 337pp. 5⅜ x 8½. T1029 Paperbound **$2.75**

LIFE HISTORIES OF NORTH AMERICAN GALLINACEOUS BIRDS. Including partridge, quail, grouse, pheasant, pigeons, doves, and others. Index. Bibliography. 93 full-page plates including 170 photographs. xiii + 490pp. 5⅜ x 8½. T1028 Paperbound **$2.75**

THE MALAY ARCHIPELAGO, Alfred Russel Wallace. The record of the explorations (8 years, 14,000 miles) of the Malay Archipelago by a great scientific observer. A contemporary of Darwin, Wallace independently arrived at the concept of evolution by natural selection, applied the new theories of evolution to later genetic discoveries, and made significant contributions to biology, zoology, and botany. This work is still one of the classics of natural history and travel. It contains the author's reports of the different native peoples of the islands, descriptions of the island groupings, his accounts of the animals, birds, and insects that flourished in this area. The reader is carried through strange lands, alien cultures, and new theories, and will share in an exciting, unrivalled travel experience. Unabridged reprint of the 1922 edition, with 62 drawings and maps. 3 appendices, one on cranial measurements. xvii + 515pp. 5⅜ x 8. T187 Paperbound **$2.00**

THE TRAVELS OF WILLIAM BARTRAM, edited by **Mark Van Doren.** This famous source-book of American anthropology, natural history, geography is the record kept by Bartram in the 1770's, on travels through the wilderness of Florida, Georgia, the Carolinas. Containing accurate and beautiful descriptions of Indians, settlers, fauna, flora, it is one of the finest pieces of Americana ever written. Introduction by Mark Van Doren. 13 original illustrations. Index. 448pp. 5⅜ x 8. T13 Paperbound **$2.00**

COMMON SPIDERS OF THE UNITED STATES, J. H. Emerton. Only non-technical, but thorough, reliable guide to spiders for the layman. Over 200 spiders from all parts of the country, arranged by scientific classification, are identified by shape and color, number of eyes, habitat and range, habits, etc. Full text, 501 line drawings and photographs, and valuable introduction explain webs, poisons, threads, capturing and preserving spiders, etc. Index. New synoptic key by S. W. Frost. xxiv + 225pp. 5⅜ x 8. T223 Paperbound **$1.45**